The Geology Companion
Essentials for Understanding the Earth

The Geology Companion
Essentials for Understanding the Earth

Gary L. Prost
Benjamin P. Prost

CRC Press
Taylor & Francis Group
Boca Raton London New York

CRC Press is an imprint of the
Taylor & Francis Group, an **informa** business

CRC Press
Taylor & Francis Group
6000 Broken Sound Parkway NW, Suite 300
Boca Raton, FL 33487-2742

First issued in paperback 2020

© 2018 by Taylor & Francis Group, LLC
CRC Press is an imprint of Taylor & Francis Group, an Informa business

No claim to original U.S. Government works

ISBN 13: 978-0-367-57264-8 (pbk)
ISBN 13: 978-1-4987-5608-2 (hbk)

Library of Congress Cataloging-in-Publication Data

Names: Prost, G. L. | Prost, Benjamin.
Title: The geology companion : essentials for understanding the Earth / Gary Prost and Benjamin Prost.
Description: Boca Raton : CRC Press, 2017. | Includes bibliographical references.
Identifiers: LCCN 2017007550 | ISBN 9781498756082 (hardback : alk. paper)
Subjects: LCSH: Geology. | Mineralogy. | Rocks--Collection and preservation.
| Minerals--Collection and preservation.
Classification: LCC QE26.3 .P76 2017 | DDC 551--dc23
LC record available at https://lccn.loc.gov/2017007550

Visit the Taylor & Francis Web site at
http://www.taylorandfrancis.com

and the CRC Press Web site at
http://www.crcpress.com

To our colleagues, who inspire us; to Nancy, for your love and support;

to Max and Nate: you are our future.

Contents

Preface

When Taylor & Francis reached out four or five years ago, their thought was to compile a comprehensive handbook of geology with contributions from world-renowned experts in various aspects of earth science. I countered that there are already lots of handbooks; what is needed is a book on earth science written for the general public, the "man-or-woman-on-the-street." In any case, I was still working full-time and didn't have the occasion to put together something that would require a large investment of time and effort.

About two years ago, I retired, and like so many retirees, I decided to write a book. The first advice you get about writing is to write what you know, so I cobbled together an outline of the book I had in mind, a book that would answer questions that have been directed at me over my career by family and friends: Do crystals have special powers? Why are you skeptical about man-made global warming? How did you know that that movie was filmed in the Canadian Rockies? How do you know that the rocks on either side of that valley were once continuous across the gap? How did you know that that fossil was from Morocco? What is "dirty oil?"

Lots of excellent geology texts exist for college students, and there are many children's books about geology, but there was nothing comprehensive *and* basic enough for the general public. This book is written at an average 11th-grade reading level. It is written at this level because we wanted it to be easily understood.

We feel strongly that science literacy is fundamental to a free society and a thriving economy. Politicians and school boards do students a disservice when they argue, as was done in Missouri in 2013, that the theory of evolution should be taught on an equal footing with the "theory" of intelligent design. It serves no useful purpose when a news anchor shows white "smoke" coming from a smokestack and calls it pollution (it's not; it's just water vapor), or when a documentary shows glaciers calving while an omniscient narrator calls it the result of global warming (it's not; all glaciers calve whether the climate is warming or not). Many consumers don't want to know that their food comes from slain livestock; similarly, they often don't want to acknowledge that the plastic in their computers, the rare earth minerals in their cell phones, and the metal in their cars come from digging in the earth. But the simple fact is that we are not living in huts and driving oxcarts *because* we have learned how to exploit the earth. *Exploit* isn't always a bad word.

So we wrote this book for the beginning student in earth science, the recreational rock collector, the incurably curious amateur scientist. I was joined in this effort by my son, who writes and edits K-12 educational books and software. Together, we set out to write in an easy style, keeping jargon to a minimum and defining uncommon or technical terms when necessary. We used everyday examples with photos and charts to illustrate concepts. With 487 figures, 24 charts, and 21 tables, this is basically a picture book with some text.

We did not shrink from tackling complex (and controversial) topics such as evolution, nuclear waste, oil sands, frac'ing, and global warming. In our experience, students and the curious want to be challenged, and they are usually eager to engage tough topics. The demanding subjects round out and complete the book.

We hope you enjoy the book, and that it answers all the questions you have.

Acknowledgments

We would like to thank the many reviewers who took time to help make this book better, more accurate, and more complete. They include Dr. Seyi Fatoke (evolution), Don Hladiuk (climate change), Melissa Newton (climate change and Earth history), Dr. David Haddad (deep time, the changing face of the earth, evolution, paleontology and fossils, Earth history, natural hazards, hydrocarbons, and climate change), and Dr. Walter Sheets (all sections). Finally, thanks to Adelle and Rick Palmer for providing photos of Mount Vesuvius and Mount Etna, to Monika Falkenberg for the photo of oil-stained feet, and to Adam Prost for photos from the Grand Canyon, Sierra Nevada, and Canadian Rockies.

Authors

 **Gary L. Prost** (primary author) obtained a BSc in geology from Northern Arizona University in 1973 and an MSc (1975) and PhD (1986) in geology at Colorado School of Mines. Over the past 40 years, Dr. Prost has worked for Norandex (mineral exploration), Shell U.S.A. (petroleum exploration worldwide), the US Geological Survey (geologic mapping, coal), the Superior Oil Company (mineral and oil exploration), Amoco Production Company (worldwide oil exploration, remote sensing, and structural geology), Gulf Canada (international new ventures), and ConocoPhillips Canada (Canadian Arctic exploration, field development, oil sands modeling, and reservoir characterization). Dr. Prost spent more than 20 years working as an image analyst/photointerpreter in the search for oil and minerals in more than 30 countries. During this time, he applied structural geology and remote sensing to mineral and oil exploration and development projects, and to environmental monitoring. His most recent work has been in petroleum exploration and field development, in leading field classes, and in public outreach.

Dr. Prost is the principal geologist in G.L. Prost GeoConsulting, which leads field trips, teaches classes, and assists companies with exploration and development projects applying the most recent technologies.

Prost has published two books, *Remote Sensing for Geoscientists: Image Analysis and Integration* (3rd edition, Taylor & Francis 2013) and *English–Spanish and Spanish–English Glossary of Geoscience Terms* (Taylor & Francis, 1997). He is a registered professional geologist in Wyoming (USA).

 **Benjamin P. Prost** (coauthor) has had an interest in geology since he tried to eat a sample of native sulfur in his parents' backyard. He was 3 years old, and the yellow crystals were appetizingly orthorhombic. The son of a geologist, he could identify pyrite, galena, gypsum, and several other minerals before he knew the alphabet.

A bit later, he graduated from Occidental College with a degree in something other than geology (a decision he regrets). For the past 11 years, he has been a writer, editor, grammar maven, and style-guide aficionado in Los Angeles and New York. He currently writes and edits online children's books. He is valued for his ability to limn scientific topics in language that any first grader can understand—which is much harder than it sounds. He is routinely called upon by coworkers to explain the difference between *weathering* and *erosion*.

1

Introduction

Although I was four years at the University, I did not take the regular course of studies, but instead picked out what I thought would be most useful to me, particularly chemistry, which opened a new world, mathematics and physics, a little Greek and Latin, botany and geology. I was far from satisfied with what I had learned, and should have stayed longer.

John Muir
Naturalist

When you have eliminated the impossible, whatever remains, however improbable, must be the truth.

Arthur Conan Doyle
Author, in *The Sign of Four*

Geology Matters

Millennium Tower is a 58-story luxury high rise in San Francisco. It cost $350 million to build and contains units that start at $1.6 million and go up to over $10 million. The builders chose to sink support pilings 24 m (80 ft) into bay fill mud and dense sand rather than the 60 m (200 ft) needed to reach bedrock. Since the building was completed in 2009, it has sunk 38 cm (16 in) and is tilting 5 cm (2 in) to the northwest (Matier and Ross 2016). This is not just a concern to the 400 residents. Taxpayers have already spent $58 million building a buttress between the tower and the adjacent Transbay Transit Center that is under construction. The subsidence is not yet considered a safety issue, but residents are concerned.

In March 2011, a subduction-related earthquake off the coast of Japan triggered a tsunami that killed nearly 16,000 coastal dwellers and destroyed the nuclear power plant at Fukushima. As a consequence, this led to the shutdown of all nuclear plants in Japan and the phaseout of nuclear plants as far away as Germany.

Do you care if your house is built on landfill, a landslide, or in a floodplain? How would you feel upon learning that the dam up the valley sits on an active fault? If you live near sea level, should you prepare for a tsunami? Can a volcano in Iceland change your European travel plans? Where can you find an active volcano? Do you hunt minerals and fossils? Where can you find them? What do minerals and fossils reveal about Earth's past? Just how old is the earth, anyway? Where did the moon come from? What is plate tectonics? Do you find yourself debating climate change with friends and coworkers? Is global warming man-made, natural, or a hoax? What is fracking, and does it really cause earthquakes and pollute our water? What controls the price of gasoline? Are we running out of oil?

If you picked up this book, those may be a few of the questions you have. This book will answer them.

Reading the Rocks

Geology is about observing, measuring, comparing, and interpreting what you see. To do these things—to "read the rocks"—you have to think like a geologist. And that starts with an understanding of **the scientific method**.

Hypothesis and Theory

When earth scientists talk about "the scientific method," we mean something very specific. We are talking about the process of observing and measuring, then coming up with an explanation, or **hypothesis**, for what we observe. We then test the hypothesis, correcting it as needed. A hypothesis that stands up after being tested again and again is called a **theory**.

The scientist's relationship with the word *theory* often clashes with the nonscientist's. In common parlance, "theory" means "best guess," a fragile idea that could crumble at the first serious shock. Not so in science. In science, a theory is so good at explaining observations that it has stood up to rigorous testing, nit-picking, battering, bludgeoning—and it still works. As Richard Dawkins (1996) writes in *The Blind Watchmaker*,

> Our present beliefs about many things may be disproved, but we can with complete confidence make a list of certain facts that will never be disproved. Evolution and the heliocentric theory weren't always among them, but they are now.

We test a theory by making predictions. If the predictions come to pass, we say the theory is a good one. However, even if a theory passes the test of prediction hundreds of times, the theory must be revised—or discarded—if it fails just once. Thus, earth scientists must always be open to the possibility that a theory could be incomplete or even wrong.

A theory, then, must be *falsifiable*—that is, there must be, conceivably, a set of circumstances that will prove it false. To quote Albert Einstein, "No amount of experimentation can ever prove me right; a single experiment can prove me wrong."

If a theory is not falsifiable, it is not scientific.

The theory of plate tectonics (Chapter 9), for example, has passed tests to the point of being almost universally accepted. It has predicted earthquakes; it has enabled us to sketch the boundaries of drifting continents; it has predicted the direction of their movement; and it has predicted the age of sediments on the ocean floor.

However, if we ever discovered that the mantle does not in fact transfer heat in convection cells—we can't see the mantle directly, after all—we would have to rethink plate tectonics. We might determine that the earth is gradually expanding, bursting continents at the seams, and that's why the coastlines of South America and Africa fit so well. This was actually a theory at one time. In 1835, Charles Darwin applied the concept of an expanding Earth to account for uplift in the Andes. Between 1889 and 1909, Roberto Montovani published his ideas that a single continent once covered the entire Earth, and that thermal expansion broke the landmass into smaller continents separated by rift zones, where the oceans now exist. The "expanding Earth" theory has since been discarded for lack of evidence and was replaced by the theory of plate tectonics and continental drift (see Chapter 9).

Facts

When it comes to reading the rocks, geologists accept several basic facts, including the following:

- The earth is unfathomably old.
- Sediments are deposited in horizontal layers.
- Older sediments appear below younger sediments (unless the rock layers have been disturbed).
- Rock layers gradually change across space and time.
- Rock layers in one place may be equivalent to rock layers in another place.
- Life forms change gradually, and their fossils can be used both to date rocks and to **correlate** (link up) rock layers separated by great distances.
- Missing (eroded) rock layers indicate the passage of vast expanses of time.

We use these facts, along with our own observations, to interpret what we see. Good observations of the surface allow you to infer what lies below the surface.

To square our observations with the known facts, we can use several approaches. One is known as **Occam's razor**, propounded by William of Ockham (1287–1347), an English friar. If there is more than one hypothesis that explains an observation, the least complicated one is preferable. In other words, the simplest explanation is usually the best. The reason that the coastlines of Africa and South America match is that the continents were once joined and gradually drifted to their present positions. Simple explanations are easier to test and falsify.

Another approach, promoted by Thomas C. Chamberlain (1843–1928), is the **Method of Multiple Working Hypotheses**. In this approach, the person explaining an observation will formulate multiple explanations, test each, and discard those that do not fit the facts. The advantage is more objectivity—making it easier to accept new data that may change a preferred explanation. When we see a trail of rocks on the ground, we can assume that either they tumbled down a mountainside or that they were deposited by glaciers. After examining the context, for example, a nearby cirque, scratches on the rocks, and polished rock surfaces, it becomes evident that the rocks were left by a glacier.

Bear in mind that observations are prone to several kinds of error. Observers may think they see something that is not actually there. Or they may see something that is there but not identify it correctly. Or they may identify something correctly, but attribute the wrong significance to it. The best interpretations come from training and experience—from considering all the alternatives critically, in context, with the ability to filter out unimportant information.

Taking Notes

If you are going into the mountains, taking a break along the side of a road, or examining a property, it is important to take good notes. These will allow you to recall, years later, what you thought was important at the time and where you saw it. Using a field notebook or tablet, you want to record the day and time, map location, perhaps the stop or sample number or site name, and any interesting observations.

Map locations can be recorded automatically on a cell phone or tablet, marked on a map, or recorded in a notebook.

Notes and observations should include sketches and photographs (Figure 1.1). Sometimes, photographs don't show everything you know is there, around a corner or in a shadow, and sketches make it easier to show these features. Later, the sketches help you understand and label what you see on the photograph.

What kind of observations should you make? Note the type of rock or soil. Note evidence of rock deformation. Deformation includes downslope creep in soils, folding in layered rocks, and signs of faulting or jointing, that is, where the rock is broken or one side of a break is obviously offset from the other. Color and composition are important. For example, "red sandstone" or "light gray limestone." If you know the layer (formation) names, use them. Describe the inclination of the layers and any evidence of fossils. Note the minerals you see, any mineral veins, and rock "alteration" such as rust staining. Good notes help you remember what you saw and where you saw it.

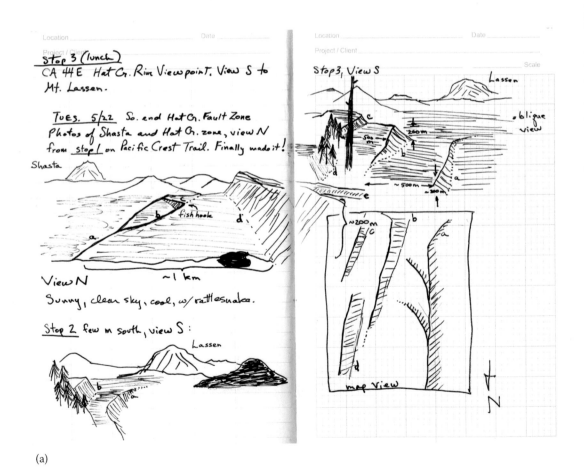

(a)

FIGURE 1.1
(a) Notebook page showing sketches of landscape and structures.   (*Continued*)

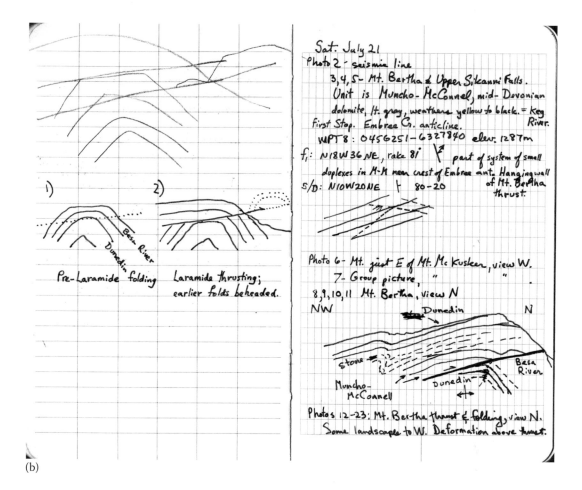

FIGURE 1.1 (CONTINUED)
(b) Notebook page with photograph references, structure sketches, and an attempt to reconstruct the events that shaped the area.

Tools of the Trade

Most geologists have a rock hammer, chisel, and safety goggles. The hammer is to break open rocks to see fresh surfaces or reveal fossils. Rock hammers come in a number of varieties (Figure 1.2). A heavier hammer (known as a single jack) is used to break larger rocks, and a sledgehammer is used to break boulders.

Besides goggles, other safety equipment includes helmets to protect you from falling rocks, safety vests to make yourself visible to oncoming traffic near road cuts, and basic first aid kits to bandage cuts or disinfect smashed thumbs.

Pickaxes are used with shovels to break up tough soil and dig into the ground.

Hand lenses (loupes, magnifying glasses) are used to help identify minerals in rock (Figure 1.3). The most common magnifications are 10× and 20×.

Muriatic acid (dilute hydrochloric acid, or 10% HCl) is used to identify carbonate minerals (limestone, dolomite) in the field. A drop of acid makes these rocks fizz.

Measuring tapes or rulers are used to measure distances or the size of objects.

(a)

(b)

FIGURE 1.2
Rock hammers. (a) Chisel-end hammers are used mostly with sedimentary rocks. (Courtesy of Wilson44691; https://commons.wikimedia.org/wiki/File:BrokenConcretion22.jpg.) (b) Pick hammers are used mainly on igneous and metamorphic rocks. (Courtesy of Alan Levine; https://commons.wikimedia.org/wiki/File: Cracked_Rocks_with_geodes.jpg.)

(Continued)

(c)

(d)

FIGURE 1.2 (CONTINUED)
Rock hammers. (c) Small sledgehammers (5 kg or 10 lb) are used with chisels or to break rocks. (Courtesy of Gomanifesto; https://commons.wikimedia.org/wiki/File:Feathers-wedges-hammer.jpg.) (d) Single-edge rock hammers are used to break large rocks. (Courtesy of FluffyBiscuit; https://commons.wikimedia.org/wiki/File:Ammonite-Fossil.jpg.)

FIGURE 1.3
Typical hand lens used by geologists. Magnification ranges from 10× to 30×. (Courtesy of Adamantios; https://commons.wikimedia.org/wiki/File:Loupe-triplet-30x-0a.jpg.)

Global positioning systems (GPS) and topographic maps (paper or electronic) help you get to and from remote locations and mark waypoints for future reference. It's now also possible to mark locations of interest on your smartphone by "dropping a pin" on Google Maps. Aerial photographs and satellite-image maps are useful, along with topographic maps, in providing a more complete picture of what is actually on the ground (Figure 1.4). Scale on these maps is given as the ratio of map distance to Earth distance. For example, if an inch on the map equals a mile on the ground, the scale is 1:63,360. The most detailed topographic maps are usually 1:24,000 (US) or 1:25,000 (international, where 1 m on the map equals 25 km on the ground). Google Earth allows you to "fly over" almost any spot on earth, viewing it from any direction, elevation, or inclination. This is a handy way to examine areas before visiting them.

Flagging tape is used to mark locations where you find samples or (as a trail of bread-crumbs) to mark your route into a remote or confusing area so you can find your way back out. Use a permanent marker to add notes.

Specimen bags are used to collect samples and label them with location and other useful information. Use a permanent marker to identify the contents and location.

Field notebooks and cameras are used to document your observations. These are slowly being replaced by laptops, tablets, and cell phone apps. However, notebooks don't need batteries, and they can get wet and still work.

Although generally not used by the nongeologist, a Brunton compass or compass cli-nometer is used to determine the orientation and inclination of rock layers and other features (Figure 1.5).

Let common sense be your guide in the field. Wear sturdy shoes with ankle support if you plan to be on steep or unstable slopes. Wear a hat and long sleeves for sun protection, and carry plenty of water when going to the desert or hot, humid tropics. Carry a satellite phone or GPS messenger like SPOT if you will be out of cell phone range. Let someone know where you are going and when you plan to be back.

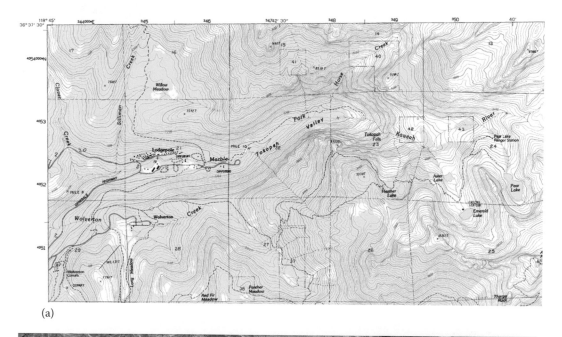

(a)

(b)

FIGURE 1.4

(a) Portion of USGS topographic map of the Lodgepole 7.5 Minute Quadrangle, California, 1993. Original map scale 1:24,000. **Topographic maps** provide information on surface elevation, streams, lakes, vegetation cover, roads, trails, and buildings. (Courtesy of the United States Department of the Interior, USGS.) (b) Google Earth satellite image of a fold northwest of Thermopolis, Wyoming. Imagery is used as maps to locate yourself, find access, trace geologic layers or formations, and map structures. (Courtesy of Google Earth.)

(a)

(b)

FIGURE 1.5
Brunton compasses. (a) Standard compass or pocket transit. (Courtesy of G. Prost.) (b) Silva Voyager compass.
(https://en.wikipedia.org/wiki/Brunton,_Inc.#/media/File:Brunton.JPG.)

"Rules of the Road" for Collecting

1. Do not collect from national parks and monuments.
2. Get permission from landowners to enter their land and collect.
3. Leave gates as you found them.
4. Leave no trash.
5. Leave no graffiti.
6. Don't endanger others (throwing rocks off cliffs, climbing above others and rolling loose rocks down a slope).
7. Record where and when you collected a sample and what it is.
8. Label samples so they can be correlated to entries in your notebook.

Why This Book?

This is a practical book for those who want to understand the earth, but don't need to become experts. If you are reading this, you are likely someone with little or no experience in geology or science, but with ample interest and curiosity. This book provides a quick, basic grounding in geological science. It explains how to apply that science in a practical way. Then, you can decide whether earthquakes or landslides or floods are worth worrying about. You will be able to determine, by observing road cuts, whether you are standing on an ancient seafloor, in a coal swamp, or next to a sand dune. You will be able to identify geologic hazards in your neighborhood, find places to look for fossils or go caving, or locate the best place to drill a water well for your vacation home. You will come to think of this book as a companion, a friendly guide to understanding the earth and geologic processes and how they affect people in their daily lives.

References

Dawkins, R. 1996. *The Blind Watchmaker*. Turtleback Books, St Louis, 400 pp.
Matier, P. and A. Ross. 2016. SF's landmark tower for rich and famous is sinking and tilting. August 1, 2016. Accessed online, August 5, 2016. http://www.sfchronicle.com/bayarea/article/SF-s -land mark-tower-for-rich-and-famous-is-8896563.php

2

Minerals

In a crystal we have the clear evidence of the existence of a formative life-principle, and though we cannot understand the life of a crystal, it is none the less a living being.

Nikola Tesla
Inventor

Diamond is just a piece of charcoal that handled stress exceptionally well.

Anonymous

The word *mineral* comes from the Latin *minera*, or mine. Until the nineteenth century, minerals were simply things that came out of mines—things that had economic value, like gold, various ores, and coal.

The modern definition is more specific. A mineral is a *homogeneous, naturally occurring inorganic substance with an orderly atomic structure*. Let's break that down:

Homogeneous means the substance is chemically distinct, uniform, and consistent. Every mineral can be expressed as a chemical formula—SiO_2 (quartz), HgS (cinnabar), $Be_3Al_2Si_6O_{18}$ (beryl), and so on. Some minerals are **native** (single) elements, such as gold, platinum, and copper. Most, however, are made up of **compounds**—two or more elements that are chemically joined.

Naturally occurring means the substance is the result of natural processes, such as magma cooling or a lake evaporating. Man-made compounds may resemble minerals in chemistry and other aspects, but a true mineral by definition must exist somewhere in nature.

Inorganic means the substance does not derive from the remains of life. Graphite and diamond, two forms of pure carbon, are minerals, but coal, a high-carbon organic rock, is not. Amber, which is fossilized tree resin, is not a mineral.

Orderly atomic structure means the substance forms crystals. As minerals precipitate out of a solution—surface water, groundwater, or magma—their atoms arrange themselves geometrically, forming crystals. Mercury, which exists in liquid form above $-39°C$, will form crystals below that temperature; mercury is therefore a mineral. Similarly, ice is a mineral, although water is not.

Every mineral has distinct physical properties, which are detailed below and in Table 2.1. It usually takes the combination of properties to identify a mineral (Hurlbut 1971; Klein and Dutrow 2007). On the other hand, if you've seen enough minerals, you can tell not only the mineral but even where it is from. Amethyst in a greenish volcanic matrix is almost always from Minas Gerais, Brazil; large crystals of rhodochrosite are almost always from the Sweet Home Mine near Alma, Colorado; the best dioptase crystals are from Tsumeb,

TABLE 2.1

Physical Properties of the Mineral Gold. Diagnostic
Properties Include Color, Hardness, and Specific Gravity

Physical Property	Description
Chemical classification	Native element
Color	Golden yellow
Streak	Golden yellow
Luster	Metallic
Transparency	Opaque
Cleavage	None
Mohs hardness	2.5 to 3.0
Specific gravity	19.3
Chemical composition	Gold, Au
Crystal system	Isometric

Namibia; barite roses come from the Garber Sandstone near Noble, Oklahoma; and classic quartz crystal clusters are from Hot Springs and Mount Ida, Arkansas (Figure 2.1).

All about Crystals

The outward appearance of a crystal always reflects its internal structure. The three-dimensional arrangement of atoms in a mineral is called a **crystal lattice**. A salt crystal, for example, has a distinct cube shape because the crystal lattice has that shape (Figure 2.2). Minerals don't have to be crystalline, but they must be capable of being crystals. A quartz crystal, for example, has a distinct hexagonal shape because the crystal lattice has that shape. Chalcedony (agate) and chert (flint) are also quartz, but their crystals are **cryptocrystalline**—so small they cannot be seen, even with an ordinary microscope. Yet, even these crystals retain the hexagonal character of quartz.

Minerals fall into one of six **crystal systems** (Figure 2.3). All the crystals in a system have the same symmetry. Here, **symmetry** refers to the arrangement of atoms or molecules about an axis (an imaginary line through the middle of a crystal). If you cut the crystal with a plane through one of the axes, the shape on one side would be a reflection of the shape on the other side. What makes each crystal system unique is, first, its total number of axes; second, the length of each axis relative to the others; and third, the angles between the axes.

These are the six crystal systems:

1. **Isometric (or cubic):** There are three axes, which are all the same length and perpendicular.
2. **Tetragonal:** There are three axes; the two horizontal ones are the same length, and the vertical axis is either longer or shorter; all three axes are perpendicular.
3. **Hexagonal:** There are four axes; three are the same length and meet at 120° angles; a fourth axis is perpendicular to the others.

(a)

(b)

FIGURE 2.1
(a) Amethyst from Minas Gerais, Brazil. Two Euro coin for scale. (Courtesy of G. Prost.) (b) Rhodocrosite from Alma, Colorado. (Courtesy of Eric Hunt; https://commons.wikimedia.org/wiki/File:The_Searchlight _Rhodochrosite_Crystal.jpg.) *(Continued)*

(c)

(d)

FIGURE 2.1 (CONTINUED)
(c) Dioptase from Tsumeb, Namibia. (Courtesy of Didier Descouens; https://commons.wikimedia.org/wiki/File:Dioptasetsumeb5.jpg.) (d) Quartz from Hot Springs, Arkansas. Two Euro coin for scale. (Courtesy of G. Prost.)

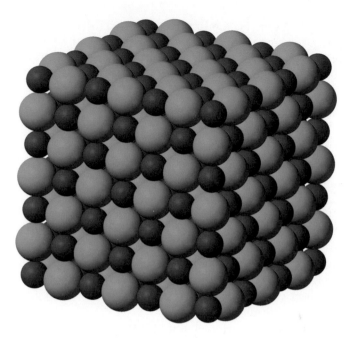

FIGURE 2.2
Model of a sodium chloride (salt) crystal lattice. Sodium atoms are purple; chlorine atoms are green. Notice that the geometry of the atoms influences the resulting crystal faces. (Courtesy of Benjah-bmm27; https://en.wikipedia.org/wiki/Lattice_energy.)

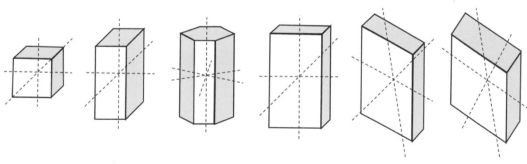

Cubic (isometric) system 3 axes at right angles; all the same length.

Tetragonal system 3 axes at right angles; 2 are the same length.

Hexagonal system 4 axes; 3 are the same length and at 120° to each other and at right angles to the 4th axis.

Orthorhombic system 3 axes at right angles and unequal lengths.

Monoclinic system 3 axes, 2 at right angles to each other and a 3rd not; all unequal lengths.

Triclinic system 3 axes, none at right angles to each other and all having unequal lengths.

FIGURE 2.3
The six crystal systems. Dashed lines are crystallographic axes.

The hexagonal system is subdivided into two "divisions"—hexagonal and rhombohedral (or trigonal). Crystals in the hexagonal division tend to form hexagonal prisms and pyramids. Crystals in the rhombohedral division tend to form triangular prisms and rhombohedrons (cube-like shapes whose faces are rhombs).

4. **Orthorhombic:** There are three perpendicular axes, as in the isometric system, but each axis is a different length.

5. **Monoclinic:** There are three axes, each a different length; two axes are perpendicular, and the third one intersects them at an oblique (nonperpendicular) angle.

6. **Triclinic:** There are three axes, each a different length, and each one intersects the other two at an oblique angle.

Crystal "**form**" describes the faces of the crystal. Common forms include prisms, pyramids, trapezohedrons, scalenohedrons, rhombohedrons, and tetrahedrons (Figure 2.4). Some minerals, such as quartz, can have more than one crystal form (called **polymorphs**) depending on the temperature and pressure of formation. Quartz is in the hexagonal crystal system and can form either hexagonal- or rhombohedral-shaped crystals. Likewise, calcite can be hexagonal or rhombohedral (Figure 2.5).

Crystal "**habit**" is the characteristic shape of a mineral crystal (Figure 2.6). Whereas *crystal form* describes the 3-D geometry of a crystal's faces, **crystal habit** describes the overall appearance of a mineral. It's like a working description based on the way a crystal habitually appears. Common crystal habits include prismatic, bladed, columnar, fibrous, acicular (needle-like), botryoidal (shaped like a cluster of grapes), platy or lenticular, druse (a coating of crystals), nodular, granular, and massive. Crystal habit is rarely diagnostic because one mineral can have several habits. Quartz, for example, can be columnar, prismatic, or druse; malachite can be botryoidal or massive.

  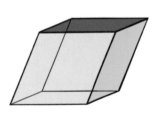 

Trapezohedron (sides made of staggered triangular facets)

Scalenohedron (faces consists of paired congruent scalene triangles)

Rhombohedron (sheared cube; faces are rhombs)

Tetrahedron: 3D shape made of four triangular faces

FIGURE 2.4
Common crystal forms.

(a)

(b)

FIGURE 2.5

(a) Hexagonal scalenohedral calcite. (Courtesy of Vassil.) (b) Red trigonal rhombohedral calcite on hematite-coated scalenohedral calcite. (Courtesy of Rob Lavinsky; https://en.wikipedia.org/wiki/Calcite.)

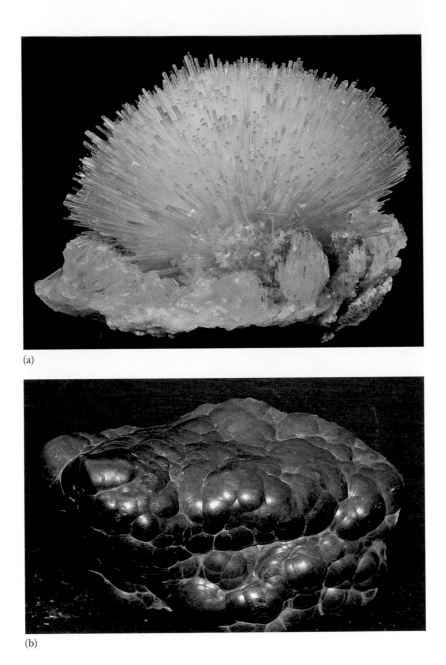

(a)

(b)

FIGURE 2.6
Various crystal habits. (a) Acicular habit, natrolite (http://www.wikiwand.com/en/Acicular_(crystal_habit)).
(b) Botryoidal habit, hematite. (Courtesy of Chris Ralph/Chris Reno; https://en.wikipedia.org/wiki/Botryoidal.)
(Continued)

(c)

(d)

FIGURE 2.6 (CONTINUED)
Various crystal habits. (c) Cubic habit, halite. (Courtesy of D. Descouens; https://en.wikipedia.org/wiki/Halite.)
(d) Dodecahedral habit, garnet (https://en.wikipedia.org/wiki/Garnet.) *(Continued)*

(e)

(f)

FIGURE 2.6 (CONTINUED)

Various crystal habits. (e) Druse habit, quartz coating on chrysocola. (Courtesy of Rob Lavinsky; https://commons
.wikimedia.org/wiki/File:Chrysocolla-Quartz-247919.jpg.) (f) Micaceous habit, lepidolite. (Courtesy of Rob
Lavinsky; https://en.wikipedia.org/wiki/Crystal_habit.)   (*Continued*)

(g)

(h)

FIGURE 2.6 (CONTINUED)
Various crystal habits. (g) Platy habit, wulfenite. (Courtesy of Eric Hunt; https://en.wikipedia.org/wiki/Crystal
_habit.) (h) Prismatic habit, tourmaline. (Courtesy of Chris Ralph/Chris Reno; https://en.wikipedia.org/wiki
/Crystal_habit.)

Mineral Properties

Color

Color may be the most obvious aspect of a mineral, but it can be misleading because several minerals can have the same color, and one mineral can have several colors. In addition to color in sunlight, some minerals display **luminescence**: they emit light of a different color. These minerals **fluoresce**, or glow various colors under ultraviolet light (e.g., fluorite, calcite, scheelite). Some specimens of a mineral may fluoresce whereas other specimens of the same mineral do not.

Streak

Streak is the color of a powdered mineral. To determine streak, one scratches a white porcelain plate with the mineral. The streak is frequently different from the color of the crystalline mineral.

Luster

Luster refers to the way a mineral's surface reflects light. There are two main types of luster, metallic and nonmetallic. Pyrite, gold, copper, and galena have a metallic luster. Nonmetallic luster is subdivided into adamantine (diamond, corundum), vitreous or glassy (quartz), pearly or waxy (biotite, quartz), silky or fibrous (asbestos), greasy (chrysocolla, smithsonite), and dull or earthy (kaolinite). Luster can be quite subjective, with different categories grading into one another. Some minerals, such as quartz, can also exhibit more than one type of luster.

Transparency

Transparency refers to the amount of light that can pass through a mineral. Some minerals are transparent (diamond, quartz), others are translucent (calcite, quartz, muscovite mica), and yet others are opaque (feldspar, biotite mica, quartz). Transparency alone is not diagnostic because the same mineral can fall into more than one category.

Cleavage

Cleavage is the tendency of a mineral to break along the planes of weakness in its crystal structure. These weaknesses are parallel to layers of atoms in the crystal that form flat, repeating surfaces. The bonds between atoms in a layer are stronger than the bonds between the layers. There are six kinds of cleavage (Figure 2.7):

- *Basal* (or pinacoidal), which is parallel to the base of a crystal (mica)
- *Cubic*, where a mineral cleaves parallel to the faces of a cube (halite)
- *Octahedral*, where a mineral will cleave parallel to one of eight crystal faces (diamond, fluorite)
- *Dodecahedral*, where a mineral will cleave parallel to 1 of 12 crystal faces (sphalerite)
- *Rhombohedral*, where a mineral cleaves parallel to the faces of a rhombohedron (calcite)
- *Prismatic*, where a mineral cleaves parallel to a vertical prism (lazulite, zircon)

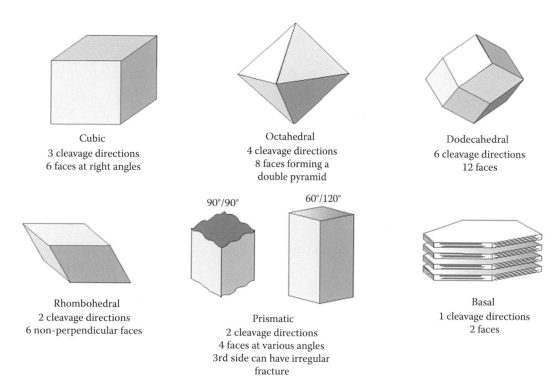

Cubic
3 cleavage directions
6 faces at right angles

Octahedral
4 cleavage directions
8 faces forming a
double pyramid

Dodecahedral
6 cleavage directions
12 faces

Rhombohedral
2 cleavage directions
6 non-perpendicular faces

Prismatic
2 cleavage directions
4 faces at various angles
3rd side can have irregular
fracture

Basal
1 cleavage directions
2 faces

FIGURE 2.7
Mineral cleavage systems.

Cleavage occurs in all minerals to some degree. Micas have well-developed cleavage that results in smooth, thin sheets. Quartz, on the other hand, exhibits cleavage only when the crystal is heated and then plunged into cold water.

Fracture

Fracture refers to breakage that does not occur along cleavage planes. In some minerals, the bonds between atoms are equally strong throughout the crystal. When these minerals break, you get an irregular surface, or fracture. The geometry of this surface can help you identify the mineral. Fractures can be conchoidal ("clam-shell," as seen in flint and obsidian), hackly or jagged (as seen in rough native metals like copper), fibrous or splintery (as seen in asbestos), uneven (a rough, irregular surface), and earthy (Figure 2.8).

Hardness

Hardness is the resistance of a mineral's surface to scratching. The most common way to gauge hardness is by using the Mohs scale, which ranges from 1 to 10. The Mohs scale is strictly relative. For example, calcite (with a Mohs value of 3) is harder than gypsum (with a Mohs value of 2), but you can't say that calcite is *twice as hard* as gypsum. Hardness refers only to a mineral's resistance to scratching—not its resistance to breaking. Diamond, which has the highest Mohs value, will still shatter if you smash it with a hammer (Table 2.2).

(a)

(b)

FIGURE 2.8
Common types of fracture. (a) Conchoidal fracture in obsidian. (Courtesy of Denis Aline; https://en.wiktionary
.org/wiki/conchoidal.) (b) Earthy fracture in limonite. (Courtesy of Tomasz Kuran aka Meteor2017; https://
en.wikipedia.org/wiki/Fracture_(mineralogy).)

TABLE 2.2

Mohs Hardness Scale

Mineral	Mohs Relative Hardness	Scratch Test
Talc	1	Scratch with fingernail
Gypsum	2	Scratch with fingernail
Calcite	3	Scratch with copper penny
Fluorite	4	Scratch with knife
Apatite	5	Scratch with knife
Orthoclase	6	Scratch with steel file
Quartz	7	Scratch with glass
Topaz	8	Will scratch quartz
Corundum	9	Will scratch topaz
Diamond	10	Will scratch corundum

Specific Gravity

Specific gravity is a measure of mineral density. It is the mineral weight relative to water, which has a specific gravity of 1.0 at 4°C. If a mineral such as quartz has a specific gravity of 2.6, then it is 2.6 times the weight of an equal volume of water (Table 2.3).

Mineral Identification

So how does mineral identification work? Let's say we find the mineral shown in Figure 2.9. The first thing we notice is the color, which we could describe as "golden." The luster is metallic. So far, it could be gold. Next, let's describe the shapes we see. The crystals belong to the isometric system and have a striated, cubic habit. Gold also belongs to the isometric system, although it's usually found as rounded nuggets.

The truly diagnostic features turn out to be hardness, density, and streak. Gold is much softer than most minerals (with a Mohs value of 2.5). This mineral has a hardness of 6.5. It has a specific gravity of about 5, whereas gold is much heavier—close to 19. Finally, we test the streak and see that this mineral has a greenish-black streak; gold is always yellow. What we've found is the mineral pyrite, also called "fool's gold" because, at first glance, it does look a lot like gold.

TABLE 2.3

Mineral Examples of Varying Specific Gravities

Specific Gravity	Relative Specific Gravity	Mineral Examples
Less than 2	Very light	Borax
2 to 2.5	Light	Gypsum, halite
2.5 to 3	Average	Quartz, calcite, feldspar, talc
3 to 4	Above average	Biotite mica
5 to 10	Heavy	Galena, hematite, magnetite
10 to 20	Very heavy	Gold, silver, uraninite
Greater than 20	Extremely heavy	Platinum, irridium

FIGURE 2.9
An unknown mineral that we would like to identify. Streak is shown. (Courtesy of Ra'ike; https://en.wikipedia
.org/wiki/Streak_(mineralogy).)

Special Powers of Crystals

Do crystals have special powers? Yes! Crystals possess certain special properties, such
as the ability to vibrate at a constant frequency, which make them useful in technology.
Quartz crystals, for example, are used to keep time in watches. When doped with special
impurities, crystals can also act as transistors in radios. Some crystals produce an electrical
charge when compressed, making them useful in precision instruments. Crystals are also
aesthetically pleasing (in other words, pretty).

Can crystal power be harnessed for emotional healing, channeling positive energy, balancing internal yin and yang, seeing the future, or warding off harmful rays? There is no
evidence to support any such claims. However, in some places, crystals do seem to have
the strange and potent power to generate New Age tourism.

Common Minerals

Most minerals you see while walking around an outcrop or a road cut reflect the local
geology. The Sierra Nevada in California, which is primarily granitic, has an abundance
of quartz, plagioclase feldspar, and micas. On the other hand, the granite at Pikes Peak,
Colorado, also contains pink-to-orange orthoclase, a feldspar more common to the interior
of the continent (Figure 2.10).

In the dark volcanic rocks of Hawaii and Iceland, you will see olivine, amphiboles, and
pyroxenes in addition to quartz and feldspars.

Areas that have been subjected to intense heat or pressure, like the Sierra Nevada foothills or the Tibetan Plateau, commonly contain garnets, serpentine, talc, chlorite, epidote,
actinolite, kyanite, staurolite, and andalusite (Figure 2.11). Less common (but more excit-
ing) minerals found in such places are corundum (rubies and sapphires) and jade.

(a)

(b)

FIGURE 2.10
Some common minerals. (a) Orthoclase feldspar. (Copyright 2016, Roger Weller.) (b) Plagioclase feldspar (large crystal and white mineral in rock). (Courtesy of Daniel Mayer; https://commons.wikimedia.org/wiki /File:Plagioclase_feldspar_phenocryst_in_Lambert_Dome-750px.jpg.) *(Continued)*

(c)

(d)

FIGURE 2.10 (CONTINUED)
Some common minerals. (c) Muscovite mica. Lincoln County, North Carolina. (Courtesy of Rob Lavinsky/iRocks
.com; https://commons.wikimedia.org/wiki/File:Muscovite-4aa51a.jpg.) (d) Biotite mica. Erongo Region, Namibia.
(Courtesy of Rob Lavinsky/iRocks.com; https://commons.wikimedia.org/wiki/File:Biotite-Orthoclase-229808.jpg.)
(Continued)

(e)

(f)

FIGURE 2.10 (CONTINUED)
Some common minerals. (e) Amphibole (Hornblende, black mineral) in plagioclase. (Copyright 2016, Roger Weller.) (f) Black pyroxene (augite) crystal. (Courtesy of Dave Tucker; https://nwgeology.wordpress.com.)

(Continued)

(g)

FIGURE 2.10 (CONTINUED)
Some common minerals. (g) Olivine (green mineral) in basalt, San Carlos Reservation, Arizona. (Courtesy of Vsmith; https://en.wikipedia.org/wiki/Olivine.)

FIGURE 2.11
Common metamorphic minerals. (Courtesy of Geology of Gems, Geologycafe.com, Roland Scal, Phil Stoffer, and Vazgen Shekoya; http://geologycafe.com/gems/chapter9.html.)

Cliffs in the Canadian Rockies are mainly limestone, so calcite, dolomite, aragonite, and siderite are common minerals. Limestone is identified by the way the calcite in it fizzes when dilute hydrochloric acid is dripped on it. Dolomite will fizz as well, but only after it has been scratched. West of Salt Lake City, you pass through the dried bed of the Great Salt Lake. The white minerals covering the ground are halite (salt) and anhydrite (Figure 2.12).

(a)

(b)

FIGURE 2.12

Some common sedimentary minerals. (a) Dolomite crystals. (Courtesy of Didier Descouens; https://en.wikipedia .org/wiki/Dolomite.) (b) Salt flats west of the Great Salt Lake, Utah. (Courtesy of G. Prost.)   (*Continued*)

(c)

(d)

FIGURE 2.12 (CONTINUED)
Some common sedimentary minerals. (c) Deformed halite (salt), Cardona salt dome, Catalonia, Spain. (Courtesy of G. Prost.) (d) Anhydrite (gypsum) twinned crystal, Green River Basin, Wyoming. (Courtesy of G. Prost.)

In contrast to rocks that are formed more or less in place, sandstone and shale reflect the geology of their source, or **provenance**. The reason is simply that these rocks are made up of grains that were transported. The white cliffs in Zion National Park in Utah, for example, contain dune sandstone of the Jurassic Navajo Formation. The Navajo Formation contains quartz, feldspar, zircon, and other minerals derived from an original source rock that may have been as far away as the Appalachian Mountains in the eastern United States. Most sandstone and shale are made up of combinations of quartz, feldspar, other rock fragments, and sometimes shell fragments.

Occasionally, sandstone will contain uncommon minerals. The winnowing of beach sand by wave action leaves smears of black sand, the accumulation of heavy minerals like magnetite. Rarely, beach sandstones contain gold, diamonds, ilmenite, and garnet. These same heavy-mineral deposits can accumulate in river channel sandstones.

Mineral Collecting: Knowing Where to Look

Minerals are intimately related to local geology. Knowing the processes that formed an area means knowing which minerals you are likely to find there.

Many fine mineral and crystal specimens come from rock cavities where they were able to grow uninterrupted. These cavities can be a gap along a fault or fracture in granitic, volcanic, or metamorphic rocks. Cavities can also form in limestone and get filled with crystals, forming geodes (Figure 2.13). Gas pockets in lava can also produce geodes.

Mine dumps have especially good mineral pickings. Consider requesting a tour of an active mine.

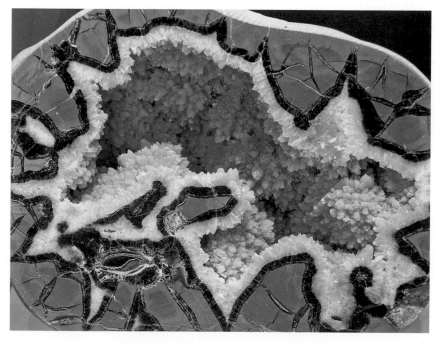

FIGURE 2.13
Calcite crystals in a septarian nodule (geode), San Juan Co., Utah. (Courtesy of Royal Ontario Museum Toronto; https://commons.wikimedia.org/wiki/File:Calcite_septarian_nodule_ROM.jpg.)

A Sampling of Mineral Collecting Field Guides

Mineral and fossil collecting guides tell you where others have found minerals. Some examples include the following:

Crow, M. 1994. *Rockhound's Guide to Texas* (and *California, Colorado, Montana, New Mexico,* and *Washington*), Falcon Press Publishing Co., 166 pp.

Dake, H. C. 1962. *Northwest Gem Trails,* Mineralogist Publishing Co., 95 pp.

Johnson, C. 1992. *Western Gem Hunters Atlas,* Cy Johnson & Son, 96 pp.

Johnson, K., and R. Troll. 2007. *Cruisin' the Fossil Freeway,* Fulcrum Publishing, 202 pp.

Johnson, L. 2014. *Rockhounding Oregon* (also *California, Colorado, Idaho, Montana, Nevada, Pennsylvania, New Jersey, Utah,* and *Wyoming*), Falcon Guides.

Mitchell, J. R. 1997. *Gem Trails of Colorado,* Gem Guides Book Co., 144 pp.

National Audubon Society. 1979. *Field Guide to Rocks and Minerals (North America),* Knopf, 856 pp.

Pough, F. 1998. *A Field Guide to Rocks and Minerals,* Houghton Mifflin Co., 349 pp.

Ransom, J. 1964. *A Range Guide to Mines and Minerals,* Harper & Row.

Romaine, G. 2009. *Gem Trails of Oregon* (also *Arizona, California, Idaho, Nevada, New Mexico, Texas, Utah,* and *Washington*), Gem Guides Book Co.

Local gem and mineral clubs and societies are also excellent sources of information on collecting locations. You can find them in a phone book or on the Internet.

References

Hurlbut, C. S. 1971. *Dana's Manual of Mineralogy,* 18th ed. John Wiley & Sons, New York, 579 pp.
Klein, C. and B. Dutrow. 2007. *Manual of Mineral Science,* 23rd ed. John Wiley & Sons, New York, 716 pp.

3

Rocks

Rocks and minerals: the oldest storytellers.

A.D. Posey
Author

Geologists have a saying: "Rocks Remember."

Neil Armstrong
Astronaut

Many people encounter rocks right in their homes. Places that sell natural stone counters call them all "granite countertops." They are not, and the difference is important. A marble countertop looks beautiful, but consists mainly of the mineral calcite, which has a Mohs hardness of 3. Granite contains quartz and feldspar, which have a hardness of 7 and 6, respectively. True granite will withstand the wear and tear of a kitchen; marble won't.

Just as elements and compounds are the building blocks of minerals, minerals are the building blocks of rocks. Some rocks consist of just one kind of mineral. Most rocks are assemblages of different minerals that have been melted, compacted, or cemented together into hard, solid materials.

The cement that holds rock together is usually precipitated from groundwater, which carries the minerals in solution owing to high temperature or pressure. Groundwater moves through the pores between mineral grains and leaves cementing minerals behind, much as tap water leaves mineral deposits on your sink and faucet. Common **mineral cements** include calcium carbonate (calcite), silica (quartz), iron oxide (limonite), anhydrite (gypsum), barite, and clays.

Rocks are divided into three main groups based on their origin: sedimentary (made up of sediments), igneous ("from fire"), and metamorphic (changed in form).

Sedimentary Rock

Sedimentary rocks form when sediments are deposited by wind or water and settle into layers. Sedimentary rocks that consist of grains are called **clastic** and include shale, siltstone, sandstone, pebble conglomerate, cobble conglomerate, and boulder conglomerate (Figures 3.1 and 3.2). Deposition and accumulation occur along beaches (both above and below the water line), in river channels and deltas, in estuaries, on river floodplains, in swamps and lakes, in deserts as sand dunes, and as valley fill deposits. The only requirement is that space is available for the sediment to accumulate.

If clastic rocks have a volcanic origin, they are **pyroclastic**. Volcanic **agglomerates** contain large, angular volcanic fragments spewed from vents, and **volcanic ash** is fine-grained

FIGURE 3.1
Dark-colored Cretaceous Mancos Shale on the lower slopes of the Book Cliffs west of Green River, Utah. Erosion-resistant sandstone forms the top of the cliff. (Courtesy of G. Prost.)

(a)

FIGURE 3.2
Some common sedimentary rocks. (a) Siltstone. (Courtesy of NASA; http://mars.nasa.gov/mer/classroom/schoolhouse/rocklibrary/source/siltstone.html.)   (*Continued*)

(b)

(c)

FIGURE 3.2 (CONTINUED)
Some common sedimentary rocks. (b) Sandstone, Antelope Canyon, Arizona. (Courtesy of Meckimac; https://commons.wikimedia.org/wiki/File:Lower_Antelope_Canyon_478.jpg.) (c) Conglomerate. (Courtesy of G. Prost.)

(*Continued*)

(d)

FIGURE 3.2 (CONTINUED)
Some common sedimentary rocks. (d) Clastic limestone, Ordovician Kope Formation near Cincinnati, Ohio. (Courtesy of Jim Stuby; https://commons.wikimedia.org/wiki/File:Limestone_etched_section_KopeFm_new.jpg.)
(Continued)

volcanic material that settles out of the air; if compacted while still hot, it melts together to form a **tuff**. Sedimentary rocks that consist mainly of shell fragments are called **coquina**.

Another type of sedimentary rock is the **chemical precipitate**. This includes **evaporites** (salt and gypsum formed by evaporation of mineral-saturated surface water), limestone and travertine (calcium carbonate that precipitates from solution in caves or hot springs; Figure 3.3), and chert (silica nodules precipitated from groundwater in limestone). Oolite shoals consist of calcite grains that precipitate in warm, calcium carbonate-saturated seawater and thus are both precipitates and clastic sediments (Figure 3.4).

Sedimentary rocks are classified on the basis of grain size, composition, and texture (Table 3.1).

Organic sediments, such as **coal** beds, form when plant material is deposited in swamps or bogs and then gets buried, dewatered, compacted, and heated to varying degrees (Figure 3.5). Organic phosphate deposits result from the accumulation of guano, mainly of seafowl and, to a lesser extent, bats. These deposits lie near coasts or on islands and may weigh up to several hundred thousand tons. **Phosphorite**, or phosphate rock, has extensive

(e)

(f)

FIGURE 3.2 (CONTINUED)
Some common sedimentary rocks. (e) Air fall tuff. (Courtesy of Maureen Feineman; https://www.e-education
.psu.edu/geosc30/node/737.) (f) Coquina, Fort Pierce, Florida. (Courtesy of James St. John; https://commons
.wikimedia.org/wiki/File:Coquinoid_quartzose_sandstone_(Anastasia_Formation,_Upper_Pleistocene_to
_lower_Holocene,_126_to_8_ka;_Indrio_Pit,_northern_side_of_the_town_of_Fort_Pierce,_southeastern
_Florida,_USA)_(15207209416).jpg.)

FIGURE 3.3
Calcite in the form of travertine, Mammoth Hot Springs, Yellowstone National Park, Wyoming. (Courtesy of I. Reid; https://en.wikipedia.org/wiki/Mammoth_Hot_Springs#/media/File:Mammoth_Hot_Springs_-_Terracing_-_August_2011.JPG.)

sedimentary layers containing at least 15% phosphate minerals (Figure 3.6). The phosphate is derived from animal bones and teeth, or from igneous mineral veins. Dissolved phosphate moves through pore spaces between sedimentary grains as part of the groundwater system before it is deposited, usually in a sandstone or limestone.

Igneous Rock

Igneous rocks cool and solidify from molten magma. If they cool underground, they are called **intrusive** or **plutonic**. If they erupt and cool at the surface, they are called **extrusive** or **volcanic**.

Igneous rocks are classified based on texture and mineralogy (Figure 3.7). Light-colored igneous rocks that consist largely of quartz and feldspar are called **felsic** (a splicing of the words *feldspar* and *silica*). Dark igneous rocks that contain iron- and magnesium-rich minerals are called **mafic** (from the words *magnesium* and *ferric*).

Volcanic (Extrusive) Rocks

Because volcanic rocks erupt as lava and cool quickly at the surface, the minerals in them do not have time to form large crystals. Most minerals in volcanic rocks are too small to see. Silica-rich lava produces light-colored rock called **rhyolite**. Rhyolite lava erupts at about 800°C and consists mainly of quartz and sodium-rich plagioclase feldspar. This silica- and gas-rich lava is viscous (sticky) and tends to erupt explosively. Because it doesn't

(a)

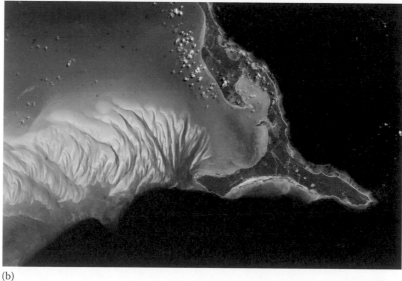

(b)

FIGURE 3.4

(a) Oolitic limestone grains precipitate from warm calcium carbonate-saturated seawater, Itaborai, Brazil. (Courtesy of Eurico Zimbres, FGEL/UERJ; https://commons.wikimedia.org/wiki/File:Calcario1Ez.jpg.) (b) Satellite image of oolite shoals (light blue) west of Eleuthera Island, Bahamas. (Courtesy of NASA.)

TABLE 3.1

Sedimentary Rock Classification Chart

	Texture	Grain Size	Composition	Comments	Rock Name
Inorganic, land-derived sedimentary rocks	Clastic (grains and rock fragments)	Pebbles, cobbles, or boulders in sand, silt, or clay	Quartz, feldspar, clay minerals, rock fragments	Rounded fragments	Conglomerate
		Sand (0.6 to 20 mm)		Angular fragments	Breccia
		Silt (0.04 to 0.6 mm)		Fine to coarse	Sandstone
		Clay (smaller than 0.04 mm)		Fine grained	Siltstone
				Very fine	Shale or mudstone
Transitional	Precipitate clastic	Sand (0.6 to 20 mm)	Calcite	Warm tropical marine setting	Oolitic sandstone
Chemically or organically formed sedimentary rocks	Crystalline precipitate	Varied	Halite	Crystals from evaporites	Halite (salt)
		Varied	Gypsum		Gypsum or anhydrite
		Varied	Calcite	Hot spring and cave deposits	Travertine
		Varied	Dolomite		Dolomite
	Bioclastic (organic fragments)	Microscopic to coarse	Calcite	Cemented shell fragments or biologic precipitates	Limestone, chalk, coquina
		Varied	Carbon	Plant remains	Coal
		Sand (0.6 to 20 mm)	Phosphate minerals in sand	Guano, bones, or hydrothermal	Phosphorite
Volcanically derived sedimentary rock	Pyroclastic	Varied	Varied	Fine ash to large bombs	Tuff
		Sand and smaller	Varied	Fine material	Ash
		Sand and larger	Varied	Coarse material	Agglomerate

FIGURE 3.5
Pennsylvanian coal at Point Aconi, Nova Scotia. (Courtesy of Michael C. Rygel via Wikimedia Commons; https://en.wikipedia.org/wiki/Coal.)

FIGURE 3.6
Phosphorite mine near Oron, Negev, Israel. (Courtesy of Wilson44691; https://commons.wikimedia.org/wiki /File:Phosphorite_Mine_Oron_Israel_070313.jpg.)

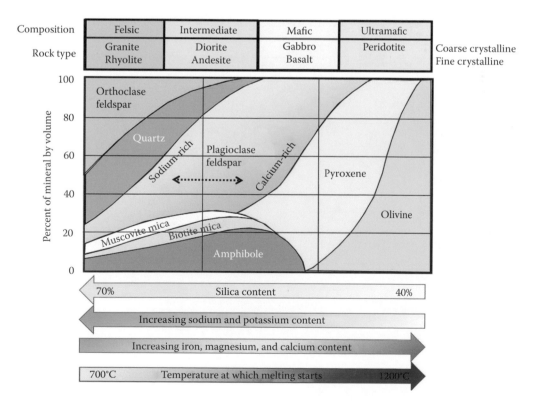

FIGURE 3.7
Igneous rock classification chart.

flow easily, rhyolite forms lava domes. Well-known rhyolite domes include Lassen Peak and Mammoth Mountain in California, Popocatépetl in Mexico, and Chaitén in Chile (Figure 3.8). When rhyolite cools with a lot of gas bubbles, it's called **pumice**, which is the only rock that will float on water. If rhyolite flows into a body of water such as a lake, it chills rapidly and forms **obsidian**, a volcanic glass. The composition of rhyolite is similar to granite, which solidifies below the surface.

(a)

(b)

FIGURE 3.8
(a) Rhyolite. (Courtesy of NASA; http://mars.nasa.gov/mer/classroom/schoolhouse/rocklibrary/index3.html.) (b) Rhyolite dome at Chaitén, Chile, is the circular feature at center. (Courtesy of NASA Goddard Space Flight Center/ Robert Simmon; https://commons.wikimedia.org/wiki/File:Chait%C3%A9n_Volcano_Lava_Dome,_Chile.jpg.)

Basalt is rich in iron and magnesium, silica-poor, and almost black (Figure 3.9). It erupts at about 1100°C and flows readily, forming low-relief shield volcanos—like the Hawaiian Islands—and extensive lava fields. Basalt contains mainly plagioclase, pyroxene, and olivine. Other examples include the Columbia River flood basalts (thick basalt that flowed over large areas), the Deccan Traps of central India ("**trap**" is another term for flood basalts),

(a)

(b)

FIGURE 3.9
(a) Basalt shield volcano. (b) Basalt flow, Craters of the Moon National Monument, Idaho. (Courtesy of G. Prost.)

and the Siberian Traps, which covered an area in Siberia roughly the size of Europe. The composition of basalt is similar to that of gabbro, which cools underground.

Lavas of intermediate composition, between that of rhyolite and that of basalt, produce the rocks **andesite** and **dacite** (Figure 3.10). Andesite contains mostly plagioclase feldspar, pyroxene, and amphibole, and is closer to rhyolite. Andesite was named after the Andes, where it is common.

Dacite has a composition closer to basalt and contains plagioclase feldspar along with biotite, amphibole, and pyroxene. Dacite was first described in Romania and is named after the former Roman province of Dacia (present-day Romania).

Another volcanic rock, **kimberlite**, originates in the Earth's mantle, at depths greater than 40 km (24 mi). When there is a weak spot in the overlying crust, the gas-charged lava erupts

(a)

(b)

FIGURE 3.10
(a) Andesite, O'Leary Peak, Arizona. (Courtesy of Jstuby at enwikipedia; https://commons.wikimedia.org/wiki/File:Olearyandesite.jpg.) (b) Dacite from Lassen Peak, California. (Courtesy of US Geological Survey; http://www.daviddarling.info/encyclopedia/D/dacite.html.)

explosively from a **diatreme** (pipe) that connects the mantle to the surface. The supersonic eruption forms a crater that is filled with chunks of rock from the mantle and crust. These eruptions are rare, and they're the source of rare minerals, notably diamond (Figure 3.11a).

An eruption of more than 50% molten carbonate is called a **carbonatite** (Figure 3.11b). These rare lavas form by separation of carbonate material from other components of a magma. The Ol Doinyo Lengai carbonatite volcano in Tanzania has the lowest temperature lava in the world, erupting at 500°C–600°C.

Plutonic (Intrusive) Rocks

Intrusive igneous rocks cool and crystallize in the subsurface. The slow cooling process allows mineral crystals to grow larger than in lavas.

Granite, the most abundant felsic rock, is similar to rhyolite in composition. Granite contains various feldspar minerals along with quartz, biotite, muscovite, and hornblende. Its characteristic pink color comes from orthoclase, a potassium-rich feldspar.

Granodiorite is often called "salt-and-pepper granite." It has less orthoclase and more biotite, hornblende, and augite than true granite. In composition, it resembles dacite (Figure 3.12).

Diorite is a little darker, having more mafic minerals. It underlies the Andes and is equivalent to Andesite in composition.

Gabbro is a dark intrusive rock with very little quartz and feldspar (Figure 3.13). Its composition resembles that of basalt. A finer-grained version of gabbro is **diabase** (North America) or **dolerite** (rest of the world).

When magma remains in place for a long time, the heavy minerals sink to the bottom of the magma chamber, and the lighter minerals rise to the top. This magmatic **differentiation** produces **ultramafic** rocks, which usually occur in the Earth's mantle at depths greater than 40 km. Ultramafic rocks are dark, rich in iron and magnesium, and low in silica and potassium. They are characterized by the minerals peridotite, pyroxene, olivine, hornblende, and calcium-rich plagioclase (anorthite).

Plutonic Structures

A **batholith** is a large body of plutonic rock that extends for more than 100 km^2. They usually consist of coarse-grained rocks like granite and granodiorite. An example is the Sierra Nevada batholith of California.

Dikes and **sills** are flat, or planar, intrusions. The difference is their orientation; dikes are nearly vertical, whereas sills are nearly horizontal. A dike usually indicates a feeder channel for a surface eruption. Sills are sheet-like intrusive rocks that are injected parallel to the surrounding rock layers. Dikes and sills can be of any composition.

Pegmatites are intrusive igneous dikes with uncommonly large crystals. The composition is usually granitic, and they can contain large gemstones. **Aplites** are intrusive igneous dikes of granitic composition with unusually small crystals.

A **laccolith** is a shallow intrusion that is injected parallel to strata, like a sill, and pushes up the overlying rock layers (Figure 3.14). Like dikes and sills, laccoliths can be of any composition. The La Sal Mountains of southeastern Utah are composed of laccoliths.

Lopoliths are rare, large, lens-shaped mafic or ultramafic intrusions that are injected roughly parallel to the surrounding rock layers. They contain important mineral deposits like nickel, copper, and platinum. At least one lopolith, the Sudbury complex in Sudbury, Ontario, is thought to be the result of a large meteorite impact.

(a)

(b)

FIGURE 3.11
(a) Kimberlite with 1.8 carat (6 mm), diamond crystal from Finsch Diamond Mine, South Africa. (Courtesy of StrangerThanKindness; https://commons.wikimedia.org/wiki/File:Diamond_-_South_Africa_-_Finsch _Mine.jpg.) (b) Carbonatite, Jacupiranga, Brazil. (Courtesy of Eurico Zimbres; https://en.wikipedia.org /wiki/Carbonatite.)

FIGURE 3.12

(a) Mt. Sinai granite, Sinai, Egypt. 2 Euro coin for scale. (Courtesy of G. Prost.) (b) Granodiorite, Yosemite National Park, California. (Courtesy of D. Monniaux; https://en.wikipedia.org/wiki/El_Capitan_Granite.)

(a)

(b)

FIGURE 3.13
(a) Diorite. (Courtesy of Siim Sepp; https://en.wikipedia.org/wiki/Diorite.) (b) Gabbro, from Thalhorn, near Fellering in Haut-Rhin, France. (Courtesy of Museum of Natural History and Ethnography in Colmar, France; Ji-Elle; https://commons.wikimedia.org/wiki/File:Gabbro-Talhorn-Mus%C3%A9e_d'histoire_naturelle_et_d'ethnographie_de_Colmar.jpg.)

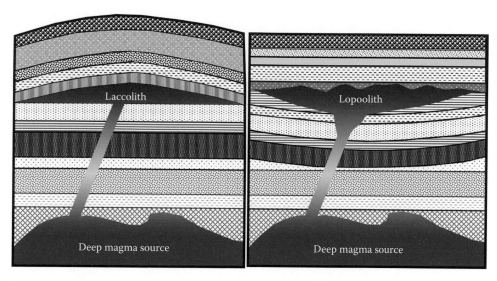

FIGURE 3.14
Difference between a laccolith and a lopolith.

Metamorphic Rock

Metamorphic rocks are any rocks—igneous, sedimentary, or already metamorphic—altered by heat, pressure, or both. When buried at great depths, rocks are subjected to heat from the earth, which is caused by radioactive decay, and to pressure from the weight of overlying rock. Tectonic stress can also pull or squeeze rocks, altering them.

Different levels of temperature and pressure produce different **metamorphic facies**. (Here, *facies* refers to an assemblage of minerals formed under similar conditions.) Metamorphic facies are classified by their composition and the amount of heat and pressure they have undergone. Most facies are named after their dominant minerals (Figure 3.15).

Ophiolites are a metamorphic assemblage caused by emplacement of oceanic crust onto continental crust or by injection of mantle rocks into continental crust. The assemblage is characterized by serpentine, diabase, chert, and pillow lava (lava erupted onto the seafloor at oceanic spreading centers). Their significance is that they reveal areas, usually in mountain belts, where ocean basins have been consumed by plate tectonic movements.

Metamorphic core complexes expose the deep crust in areas of extension. They are thought to form in areas with thick crust that is extending, as in the Basin-and-Range province of western North America. As the upper crust extends along low-angle faults, the middle and lower crust is progressively unroofed and rises buoyantly (by **isostatic rebound**). The lower crustal rocks are characterized by high-grade metamorphism (eclogite, granulite, and amphibolite facies), and the fault zones are characterized by mylonite and ductile deformation. **Mylonite** is a fine-grained, often banded, recrystallized metamorphic rock caused by crushing and grinding of rock along fault zones.

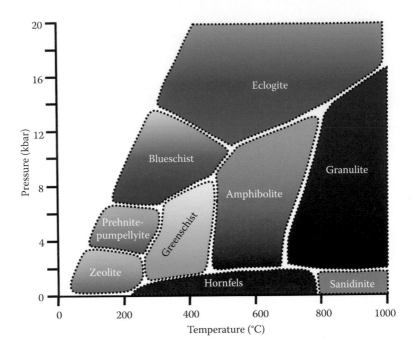

FIGURE 3.15
Metamorphic facies chart. Facies are a function of temperature and pressure. One kbar = 100 MPa = 14,500 psi.

Metamorphic rocks (as opposed to metamorphic facies) are classified by composition and texture. Rocks altered more by heat than by pressure have a recrystallized, uniform texture. Rocks altered more by pressure than by heat have a **foliated** (layered or striped) texture (Table 3.2).

Foliation refers to the alignment of platy minerals, such as micas, in a metamorphic rock (Figure 3.16). Foliation indicates changes in the degree or direction of pressure as the rock was being altered; it tends to occur along cleavage planes. Rock **cleavage** (as opposed to mineral cleavage) is a metamorphic fabric consisting of planar partings (Figure 3.17).

Pressure solution involves dissolving minerals at grain-to-grain contacts, which are areas of high stress. **Stylolites**, serrated partings, form where material was dissolved (Figure 3.18). **Axial planar cleavage** is a metamorphic fabric formed parallel to the axis of folds as a result of pressure solution (Figure 3.19).

Where metamorphic rocks occur over a large area, it is described as **regional metamorphism**. Isolated metamorphic rocks suggest **contact metamorphism**, where local rock came into contact with magma or hot fluids (Figure 3.20). The contact zone, or **metamorphic aureole**, is a few meters to a few hundreds of meters wide, and it can contain important mineral deposits known as skarn deposits (see Chapter 15).

In **prograde metamorphism**, minerals lose "volatile" components such as water and carbon dioxide to become more stable at high temperature and pressure. For example, clay minerals progressively transform into micas and ultimately garnets. **Retrograde metamorphism** is the opposite. Metamorphic rock absorbs water in low-temperature, low-pressure environments and minerals such as garnet turn into micas and ultimately clay as temperature and pressure drops.

TABLE 3.2

Metamorphic Rock Classification Chart

Texture		Grain size	Composition	Type of metamorphism	Comments	Rock name
Foliated	Mineral alignment	Fine	Mica / Quartz / Feldspar / Amphibole / Garnet / Pyroxene	Regional (Heat and pressure increase with depth of burial)	Low-grade metamorphism of shale	Slate
		Fine to medium			Foliation surfaces glitter due to mica crystals	Phyllite
					Visible mica crystals	Schist
	Banding	Medium to coarse			High-grade metamorphism, mineral segregation banding	Gneiss
Recrystallized (not foliated)		Fine	Variable	Contact (heat)		Hornfels
		Fine to coarse	Quartz	Regional or contact		Quartzite
			Calcite and/or dolomite			Marble
		Coarse	Various minerals			Metaconglomerate

FIGURE 3.16

Foliation in gneiss as a result of compositional layering. Prague, Czech Republic. (Courtesy of Huhulenik; https://upload.wikimedia.org/wikipedia/commons/a/af/Orthogneiss_Geopark.jpg.)

FIGURE 3.17
Pencil cleavage in limestone. (Courtesy of Colinlangford; https://commons.wikimedia.org/wiki/File:Pencil
_Cleavage.JPG.)

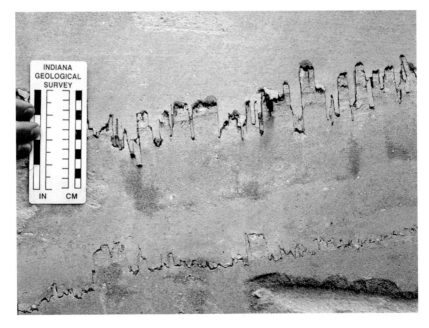

FIGURE 3.18
Stylolites in Mississippian Salem Limestone, Bloomington, Indiana. (Courtesy of Michael Rygel; https://
commons.wikimedia.org/wiki/File:Stylolites_mcr1.jpg.)

FIGURE 3.19
Near-vertical axial plane cleavage in the Devonian Nisku Formation, Allstones Lake, Alberta. (Courtesy of G. Prost.) Bedding dips about 60° to the left.

FIGURE 3.20
Mount Gould, Glacier Park, Montana. Contact metamorphism is the light rim around the dike (dark band on cliffs). (Courtesy of G. Prost.)

Metamorphic Rock Types

Quartzite is recrystallized, quartz-rich sandstone used in countertops and buildings (Figure 3.21a). It resembles sandstone, but there's an easy way to tell them apart: when you hit sandstone with a hammer, it goes "tunk"; when you hit quartzite with a hammer, it goes "tink."

Marble, a recrystallized limestone or dolomite, is used in countertops, in sculptures, and as facing stone (Figure 3.21b). While attractive, marble is soft rock, and real marble countertops *will* scratch.

With increasing metamorphism, shale, a sedimentary rock, turns into **slate** (Figure 3.21c), **phyllite** (Figure 3.21d), **schist** (Figure 3.22a), and ultimately **gneiss** (Figure 3.22b). Slate is used in roofing, in flooring, and as paving stones. Gneiss is both recrystallized and foliated: the original rock may have been sedimentary, igneous, or metamorphic. **Granite gneiss** has been heated to the point where it starts melting (Figure 3.22c).

Serpentinite is a low-temperature metamorphic rock formed by the hydration of mafic and ultramafic rocks such as peridotite (Figure 3.22d). Abundant in the mineral serpentine (named for its green, serpent-like color), serpentinite is an important source of asbestos, chrome, and nickel.

Soapstone is formed by the retrograde metamorphism of peridotite, dunnite, or serpentinite. It consists mainly of talc, from which it gets its "soapy" feel. Soft and easy to carve, soapstone is used in arctic aboriginal art (Figure 3.22e).

Novaculite is a nonfoliated metamorphic rock composed almost entirely of silica derived from the shells of **diatoms**, a kind of single-celled marine algae.

Amphibolite is a dark, coarse-grained, nonfoliated metamorphic rock consisting mainly of amphibole (hornblende) and plagioclase feldspar (Figure 3.23a). It derives from the metamorphism of basalt, gabbro, or clay-rich sedimentary rocks. Amphibolite is used as aggregate or building stone.

Hornfels is a fine-grained, nonfoliated metamorphic rock formed by contact metamorphism. Light to dark in color, it has no characteristic composition (Figure 3.23b). It is used mainly as aggregate.

What Rocks Tell You

You can learn a lot from rocks. The composition of sedimentary rocks tells you how and where they were formed, and what their original source might have been. Their texture, grain size, grain shape, and grain sorting tell you about their depositional history. For example, conglomerates with large, angular cobbles rarely occur far from their source, whereas fine siltstone and mudstone can be found thousands of kilometers from their source. This is because the particles get ground, crushed, and broken into progressively smaller grains as they travel. A quartz sandstone with frosted round grains suggests strongly that the rock is a windblown deposit: the frosting is from grains bumping into each other. A rock containing a large assortment of grain compositions and grain sizes, both rounded and angular, suggests a flood or landslide deposit that mashes everything together.

Pore spaces in sedimentary rocks contain traces of fluids or minerals, revealing the history of fluids that moved through the rocks (hot water, oil, gas) and the temperature and

FIGURE 3.21

Moderately metamorphosed rocks. (a) Quartzite. Note how the rock breaks through the grains, not around them. Near Loch Assynt, Scotland. (Courtesy of Lysippos; https://commons.wikimedia.org/wiki/File:Quartzite ,_assyn_u.jpg.) (b) Marble. (Courtesy of NASA; http://mars.nasa.gov/mer/classroom/schoolhouse/rocklibrary /index3.html.) *(Continued)*

(c)

FIGURE 3.21 (CONTINUED)
Moderately metamorphosed rocks. (c) Slate. (Courtesy of Vincent Anciaux; https://commons.wikimedia.org
/wiki/File:Fumayschistewiki.jpg.)   *(Continued)*

pressure history of the rock. For example, pores can be filled with silica that was dissolved
from grains in a high-temperature and high-pressure environment and redeposited in
areas of lower temperature and pressure. Pores can be destroyed by pressure that causes
interlocking grains. New pores can be formed by dissolving minerals.

Likewise, the composition, texture, and grain size of igneous and metamorphic rocks
tell their cooling or deformation history. Coarse-grained rocks cooled slowly; fine-grained
rocks cooled quickly. Rocks with compositional banding or alignment of platy minerals
formed under high temperatures and pressures. Rocks containing minerals like staurolite, kyanite, and sillimanite indicate medium-pressure, high-temperature metamorphism. Granite-gneiss indicates high-temperature metamorphism to the point of melting.
Stylolites in limestone indicate low temperature but high pressure that caused pressure
solution of calcite minerals.

Rocks are found in the earth either as layers (sedimentary or volcanic strata) or in massive igneous or metamorphic bodies. The study of rock layers is called **stratigraphy**, and in
order to understand what the layers mean, we will review some basic concepts (Chapter 4).

(d)

FIGURE 3.21 (CONTINUED)

Moderately metamorphosed rocks. (d) Phyllite. From Prague, Czech Republic. (Courtesy of Chmee2; https://commons.wikimedia.org/wiki/File:Phyllite_in_Geopark_on_Albertov.JPG.)

A Sampling of Geologic Guidebooks

There are a number of guidebooks to the geology of parks and regions. Examples in North America include the Roadside Geology series and U.S. Geological Survey Bulletins. A few of these are listed below. Check Amazon.com for more.

Alt, D., and D. W. Hyndman. 1986. *Roadside Geology of Montana.* Mountain Press Publishing Co., Missoula, MT, 427 pp.

Alt, D., and D. W. Hyndman. 2000. *Roadside Geology of Northern and Central California.* Mountain Press Publishing Co., Missoula, MT, 369 pp.

Chronic, H. 1983. *Roadside Geology of Arizona.* Mountain Press Publishing Co., Missoula, MT, 314 pp.

Chronic, H., and F. Williams. 2002. *Roadside Geology of Colorado.* Mountain Press Publishing Co., Missoula, MT, 398 pp.

Eyles, N., and A. Miall. 2007. *Canada Rocks—The Geologic Journey.* Fitzhenry and Whiteside, Markham, ON, 512 pp.

Gadd, B. 2008. *Canadian Rockies Geology Road Tours.* Corax Press, 576 pp.

Hansen, W. R. 1969. *The Geologic Story of the Uinta Mountains.* U.S. Geological Survey Bulletin 1291, 144 pp.

(a)

(b)

FIGURE 3.22
Highly metamorphosed rocks. (a) Garnet schist from Syros, Greece. Euro coin for scale. (Courtesy of Graeme Churchard (GOC53); https://commons.wikimedia.org/wiki/File:Garnet_Mica_Schist_Syros_Greece.jpg.) (b) Folds in Gneiss. (Courtesy of Anne Burgess; https://commons.wikimedia.org/wiki/File:Folds_in_Gneiss_-_geograph.org.uk_-_1370873.jpg.)   (*Continued*)

(c)

(d)

FIGURE 3.22 (CONTINUED)

Highly metamorphosed rocks. (c) Port Deposit granite gneiss, Maryland. Width is approximately 10.7 cm (https://commons.wikimedia.org/wiki/File:Foliated_granite_PlateVIII_MD_Geological_Survey_Volume_2 .jpg). (d) Serpentinite (8.6 cm across) from Tasmania. Greenish = serpentine; purplish = stichtite. (Courtesy of James St. John; https://commons.wikimedia.org/wiki/File:Stichtitic_serpentinite,_Dundas_Ultramafic _Complex.jpg.) *(Continued)*

(e)

FIGURE 3.22 (CONTINUED)
Highly metamorphosed rocks. (e) Inuit soapstone carving. (Courtesy of G. Prost.)

(a)

FIGURE 3.23
Nonfoliated metamorphic rocks. (a) Amphibolite, Val di Fleres, Italy. (Courtesy of Bernabè Egon; https://en
.wikipedia.org/wiki/Amphibolite.)
                                                                                (Continued)

(b)

FIGURE 3.23 (CONTINUED)

Nonfoliated metamorphic rocks. (b) Hornfels from Novosibirsk, Russia. Banding is due to original bedding. (Courtesy of Fed; https://en.wikipedia.org/wiki/Hornfels.)

Keefer, W. R. 1971. *The Geologic Story of Yellowstone National Park*. U.S. Geological Survey Bulletin 1347, 92 pp.

Lohman, S. W. 1974. *The Geologic Story of Canyonlands National Park*. U.S. Geological Survey Bulletin 1327, 126 pp.

Lohman, S. W. 1975. *The Geologic Story of Arches National Park*. U.S. Geological Survey Bulletin 1393, 113 pp.

Purkey, B. W., E. M. Duebendorfer, E. I. Smith, J. G. Price, and S. B. Castor. 1994. *Geologic Tours in the Las Vegas Area*. Nevada Bureau of Mines and Geology Special Publication 16, 156 pp.

Maps

Geologic road maps and national park geologic maps present geology graphically. Some examples are as follows:

American Association of Petroleum Geologists. 1972. Geological Highway Map Northern Rocky Mountain Region—Idaho–Montana–Wyoming.

American Association of Petroleum Geologists. 1973. Geological Highway Map Pacific Northwest Region—Washington–Oregon (Idaho in Part).

American Association of Petroleum Geologists. 1974. Geological Highway Map Alaska and Hawaii.

American Association of Petroleum Geologists. 1986. Geological Highway Map Mid-Continent Region—Kansas–Missouri–Oklahoma–Arkansas.

Cooper, M. (ed.). 2010. Geological Highway Map of Alberta. The Canadian Society of Petroleum Geologists.

Hintze, L. F. 1997. Utah Geologic Highway Map. Department of Geology, Brigham Young University.

Huntoon, P. W., G. H. Billingsley, Jr., W. J. Breed, J. W. Sears, T. D. Ford, M. D. Clark, R. S. Babcock, and E. H. Brown. 1986. Geologic Map of the Eastern Park of the Grand Canyon National Park, Arizona. Grand Canyon Natural History Association and the Museum of Northern Arizona.

Wyoming Geological Survey. 2006. Wyoming Geologic Highway Map. CTR Mapping.

4

Geologic Principles

From whatever I have been able to observe up to this time the series of strata which form the visible crust of the earth appear to me classified in four general and successive orders. These four orders can be conceived to be four very large strata, as they really are, so that wherever they are exposed, they are disposed one above the other, always in the same order.

Giovanni Arduino
Geologist

No collateral science had profited so much by palæontology as that which teaches the structure and mode of formation of the earth's crust, with the relative position, time, and order of formation of its constituent stratified and unstratified parts.

Sir Richard Owen
Paleontologist

A distinct group of sedimentary layers that can be mapped is called a **formation**. It's the basic unit of stratigraphy. Every formation has a name, one that includes either the word *formation* or the dominant type of rock you'll find in it—*Morrison Formation*, for example, or *Redwall Limestone*. A stack of formations in a given spot is called a **stratigraphic section** or **stratigraphic column** (Figure 4.1). An assemblage of non-sedimentary rock, which is called a **massive unit**, also gets a name, such as *Pikes Peak granite* or *Idaho Springs Formation gneiss*.

Where sedimentary formations have been folded, faulted, eroded, overturned, or what have you, or where they butt up against massive units, you can apply some basic principles to "read the rocks"—to glean the geological processes at work. Using these principles, you can determine the relative ages of the rocks, tell "which way is up" (which rocks are younger, which are older), see which formations taper and disappear, and understand why some formations change rock type from one area to the next.

Principle of Original Horizontality

Sediments are always deposited parallel to the earth's surface—in other words, horizontally. (An exception might be wind-sculpted sand dunes, where sediments can be inclined up to 34°, but these are rare.) Sedimentary rocks, then, reflect this horizontality when they form. The corollary is this: any sedimentary rock layers that are *not* horizontal must have been tilted or folded after deposition.

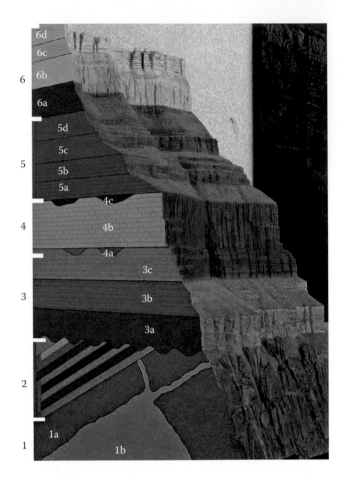

FIGURE 4.1
Stratigraphic column for the Grand Canyon, Arizona. 6d, Kaibab Limestone; 6c, Toroweap Formation; 6b, Coconino Sandstone; 6a, Hermit Shale. 5, Supai Group; 5d, Esplanade Formation; 5c, Wescogame Formation; 5b, Manakacha Formation; 5a, Watahomigi Formation. 4c, Surprise Canyon Formation; 4b, Redwall Limestone; 4a, Temple Butte Limestone. 3c, Muav Limestone; 3b, Bright Angel Shale; 3a, Tapeats Sandstone. 2, Grand Canyon Supergroup. 1, Vishnu Group; 1b, Zoroaster Granite; 1a, Vishnu Schist. (Courtesy of US National Park Service and US Geological Survey, Daniel Mayer; https://commons.wikimedia.org/wiki/File: Grand_Canyon_geologic_column.jpg.)

Principle of Superposition

In any sequence of sedimentary rocks, the layers at the bottom were deposited first. In other words, the bottom layer is the oldest. The layers get younger and younger as you go up. On the basis of this principle, you can determine the relative age of any rock layer in any rock sequence. (This principle, however, applies only to rocks that haven't been deformed or displaced. It's possible to see older layers on top of younger layers, but in those cases, the rocks are upside down, or old rocks have been faulted over younger rocks.)

It follows that fossils in the bottom layer of a sequence are older than the fossils in overlying layers. The principle of superposition, then, is also useful for determining the relative ages of life forms.

Unconformities

In 1788, the Scottish naturalist James Hutton found a curious outcrop at Siccar Point, on the east coast of Scotland. There, he saw gently inclined beds of red sandstone overlying nearly vertical beds of a different sandstone. The contact surface between the two formations, Hutton surmised, represented a gap in deposition. Below the gap, the originally horizontal sandstone must have been tilted, eroded, and then submerged in a sea. Then, new sands were deposited on the erosion surface. The new sands became the red sandstone that Hutton saw above the gap.

Such a gap is called an **unconformity**. It indicates that rock is either missing—eroded away before new sediments were deposited—or that deposition stopped for a period of time.

The surface of an unconformity is usually undulating or irregular. It may extend only a short distance, or it may extend over entire regions. The same unconformity can represent a longer or shorter gap in time depending on how long erosion occurred in any given area. Likewise, the amount of erosion can be greater or smaller as you move along the unconformity.

Types of Unconformities

If the rocks above and below the unconformity are parallel, it's called a **paraconformity**, which suggests a period of nondeposition with little or no erosion (Figure 4.2a). Paraconformities can be hard to see, because the surface of the unconformity looks like an ordinary bedding plane.

If the layers above and below the unconformity are parallel, but the layers lying just below the unconformity are truncated, it's called a **disconformity** (Figure 4.2b). A disconformity suggests that substantial erosion took place before deposition resumed.

If the layers below the unconformity are inclined and the layers above it are not, it's called an **angular unconformity** (Figures 4.2c and 4.3). Here, the layers below the unconformity were tilted and eroded before the layers above were deposited. This type of unconformity is what Hutton saw at Siccar Point.

If rock layers are deposited on top of a massive unit, such as an eroded granite, it's called a **nonconformity** (Figures 4.2d and 4.4).

Principle of Stratigraphic Correlation and Lateral Continuity

A stratigraphic section shows the sequence of rock layers in one place. For example, an outcrop may have a layer of sandy shale on top, a layer of mudstone in the middle, and a layer of limestone at the bottom. Some distance away, another outcrop might reveal the same sequence of rocks. Even farther away, an outcrop might reveal a thin layer of sandy shale directly on top of a thick layer of limestone. By drawing a stratigraphic section of each outcrop, a geologist can **correlate** these rock layers. **Correlation** is the way geologists determine that a layer of rock in one place is the same **lithology** (rock composition, color, grain size) as a layer of rock in another place.

When naturalists first began looking at rock layers, they noticed that many layers could be traced across Europe. They figured that any given rock layer must have been deposited at the same time everywhere on Earth. Then, the next layer was deposited everywhere, and

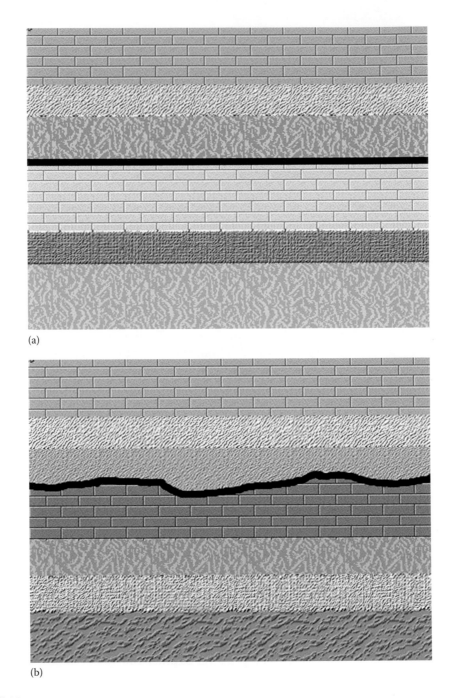

(a)

(b)

FIGURE 4.2
Types of unconformities indicated by the heavy black lines. (a) Paraconformity. (b) Disconformity. This is usually an erosion surface.   (*Continued*)

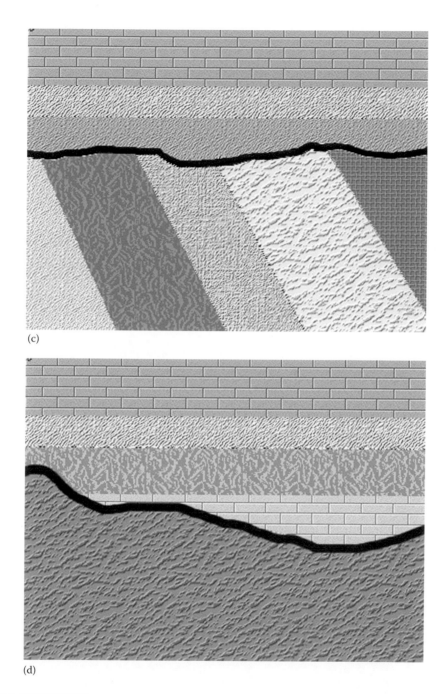

(c)

(d)

FIGURE 4.2 (CONTINUED)
Types of unconformities indicated by the heavy black lines. (c) Angular unconformity. (d) Nonconformity (https://en.wikipedia.org/wiki/Unconformity).

FIGURE 4.3
The Great Unconformity in the Grand Canyon, Arizona, separates the northeast-tilted Chuar Group from flat-lying Tapeats Sandstone. Depending on where you are along this angular unconformity (dotted line), the time interval of missing rocks varies from 175 million to 1.6 billion years. (Courtesy of G. Prost.)

FIGURE 4.4
Four-hundred-million-year-old Fremont Canyon Sandstone over 2.6-billion-year-old granite, Fremont Canyon, Wyoming. This nonconformity represents a 2.2-billion-year gap (dotted line), making it one of the largest unconformities in North America. (Courtesy of G. Prost.)

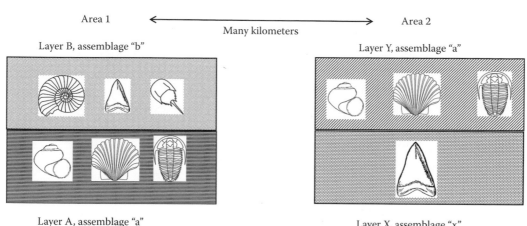

FIGURE 4.5
Concept of correlation. Layer A has the same fossil assemblage as Layer Y. They are assumed to be deposited at the same time and correlate to one another.

so on. Thus, a red sandstone in England was the same as the red sandstone in Germany, and both were the same age.

Upon closer examination, it was found that most formations thinned out to nothing and were replaced by other rock layers. That is, the formations changed sedimentary **facies** (rock characteristics that reflect its origin) over long distances. If you find a specific type of trilobite fossil in a layer of shale, and some distance away you find the same type of trilobite fossil in a siltstone, then you can assume that those two rocks—the shale and the siltstone—were deposited at the same time. In other words, if the same fossils occur in rock layers that are far apart, you can surmise that the two layers formed at the same time: they correlate in age but not lithology (Figure 4.5).

Geologists have found that lithology and age are not necessarily linked. A formation may consist of the same rock type (lithology) over a large area, but the age of that layer changes across the region. For example, you may be able to trace a layer of sandstone from one outcrop to another outcrop many kilometers away. Both outcrops reveal the same sandstone. However, the sandstone at Outcrop A may be younger or older than the sandstone at Outcrop B.

The reason is that depositional environments change over time. For example, a slowly falling sea level will cause a beach to migrate across a region—from, say, east to west. As the beach migrates west, it contains the same kind of sand, but the sand will be younger. Over time, the beach sand turns into sandstone. Eventually, this sandstone reveals itself in widely separated outcrops. You can correlate the sandstone layer from one outcrop to another—the lithology is the same—but you can't say that the sandstone is the same age across its extent. The same sandstone "crosses" time; it is said to be **diachronous**, or **time-transgressive** (McKee and Resser 1945).

Principle of Faunal Succession

William Smith (1769–1839) worked in coal mines in Somerset, England, and helped survey the Somerset Coal Canal. He noticed a specific sequence of rock layers that always occurred in the

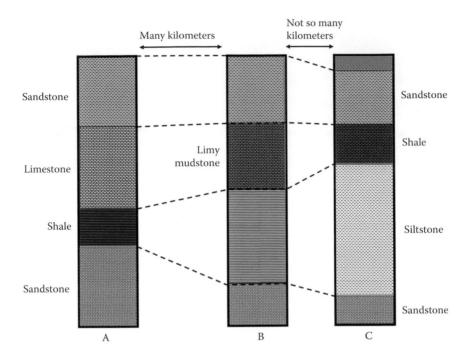

FIGURE 4.6

Stratigraphic correlation. The units in column A are correlated to the units in columns B and C many kilometers away. Each formation changes thickness, and the green and blue formations change rock type gradually across the region.

same relative position. Also, distinct fossils could always be found in each layer. Smith predicted that this would always be the case, and he confirmed this prediction during his travels in England, Wales, and Scotland. By correlating similar outcrops across these countries, he was able to draw and publish the first geologic map of Britain, in 1815 (Winchester 2001). Geologists have been using the technique of correlation to make geologic maps ever since (Figure 4.6).

Smith's discovery that the same groups of fossils are always found in the same vertical sequence is called the **principle of faunal succession**. Rocks containing fossils, both animals and plants, succeed each other vertically in the rocks, that is, over periods of time, in a reliable, specific order that can be recognized over wide areas.

Principle of Cross-Cutting Relationships

This principle states that if one geologic feature cuts across another, the one that is cut is older. If a dike cuts across a granite, the granite is older than the dike. (In other words, for the dike to cut through the granite, the granite already had to be there.) If a fault offsets a sandstone, the fault is younger than the sandstone. If an unconformity truncates a shale, the unconformity is younger than the shale.

You may see multiple cross-cutting features in the same unit, as when a sandstone is faulted, and the fault is truncated at an unconformity (Figure 4.7). In this case, the unconformity would

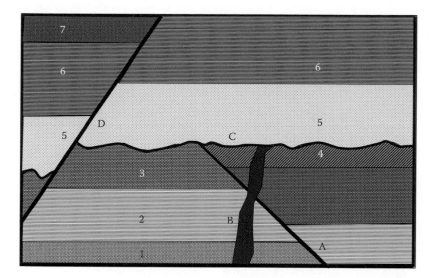

FIGURE 4.7
Cross-cutting relationships. Fault A cuts layers 1 to 4, so those layers are older than A. Dike B cuts fault A but terminates at unconformity C, so it is younger than A and older than C. Unconformity C terminates layers 3 and 4, so it is younger than those layers. Fault D cuts through all layers, so it is younger than layer 7.

be younger than the fault, and the fault would be younger than the sandstone. Cross-cutting relationships help you reconstruct the sequence of events in the geologic history of an area.

Principle of Inclusions

When one rock contains fragments (inclusions) of another rock, the fragments had to exist before the rock that contains them. Therefore, the inclusions are older than the rock that contains them (Figure 4.8). If a sandstone contains inclusions of basalt, the basalt had to exist before the sandstone. If you see schist inclusions in a granite, the schist had to exist before the granite.

Principle of Uniformitarianism

In 1788, James Hutton published a paper in which he stated, "In examining things present, we have data from which to reason with regard to what has been; and, from what has actually been, we have data for concluding with regard to that which is to happen here after." Charles Lyell, another Scottish geologist, later condensed this idea into a short axiom: "The present is the key to the past." We can explain what happened in the past by observing processes at work around us today, because those processes have been essentially uniform throughout Earth's history.

Underlying this principle is the assumption that geologic processes are slow and continuous, acting over vast amounts of time. Erosion may be imperceptibly slow; an earthquake

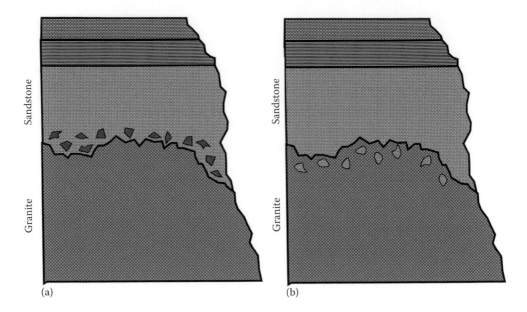

FIGURE 4.8
Principle of inclusions. (a) Granite inclusions in the overlying sandstone indicate that the granite is older.
(b) Sandstone inclusions in the granite indicate that the sandstone is older and was intruded by the granite.

may cause only a few millimeters of offset, but the cumulative effects of these processes over long time spans are great. This theory, **uniformitarianism**, emerged in the late eighteenth and early nineteenth centuries. It contradicted the prevailing theory at the time, which proposed that sudden, catastrophic events shaped the earth. That theory is known as **catastrophism**.

Geologists now agree that most processes are slow and continuous, punctuated now and then by rare catastrophic events like asteroid impacts, massive floods, or super-volcano eruptions that can have major effects on geologic history.

References

McKee, E. D., and Resser, C. E. 1945. *Cambrian History of the Grand Canyon Region*. Carnegie Inst. Washington Pub. 563, 232 pp.
Winchester, S. 2001. *The Map that Changed the World—William Smith and the Birth of Modern Geology*. Harper Collins Publishers, New York, 329 pp.

5

Deep Time

I incline to this opinion, that from the evening ushering in the first day of the World, to that midnight which began the first day of the Christian era, there was 4003 years, seventy days, and six temporarie hours…

James Ussher
Anglican Archbishop of Armagh

Rocks are records of events that took place at the time they formed. They are books. They have a different vocabulary, a different alphabet, but you learn how to read them.

John McPhee
In Suspect Terrain

One of the hardest concepts to grasp, for geologists as well as non-geologists, is that of **deep time**. These are depths of time so far beyond our everyday experience that it is almost impossible to comprehend. In his book *Basin and Range* (1982), John McPhee tried to describe deep time:

> Consider the Earth's history as the old measure of the English yard, the distance from the King's nose to the tip of his outstretched hand. One stroke of a nail file on his middle finger erases human history.

Consider how long it must take for tiny sea shells to accumulate, one upon another, to form the limestone layers thousands of meters thick seen in Banff National Park (Figure 5.1). Consider how long it takes to deposit layer upon layer of sand, one grain at a time, to get the sandstone cliffs seen in the Grand Canyon. And yet we see thousands of meters of sedimentary layers everywhere we look, most with rock missing at unconformities because of erosion, indicating that formerly there had been even thicker layers.

The Concept of Deep Time

James Hutton (1726–1797) was a Scottish naturalist and farmer. He noticed that, every year, his fields lost some soil to erosion and that the local rivers carried the sediment to the sea where it was deposited as sand and mud. He figured it would take a long, long time to build the thick rock layers of sandstone and mudstone he saw on the sea cliffs. At Siccar Point, he noticed the angular unconformity mentioned earlier in our discussion of unconformities. Not only was erosion and deposition a slow process that took large amounts of time, the sediments then had to become consolidated (turned into rock), be tilted and uplifted above the sea, be eroded again, and then dropped below sea level and have more layers of sediment deposited above the erosion surface (Figures 5.2 and 5.3). In 1788, he wrote that the history of the earth is so long

FIGURE 5.1
Mt. Rundle, Banff National Park, Canada, consists of limestone layers hundreds of meters thick. (Courtesy of G. Prost.)

FIGURE 5.2
Hutton's unconformity at Siccar Point, Scotland. (Courtesy of Dave Souza; https://commons.wikimedia.org /wiki/File:Siccar_Point_red_capstone_closeup.jpg.)

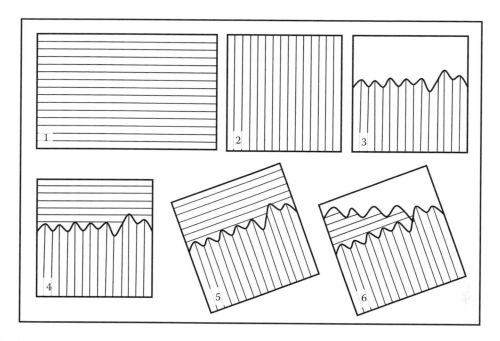

FIGURE 5.3
Schematic sequence of events leading up to the outcrops seen at Siccar Point. (1) Deposition, burial, and consolidation. (2) Rotation to vertical during a mountain-building event. (3) Erosion. (4) Subsidence and deposition of new layers. (5) Tilting during a second mountain-building event. (6) Present-day erosion.

that "we find no vestige of a beginning, no prospect of an end." In other words, the lengths of time involved are so large they are beyond comprehension.

Before Hutton, most people in Europe and the Middle East assumed Earth was only around 6000 years old based on chronologies in the Bible. In fact, in 1650, Bishop Ussher of Trinity College, Dublin, worked backward through the lives of the patriarchs and determined that Earth was created on the evening of October 22, 4004 BCE. Again, based on the Bible, most Europeans believed not only that Earth was young, but also that all rocks had formed at the same time Earth was created.

Hutton argued that if the processes we see today have always been at work, then the earth must be much older than 6000 years. But how much older? Two systems of dating were developed by these early scientists: relative dating based on fossils and cross-cutting relationships, and absolute dating based on experimentation and measurement. Both of these dating methods suggested that the earth was older than previously thought, but it was not until the development of radiometric dating that the age was determined to be around 4.5 billion years old.

How to Tell the Age of Rocks: Relative and Absolute Dating

Relative age dating means figuring out which rocks are older or younger than other rocks. If we know that Rock Layer A is always below Rock Layer B, then, by the principle of superposition, A should be older than B. We notice that Layer A has fossils a1, a2, and a3 (the "a" assemblage) and Rock Layer B has fossils b1, b2, and b3 (the "b" assemblage).

So any rocks anywhere in the world with the "a" assemblage will be older than rocks anywhere with the "b" assemblage (Figure 5.4). This is relative dating: even though we do not know the absolute age of Layers A and B, we know that Layer A is older than Layer B.

Upon closer examination, we also notice that these units are separated by a distinctive volcanic ash layer. That layer is a point-in-time marker, since eruptions occur in a geologic instant. Such a marker should be the same age everywhere.

It turns out that observing the gradual change in fossil **morphology** (shape) and **assemblages** (groups) as rocks get younger provides fundamental evidence for the theory of evolution, the gradual change in plants and animals over time. Because fossil shapes evolve over geologic time, generally from simple to more complex forms, one can use the distinctive shape of fossils in a sedimentary layer to determine its age relative to another fossil-bearing sedimentary layer.

Absolute age dating assigns an age in years to a rock layer. A plus–minus error is usually given. Taking Layer A as an example, suppose it is intruded by a dike that can be dated at 240 million years ± 5 million years. (Geologists refer to a million years as a mega-annum, or Ma.) This must mean that Layer A is older than about 240 Ma. Now, suppose Layer A also lies above an ash layer that has been dated as 265 Ma ± 5 million years. This means Layer A is younger than about 265 Ma. Thus, we have determined that Layer A is between 240 and 265 million years old (Figure 5.5).

Radiometric age dating is an absolute dating technique that uses radioactive elements such as uranium that decay at a known rate. Radioactive elements are special in that they are not physically stable and yet want to be. In order to reach a stable state, they must go through a predictable series of transformations from unstable parent to more stable daughter products. Depending on the parent element, this can take thousands of years to hundreds of millions of years. This transformation process is called **radioactive decay**.

The more time that has passed since a rock solidified from a molten state, the less of the parent and more of the daughter product are in the rock. By measuring the amounts and determining the ratio of parent to daughter, we can determine the age of the mineral or rock (Figure 5.6). This process is carefully measured, is repeatable, and the range of uncertainty is understood.

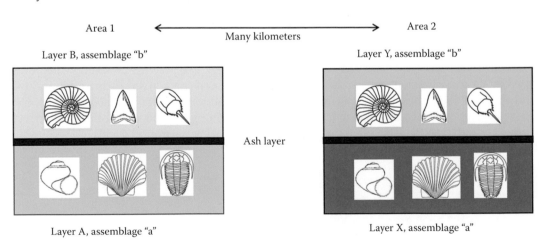

FIGURE 5.4
Relative age dating using fossil groups. Layers A and B are in our map locality, and Layers X and Y of different rock types are farther away. Using the fossil assemblages, we know that Layer X is the same age as Layer A, and Layer Y is the same age as Layer B. We also have an ash layer that we know is the same age everywhere in the region.

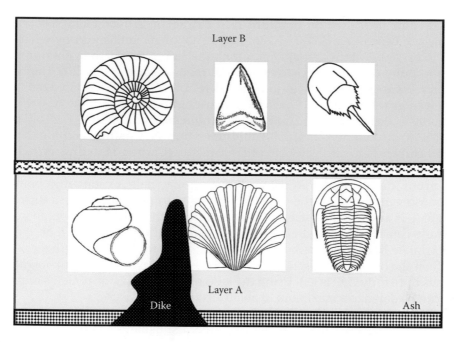

FIGURE 5.5

The dike is dated at 240 million years, and the ash layer is 265 million years. Using the Principle of Crosscutting Relationships, we see that Layer A is older than the dike. Using the Principle of Superposition, we see that Layer A is younger than the ash bed. We use the absolute ages of the volcanic ash layer and the dike to determine that the age of Layer A is between 240 and 265 million years.

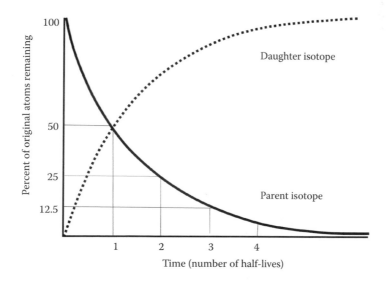

FIGURE 5.6

Principles of radiometric dating. The ratio of parent-to-daughter product gives the time since crystallizing from a melt.

Zircon is commonly used for **uranium–lead dating**: once the mineral has crystallized from a melt, it becomes a closed system and is no longer influenced by what happens outside the crystal. Within that system, uranium-238 breaks down to thorium; thorium decays to radium; radium decays to radon; and radon decays to lead-206. Half the U-238 decays to Pb-206 in 4.5 billion years (the "**half-life**" of U-238). This is called the **uranium–thorium–lead geochronometer**. Other radiometric age dating techniques use the decay of potassium-40 to argon-40 in mica (**potassium–argon method**; half-life of 1.3 billion years), rubidium-87 to strontium-87 in mica (**rubidium–strontium dating**; half-life of 50 billion years), and carbon-14 to nitrogen-14 in organic matter (**Carbon-14 dating**, or **radiocarbon method**; half-life of 5700 years).

Radiocarbon dating works on relatively young materials. Cosmic rays bombard the upper atmosphere and cause the N-14 to combine with neutrons to form unstable C-14. The ratio of C-14 and C-12 (the stable carbon isotope) in the atmosphere, oceans, and lakes is essentially constant. Living organisms don't care which isotope of carbon they use, so they incorporate both of the carbon isotopes equally into their tissues, bones, shells, or wood. When the organism dies, the C-14 continues to decay to N-14 at a known rate. The C-14 decays until only C-12 is left. Using the ratio of C-14 to C-12, the remains of plants and animals can be dated to a maximum of about 70,000 years.

Another type of absolute age dating uses fission tracks. **Fission track dating** counts the number of tracks produced in certain minerals, usually apatite or zircon, by the splitting nuclei of U-238 (Figure 5.7). Since U-238 decays at a known rate, the more tracks there are, the older the mineral is. For igneous rocks, this tells when the mineral crystallized from a melt. In the case of metamorphic rocks, it tells when the clock was set by recrystallization of a mineral at a certain temperature. These methods do not tell the age of sedimentary rock: rather, they tell when the constituent grains crystallized from a melt.

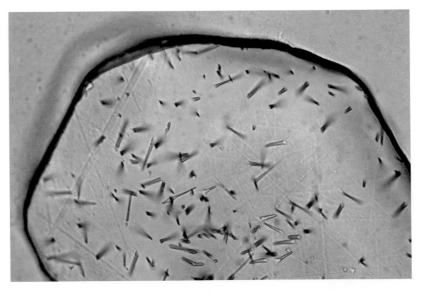

FIGURE 5.7
Fission tracks in an apatite crystal. (Courtesy of Dr. Phillip Armstrong; http://earthsci.fullerton.edu/parmstrong /research.htm.)

Age of the Earth and the Solar System (and How Long Earth Will Last)

Early attempts at estimating the absolute age of the earth considered sedimentation rates, the rate of cooling from an originally molten state, and the rate of salt accumulation in the ocean.

Around 450 BCE, the Greek **Herodotus** observed that the Nile deposited a thin layer of sediment each year and concluded that the building of the Nile delta must have taken several thousand years. Some scientists in the 1800s felt that, if one could find the thickness of all rocks ever deposited, and determine the rate of deposition through time, then the age of the earth could be calculated. Unfortunately, there is no complete sedimentary section anywhere (rocks are always missing), and below a certain depth, sedimentary rocks end and igneous or metamorphic rocks begin. Also, there is no constant rate of sedimentation: some thick sections are deposited instantaneously (like landslides), and some thin sections (such as clays in lakes) can take thousands of years to accumulate.

The French naturalist **Georges-Louis Leclerc, Comte de Buffon** (1707–1788) proposed that a comet had struck the sun and had broken off debris that became the planets of the solar system. He figured that the Earth started molten and has gradually cooled to form dry land, with rain condensing in the atmosphere to form oceans. Buffon experimented with heated iron spheres to see how long they took to cool and then estimated how long it would take for an Earth-sized object. He projected that Earth cooled from a molten state in about 75,000 years. The physicist **William Thompson** (**Lord Kelvin**, 1824–1907) followed this line of thought to calculate the age of the earth by assuming the sun and Earth are cooling at a constant rate from an original molten state. In 1862, he estimated that Earth was between 20 and 400 million years old. Over time, he refined this until, by the 1880s, he settled on an age of 20 million years.

John Joly (1857–1933), an Irish geologist, in 1899 calculated the age of the earth by looking at the salinity of ocean water. His reasoning was that rain falls as fresh water, but as it runs into rivers, it picks up minerals, some of which dissolve. By the time the water reaches the ocean, it contains a large amount of dissolved minerals, including salt. The ocean is salty because evaporating ocean water leaves dissolved salt behind. Joly calculated the rate at which the oceans should have accumulated salt and determined that the oceans are between 80 and 150 million years old. He incorrectly assumed, among other things, that the rate of salt input to the oceans is constant and that the amount of salt in the ocean has been increasing steadily though time. In fact, the oceans are in a state of salinity equilibrium, the input being balanced by the removal of salt (as in layered salt formations).

The discovery of radioactivity by **Ernst Rutherford** around 1902 allowed a new way to estimate age. In 1907, **Bertram Boltwood** used the known decay rate of radium to estimate the age of the earth as between 400 million and 2.2 billion years. In 1913, **Arthur Holmes** looked at the relationship of uranium to lead and estimated that Earth was about 1.6 billion years old. **Claire Patterson** developed the currently used uranium–lead dating method. He looked at lead in iron meteorites because it was the kind of lead expected in the early solar system. In 1956, he published his estimate of 4.55 billion years, a value that still stands.

By the way, cosmologists figure that the **big bang** that created the universe and time happened 13.8 billion years ago. They figured this out by knowing the current speeds and distances of galaxies and the rate the universe is expanding. They worked backward to the time when everything would have collapsed to one point. Another technique involves measuring the distance in light years to the oldest stars, those that are farthest away from

FIGURE 5.8
This gneiss from near the Acasta River, Northwest Territories, Canada, is 4.03 billion years old. (Courtesy of Mike Beauregard, Nunavut, Canada; https://en.wikipedia.org/wiki/Oldest_dated_rocks.)

Earth. Since the universe is thought to be expanding equally in all directions, that distance provides the time that light from those most distant stars has been traveling toward Earth and thus the age of the universe. These two methods give the same age.

Earth is thought to have formed by collision and accretion of small dust and rock-sized particles in the interstellar dust cloud that also formed the sun and solar system. The **oldest rocks**, "faux-amphibolites," from northern Quebec, Canada, are dated at about 4.28 billion years, and rocks from the Northwest Territories are 4.03 billion years old (Figure 5.8). The oldest minerals are 4.4-billion-year-old zircons from Australia's Jack Hills. Earth is now estimated to be about 4.55 billion years ± 50 million years. This is based at least partly on the 4.5-billion-year age of the oldest moon rocks.

Over the next 4 billion years, as the sun consumes itself in fusion reactions, the amount of solar radiation reaching Earth will increase. As the sun's hydrogen is depleted (the hydrogen fuses to helium and heavier elements), the density of the core increases, raising the pressure at the surface of the core. This causes the nuclear reactions to run hotter and the sun to get brighter. That process is slow at first, but eventually accelerates. It is estimated that in about 1.1 billion years, solar luminosity will be 10% higher than now. This will cause the oceans to boil off. Four billion years from now, the increase in Earth's surface temperature will cause the extinction of all life on the surface. In about 7.5 billion years, the planet will be absorbed by the sun, which will have become a red giant and expanded across Earth's orbit.

References

McPhee, J. 1982. *Basin and Range*. Farrar, Straus and Giroux, New York, 224 pp.
McPhee, J. 1984. *In Suspect Terrain*. Farar, Strauss and Giroux, New York, 224 pp.

6

Earth History

Time is a sort of river of passing events, and strong is its current; no sooner is a thing brought to sight than it is swept by and another takes its place, and this too will be swept away.

Marcus Aurelius
Roman Emperor and stoic philosopher

Reunite Gondwana!

Anonymous

The Geologic Time Scale

The T-Rex, velociraptors, and most other dinosaurs featured in *Jurassic Park* are in the wrong place. These animals are from the late Cretaceous, 45 to 70 million years after the Jurassic. They belong in Cretaceous Park.

The **geologic time scale** refers to sequential periods of *time* (not rocks) as defined by the fossil succession found in rocks and described by European geologists in the 1700s to 1800s. The major divisions (**Eons**) represent phases in development of the earth. The time before the first fossils was originally called the Precambrian and is now subdivided into **Hadean** (from Hades, the hellish conditions on early Earth), **Archaean** (ancient—no life), **Proterozoic** (earlier life), and **Phanerozoic** (visible life). The Phanerozoic is separated into **Eras** on the basis of fossils and includes the **Paleozoic** (early life), **Mesozoic** (middle life), and **Cenozoic** (new life). Eras are in turn divided into Periods and Epochs (Figure 6.1).

Time designations come from locations where rocks of that age were first described ("Devonian" from Devon in England; "Jurassic" from the Jura Mountains on the French–Swiss border), ancient tribes that lived nearby ("Ordovician" after the Ordovice in Wales), rock type in the area ("Cretaceous" from the Latin *creta*, or "chalk"), or a sequence of rocks ("Triassic" from three distinct layers found throughout Germany). The original relative positions on this chart have been refined and updated by new fossil relationships and by absolute radiometric age dating of interlayered igneous rocks. There are slight name differences from place to place, such as "Carboniferous" in Europe that is the same interval as the "Mississippian" plus "Pennsylvanian" in the United States.

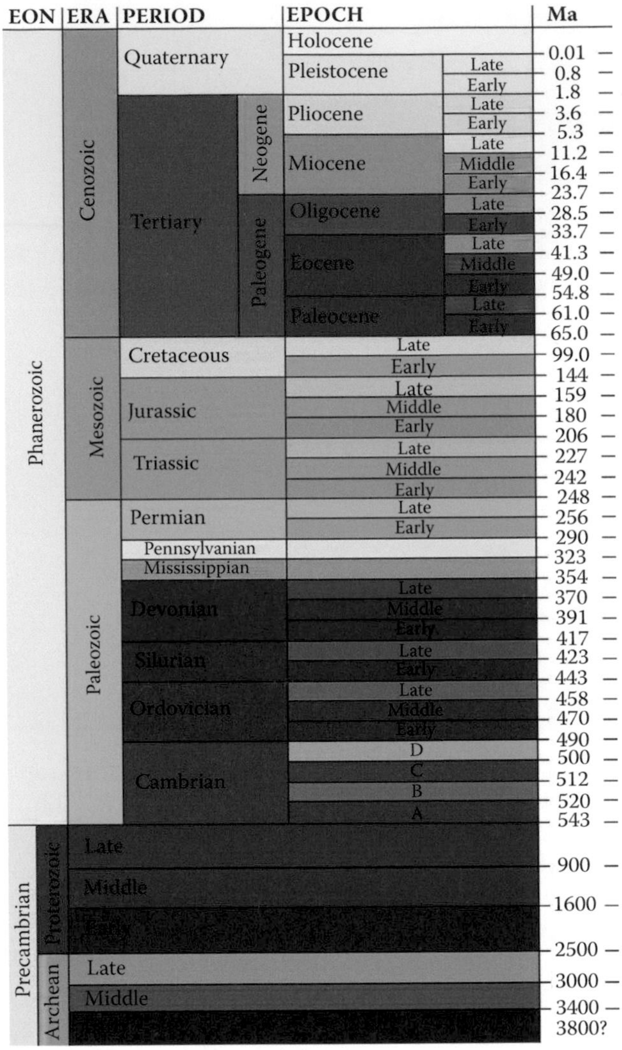

EON	ERA	PERIOD		EPOCH		Ma
Phanerozoic	Cenozoic	Quaternary		Holocene		0.01 —
				Pleistocene	Late	0.8 —
					Early	1.8 —
		Tertiary	Neogene	Pliocene	Late	3.6 —
					Early	5.3 —
				Miocene	Late	11.2 —
					Middle	16.4 —
					Early	23.7 —
			Paleogene	Oligocene	Late	28.5 —
					Early	33.7 —
				Eocene	Late	41.3 —
					Middle	49.0 —
					Early	54.8 —
				Paleocene	Late	61.0 —
					Early	65.0 —
	Mesozoic	Cretaceous		Late		99.0 —
				Early		144 —
		Jurassic		Late		159 —
				Middle		180 —
				Early		206 —
		Triassic		Late		227 —
				Middle		242 —
				Early		248 —
	Paleozoic	Permian		Late		256 —
				Early		290 —
		Pennsylvanian				323 —
		Mississippian				354 —
		Devonian		Late		370 —
				Middle		391 —
				Early		417 —
		Silurian		Late		423 —
				Early		443 —
		Ordovician		Late		458 —
				Middle		470 —
				Early		490 —
		Cambrian		D		500 —
				C		512 —
				B		520 —
				A		543 —
Precambrian	Proterozoic	Late				900 —
		Middle				1600 —
		Early				2500 —
	Archean	Late				3000 —
		Middle				3400 —
						3800?

FIGURE 6.1
Simplified geologic time scale. "Ma" is used to indicate millions of years (megaannum) before present. Note that the Hadean (4550 to about 4000 Ma) is not on this scale, and that the "scale" is not linear. (Courtesy of US Geological Survey.)

Key Events of the Precambrian Eon

The Precambrian covers the first 4 billion years of Earth's 4.5-billion-year history. Most of geologic time occurred before there were any obvious signs of life. This huge span of time has been divided into the Hadean, Archean, and Proterozoic eons.

The Hadean: Formation of the Earth, Moon, and Atmosphere

The **Hadean** is when Earth was formed by the accretion of space debris and was mostly molten. The heat came from constant impacts blasting the surface and from an abundance of radioactive elements (uranium, thorium, and potassium). Lasting from the formation of

FIGURE 6.2
Cathodoluminescence image of a 4.374-billion-year-old old zircon crystal from Jack Hills, Australia. Crystal is 400 microns in size. (Courtesy of Dr. John Valley, University Wisconsin, Madison; http://geoscience.wisc.edu /geoscience/people/faculty/john-valley/john-valley-incle-on-zircons/.)

Earth 4.55 billion years ago to around 4.0 billion years ago, there are only a few locations in western Greenland, northwestern Canada, and western Australia with minerals (zircons) this old (Figure 6.2).

The **giant impact hypothesis** claims that, about 50 million years after Earth reached a size close to its present mass (about 4.5 billion years ago), it was struck by an object about a third the mass of Earth at a shallow angle, what Robert Hazen calls "the big thwack" (Hazen 2012). The collision vaporized some of Earth's outer layers and melted both bodies. Some mantle material was ejected into space. The ejecta in orbit around Earth coalesced under the influence of gravity into a single, spherical body, the **Moon** (Figure 6.3).

Gases released by volcanoes created a **primitive atmosphere** rich in water vapor, carbon dioxide (CO_2), sulfur dioxide, and ammonia, with little or no free oxygen. Some surface water is thought to have existed, despite the high temperatures, owing to the high atmospheric pressure of the CO_2-rich atmosphere.

The Archaean: Beginning of Life and the Oldest Rocks

By the **Archean**, 4 billion years ago, Earth's crust had solidified and was cooling. There is an ocean derived from condensed atmospheric water vapor. The **"late heavy bombardment"** was a time of large impacts between 3.8 and 4.1 billion years ago. Many of the Moon's craters record this event. There are no craters of this age on Earth because erosion and plate tectonics have recycled the rocks. By 4 billion years ago, continents exist and plate tectonics, the slow movement of Earth's crustal plates, has begun. Surface water erodes and deposits the

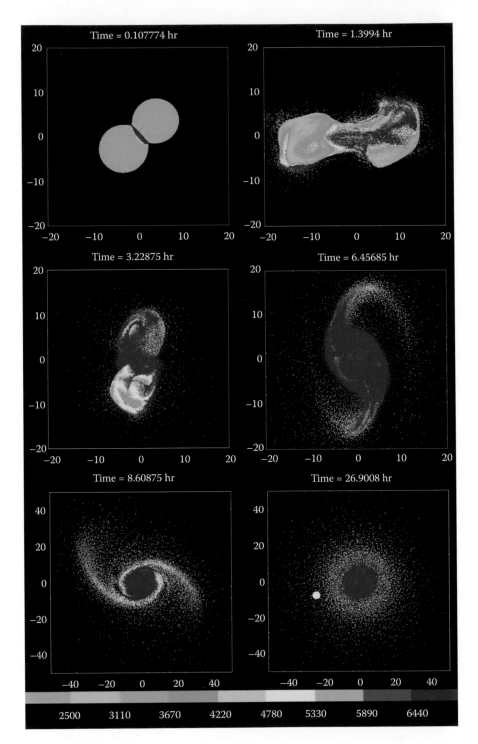

FIGURE 6.3
Model simulation for formation of the Moon by a large impact at a grazing angle. (Modified after NASA and Dr. R. Canup of Southwest Research Institute; https://sservi.nasa.gov/articles/nasa-lunar-scientists-produce -new-model-for-earthmoon-formation/.)

first sedimentary rocks, and metamorphic rocks form by deep burial and heating. Archean rocks have been found in Greenland, Canada, the US (Wyoming and Montana), the Baltic Shield province in Scandinavia and Russia, the Ukranian Shield in Ukraine and Russia, Scotland, India, Brazil, Western Australia, and southern Africa. The Archean atmosphere has evolved and contains CO_2, nitrogen, ammonia, methane, and traces of free oxygen. Life probably started near the beginning of the Archean and evolved ever so slowly into simple single-celled organisms (archaea and bacteria—see Chapter 8, Evolution).

The Proterozoic: Photosynthesis, the Great Oxygenation Event, and Snowball Earth

At the beginning of **Proterozoic** time about 2.5 billion years ago, single-celled organisms had evolved into primitive photosynthetic bacteria. Breathing carbon dioxide and gener- ating free oxygen over a 100-million-year period, from 2.35 to 2.45 billion years ago, they depleted the atmosphere's carbon dioxide enough to plunge the earth into several ice ages during that interval. By 2.22 billion years, we have definite evidence of an oxygen-rich atmosphere: the extensive manganese oxide deposits in the Kalahari field of South Africa could only have formed in an oxidizing environment. By this time, the first eukaryotes, cells with a nucleus containing mitochondria in animals and chloroplasts in plants, had evolved. Chloroplasts use the energy from sunlight to convert atmospheric carbon and mineral nutrients into sugars in the process called **photosynthesis**. Photosynthesis was perfected by cyanobacteria (formerly called "blue-green algae"). Extracting the carbon from carbon dioxide generates oxygen as a waste product. The first oxygen produced by bacteria oxidized iron and other minerals in the oceans. This caused precipitation of iron- rich sediments that became **banded iron formations**, rocks that contain important iron deposits in Minnesota, South Africa, and Western Australia (Figure 6.4). About 2.4 bil- lion years ago, free oxygen began to accumulate in the atmosphere. The conversion to an oxygen-rich atmosphere around 2.2 billion years ago is called the "**great oxygenation event**." Ozone (O_3) could not form until there was free oxygen in the atmosphere. Ozone in

FIGURE 6.4
Two-billion-year-old banded iron formation, Karijini National Park, Western Australia. (Courtesy of G. Churchard; https://simple.wikipedia.org/wiki/Banded_iron_formation.)

the atmosphere protects surface life from the harmful effects of solar ultraviolet radiation. Since humans depend on oxygen and ozone, we owe our very existence to lowly bacteria.

The billion years between the great oxygenation event and the appearance of multicellular life around 1 billion years ago has been called the "boring billion." It was thought that not much happened during this interval. And yet we see that 2 billion years ago, only single-celled eukaryotes were common, mostly amoeba, paramecia, and their ilk. Multicellular life at this time consisted of slime molds and colonies of single-celled algae that formed layered mounds called **stromatolites**. *Grypania*, spiral-shaped strings of cells held together by membranes, developed about 1870 Ma. Multicellular plants, probably similar to the green and red algae found in shallow marine settings today, evolved slightly over a billion years ago. There is evidence that single-celled plants may have spread extensively on land starting around 750 Ma. A lot was happening, albeit slowly, over these billion years.

Animals are multicellular life with specialized cells that work together to perform complex behaviors such as movement, respiration, circulation, and digestion. Animals didn't appear in abundance until around 600 Ma. Complex multicellular organisms require a lot of energy to communicate between cells. Oxygen provides that energy. The newly oxygen-rich atmosphere of 2.2 billion years ago allowed development of complex multicellular organisms. (The composition of the **atmosphere** has continued to change through geologic time: it is currently 21% oxygen, 78% nitrogen, 0.9% argon, and 0.04% CO_2.)

Organisms multiplied and diversified. Their soft-bodied remains in rocks as old as 570 Ma are seen in the Ediacara fauna of South Australia, Newfoundland, and the Northwest Territories in Canada (see Chapter 7, Fossils).

There is evidence for at least three global glaciations between 740 and 580 Ma that effectively put a damper on life for a while, particularly life dependent on sunlight. This **"Snowball Earth" hypothesis**, supported by glacial deposits and an increase in light carbon isotopes in limestones, asserts that, during the **Cryogenian** period (635–720 Ma), life barely survived in cold, ice-covered seas. The reasons for the cooling are speculative but may have to do with rock weathering and massive cyanobacteria blooms removing large amounts of carbon dioxide, a greenhouse gas, from the atmosphere. Life would have survived only where the ice was missing, either near the equator or around volcanic thermal vents. Warming at the end of this period, possibly attributed to a buildup of atmospheric carbon dioxide from volcanism, provided fresh opportunities for diversification. Life again flourished. Body size and complexity increased as a result of evolutionary pressure (climate change and competition), and creatures generated the first skeletons and shells. Because hard parts are easier to preserve as fossils, it looked like many different life forms emerged suddenly out of nowhere.

Key Events of the Paleozoic Era

An Explosion of Life

The **Cambrian explosion** describes the sudden appearance of diverse fauna at the beginning of the Paleozoic Era about 543 million years ago. The base of the Cambrian is recognized by tracks and burrows of moving animals. The oldest animal fossils are tracks of 525-million-year-old wormlike creatures found in the Chengjiang Shale, China. The

Cambrian (543–490 Ma) is when most of the major animal phyla appear in the fossil record. All animal life at this time is still in the ocean, and the seas are dominated by trilobites.

Maps that show what Earth looked like in the past are called **paleogeographic reconstructions**. Ancient geography is derived from the age and type of rock (beach sand, sand dunes, coral reefs) and fossils (leaves, seashells, bones) that provide clues to the environment.

Reconstructions suggest that the Paleozoic era began with a large southern continent, called **Gondwana**, and several smaller landmasses (Figure 6.5). During the **Avalonian orogeny**, a period of mountain building, a mountain chain developed from the Avalon Peninsula in Newfoundland through New England to the eastern Appalachians.

During the **Ordovician** (490–443 Ma), the **Taconic orogeny** raised up mountains in Vermont (Figure 6.6). Glaciers covered parts of Africa and adjacent South America, and global cooling is credited with causing an end-Ordovician mass extinction.

In the **Silurian** (443–417 Ma), ancestral North America–Greenland collided with proto-Europe during the **Caledonian orogeny**, creating the Caledonian Mountains that extend from Scandinavia through Britain, Ireland, and Greenland to the Appalachians of New York. The northern microcontinents began to collide to form a northern continent called **Laurasia** or **Euramerica** (Figure 6.7).

The **Devonian** (417–354 Ma), saw the **Acadian orogeny** build the Appalachians from the Gaspe Region of Canada to Alabama, and possibly extending as far as west Texas and northern Mexico. It was caused by the collision of tectonic plates carrying Avalonian continental fragments and Laurasia. The Acadian orogeny was coincident with the **Antler orogeny** mountain building event in Nevada and Utah, and with the early **Variscan orogeny** in Europe (Figure 6.8).

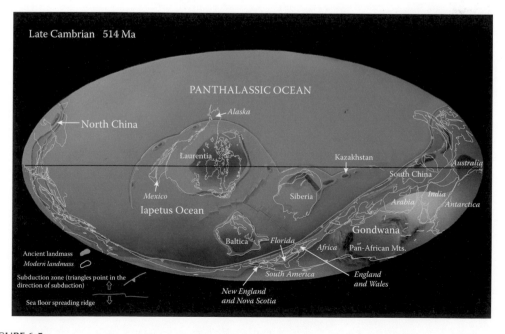

FIGURE 6.5
Late Cambrian geography (514 Ma). (From Scotese, C. R. 2001. *Atlas of Earth History, Volume 1, Paleogeography*, PALEOMAP Project, Arlington, Texas, 52 pp., with permission; http://www.scotese.com/earth.htm.)

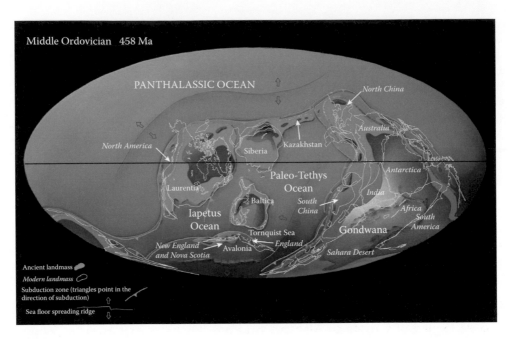

FIGURE 6.6
Middle Ordovician geography. (From Scotese, C. R. 2001. *Atlas of Earth History, Volume 1, Paleogeography*, PALEOMAP Project, Arlington, Texas, 52 pp., with permission; http://www.scotese.com/earth.htm.)

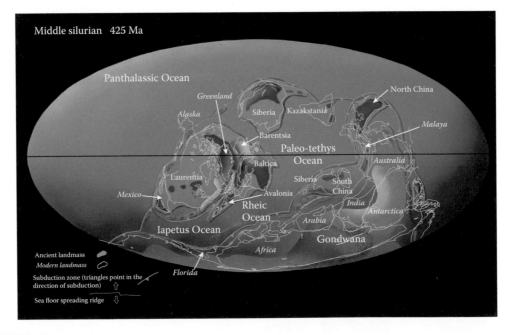

FIGURE 6.7
Middle Silurian geography. (From Scotese, C. R. 2001. *Atlas of Earth History, Volume 1, Paleogeography*, PALEOMAP Project, Arlington, Texas, 52 pp., with permission; http://www.scotese.com/earth.htm.)

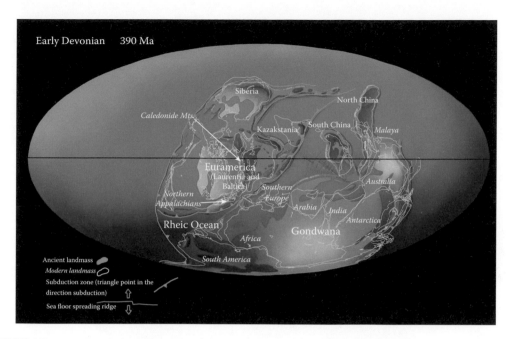

FIGURE 6.8
Early Devonian geography. (From Scotese, C. R. 2001. *Atlas of Earth History, Volume 1, Paleogeography*, PALEOMAP Project, Arlington, Texas, 52 pp., with permission; http://www.scotese.com/earth.htm.)

During the Devonian, **Placoderms** (armored fish) became dominant. The first coral reefs were built during the Devonian and cartilaginous fish such as sharks became common. A Late Devonian mass extinction affected mainly marine life and may have been caused by global warming, the first "**greenhouse extinction**."

First Land Plants

True **land plants** evolved perhaps as long ago as 510 Ma (during Cambrian time) from single-celled green algae living on land and in shallow freshwater pools. Fossil spores indicate that land plants existed in the Ordovician, about 450 Ma. Until then, the oceans had been teeming with life, but the land was largely barren. Land plants required development of not only a cuticle covering to prevent drying but also holes (stomata) to allow CO_2 and O_2 to pass through the cover. The first land plants were low-lying mosses and fungi. Standing upright allowed access to more light but required incorporating stiffening agents and development of roots to tie the plant to the ground. Land plants began to diversify about 430 Ma, in the Late Silurian (Figures 6.9 and 6.10). The oldest land plant, **Cooksonia** (about 425 Ma), had a spikey stem without leaves. Leafy plants were not common until around 360 Ma. Leaves, which capture yet more light, required lots of stomata and relatively low temperatures to prevent drying out the plant. Deep roots led to deeper weathering of bedrock, to reduced erosion, and to extensive soil development on land. **Ferns** became abundant, as well as plants with leaves and woody stalks. By late Devonian (360–380 Ma), there were forests of woody plants with roots and leaves. The oldest fossil **conifer** dates from around 300 Ma during the Carboniferous.

(a)

FIGURE 6.9
Some of the earliest land plants. (a) Silurian *Cooksonia pertoni* from Shrewsbury, England. (Courtesy of Hans
Steur, Hans' Paleobotany Pages; https://steurh.home.xs4all.nl/eng/old1.html.)   *(Continued)*

First Land Animals

Moving out of the sea required development of a respiratory system that stayed moist.
Scorpions and spiders (**arthropods**) developed these first, around 430 Ma, followed by mil-
lipedes and, around 410 Ma, early **insects**. The best fossils come from the **Rhynie Chert** of
Scotland. Insects that we recognize today didn't develop until around 330 Ma.

Arthropods' external skeleton kept their insides from drying out, but required holes to
let in oxygen and membranes to let the gas into their blood. Their gills become modified
into a set of flat membranes arranged like pages in a book, so-called "book lungs." These
passive lungs did not require pumping of air over the membranes. Insects developed a
different passive system that involved tube-like trachea that let oxygen into their blood.

Tracks in 490 Ma sand dunes suggest that amphibious arthropods were already on land
at the end of Cambrian time (Figure 6.11). The oldest known fossil of a land animal is a
Middle Silurian millipede, about 428 Ma.

The emergence of vertebrates onto land required relatively high oxygen levels for easier
breathing, modification of fins to fleshy lobes for movement out of water, and lungs. The
oldest bones are dated around 360 Ma, but there are tracks as old as 400 Ma. Gills evolved

(b)

FIGURE 6.9 (CONTINUED)
Some of the earliest land plants. (b) Devonian *Sawdonia ornata*, Glasgow, Scotland. (Courtesy of Hans Steur, Hans' Paleobotany Pages; https://steurh.home.xs4all.nl/eng/old2.html.)

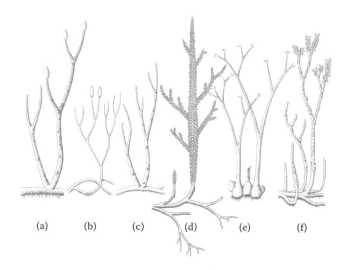

FIGURE 6.10
Lower Devonian (408–360 Ma) terrestrial plants from the Rhynie chert. (a) *Rhynia gwynne-vaughanii.* (b) *Aglaophyton major.* (c) *Ventarura lyonii.* (d) *Asteroxylon mackiei.* (e) *Horneophyton lignieri.* (f) *Nothia aphylla.* (Drawing by Falconaumanni; https://commons.wikimedia.org/wiki/File:Rhynie_Chert_flora.jpg.)

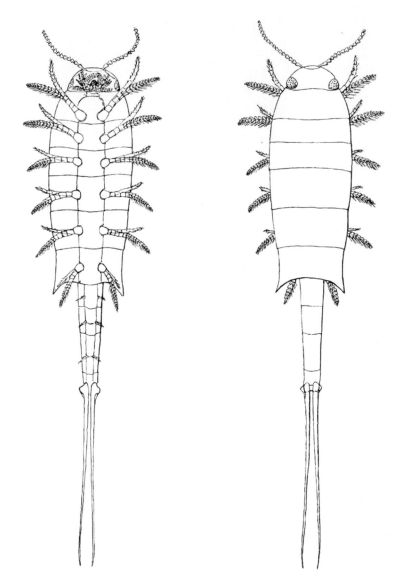

FIGURE 6.11
Euthycarcinoid arthropod that left 490 million year old tracks. (Courtesy of A. Handlirsch; https://en.wikipedia
.org/wiki/Euthycarcinoidea.)

to lungs that could breathe air. The rib cage, wrist, ankle, backbone, shoulders, and pelvic bones had to evolve to support weight out of water. These modifications were not in place until 330–340 Ma.

The first land vertebrates were **Tetrapods** (four limbed creatures) that evolved from lobe-finned fish in the late Devonian. **Tiktaallik**, a lobe-finned fish, is thought to have lived about 375 Ma in oxygen-poor shallow waters and may have been the transitional form to terrestrial amphibians (Figures 6.12 and 6.13). The first **amphibians**, animals like frogs and salamanders that live on land or in freshwater and need water to reproduce, are found in Devonian rocks from 360 to 370 million years old.

FIGURE 6.12
Devonian Tiktaallik skeleton from Ellesmere Island, Nunavut, Canada. (Courtesy of Eduard Solà, Field Museum, Chicago; https://en.wikipedia.org/wiki/Tiktaalik.)

FIGURE 6.13
Reconstruction of Tiktaallik. (Courtesy of Obsidian Soul; https://en.wikipedia.org/wiki/Tiktaalik.)

The **Carboniferous** (carbon-bearing) is named for the abundant coal beds derived from swamps and forests that existed 355–290 million years ago. The first known **reptiles** are from this time (Figure 6.14). The collision of the northern and southern continents during the Carboniferous created the supercontinent **Pangaea**. The single continent was surrounded by the **Panthalassa Sea**. Animals and plants were able to move more or less freely across this continent. The Ancestral Rockies formed in western North America.

FIGURE 6.14
Reconstruction of *Hylonomus*, an early Carboniferous reptile. (Courtesy of Nobu Tamura; https://en.wikipedia
.org/wiki/Evolution_of_reptiles.)

The Ouachita Mountains of Oklahoma and the Allegheny and southern Appalachian
Mountains of the eastern United States were raised up during the **Alleghenian orogeny**,
when Africa collided with North America (Figures 6.15 and 6.16). The same event raised
the anti-Atlas mountains in northeast Africa. The **Hercynian orogeny** built mountains
that extend from Ireland to southeastern Germany. The Great Dividing Range of Australia
was formed during the Carboniferous when Australia collided with South America–New
Zealand. Glaciers covered parts of present-day Antarctica, Australia, southern South
America, Africa, and India.

The **Permian**, lasting from 290 to 248 Ma, saw the main uplift of the Ural Mountains
when Europe and Asia collided. The Hercynian orogeny continued in Europe, and the
southern Appalachians continued to rise as Africa–South America crashed into North
America (Figure 6.17).

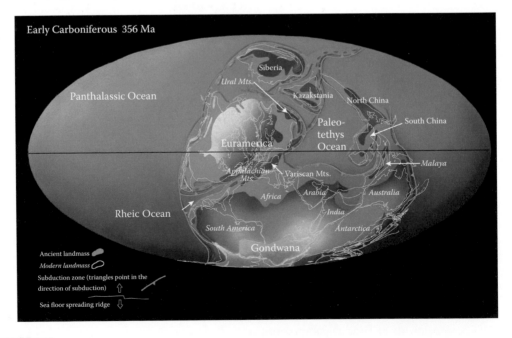

FIGURE 6.15
Early Carboniferous (Mississippian) geography. (From Scotese, C. R. 2001. *Atlas of Earth History, Volume 1,
Paleogeography*, PALEOMAP Project, Arlington, Texas, 52 pp., with permission; http://www.scotese.com/earth.htm.)

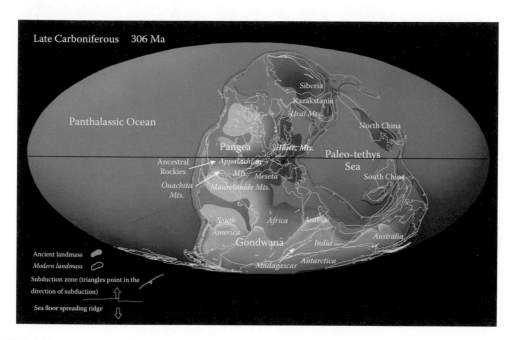

FIGURE 6.16
Late Carboniferous (Pennsylvanian) geography. (From Scotese, C. R. 2001. *Atlas of Earth History, Volume 1, Paleogeography*, PALEOMAP Project, Arlington, Texas, 52 pp., with permission; http://www.scotese.com/earth.htm.)

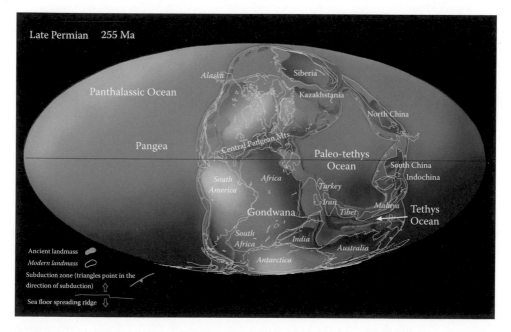

FIGURE 6.17
Late Permian geography. All previous continents have combined into one supercontinent, Pangaea. (From Scotese, C. R. 2001. *Atlas of Earth History, Volume 1, Paleogeography*, PALEOMAP Project, Arlington, Texas, 52 pp., with permission; http://www.scotese.com/earth.htm.)

FIGURE 6.18
Permian Dimetrodon skeleton from Texas. (Courtesy of H. Zell, Staatliches Museum für Naturkunde Karlsruhe; https://en.wikipedia.org/wiki/Dimetrodon.)

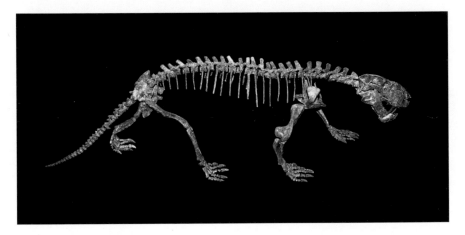

FIGURE 6.19
Upper Permian Gorgonopsid from the Ruhuhu Valley, Tanzania. (Courtesy of H. Zell, Staatliches Museum für Naturkunde Karlsruhe; https://en.wikipedia.org/wiki/Gorgonopsia.)

The iconic reptile **Dimetrodon** lived during the Permian (Figure 6.18). **Gorgonopsids** were a group of reptiles that included the ancestors of mammals. During the Permian, they evolved from dog-sized to rhinoceros-sized dominant predators (Figure 6.19).

The Great Extinction

Glaciers covered much of Gondwana. Toward the end of Permian time, massive flood basalt eruptions in Siberia may have contributed to the greatest mass extinction in history, wiping out nearly 95% of all marine species and 70% of terrestrial vertebrates. Flood basalts are associated with vast outpourings of carbon dioxide and thus global warming.

There is evidence from oxygen isotopes that seawater was in excess of 40°C (104°F) and that continents near the equator had temperatures in the range of 60°C (140°F) during the first million or so years after the extinction. Other hypotheses include a large impact (no evidence has been found) and the suggestion that a low oxygen atmosphere combined with rotting of abundant organic matter released massive amounts of poisonous hydrogen sulfide into the atmosphere and oceans over several pulses.

Key Events of the Mesozoic Era

Dinosaurs

The Mesozoic is when dinosaurs became the dominant life form on Earth. It is also the time when the first bird and mammal fossils appear. Fossils indicate that **deciduous trees** and **flowering plants** (**angiosperms**) first appeared and diversified rapidly during the Cretaceous. Pangea broke into the northern continent of **Laurasia** and a southern continent of **Gondwana**, and these then broke into several subcontinents. This dispersion of continents led to a divergence of species.

After the Permian extinction, it took almost 30 million years to re-establish stable ecosystems. The extinction of **rugose** and **tabulate corals** (see Corals, Chapter 7) led to the rapid spread of **hexacorals**, the coral we know today. The extinction of the Gorgonopsids allowed **crocodilians** and **dinosaurs** to became the dominant predators starting in the **Triassic** (248–206 Ma). This drove the ancestors of mammals to become burrowing nocturnal insect eaters. Key evolutionary changes in mammals at that time include differentiation of their teeth, better thermal regulation for nocturnal life, and an improved sense of hearing and smell. The oldest group of mammals are egg-laying **monotremes** that today include the platypus and echidna. Mammals split into monotremes, **marsupials** (those with a pouch for their young), and **placentals** (mammals with a placenta, including us) during the Jurassic.

Birth of the Atlantic Ocean

During the Triassic, the **Atlantic Ocean** began to form by rifting in the Caribbean and between North America and North Africa (Figure 6.20). Flood basalts and associated volcanic gases of the rift-related Central Atlantic Magmatic Province have been cited as a cause for the end-Triassic mass extinction about 200 Ma. About 250 million years ago, the **Tethys Sea** began as a rift in southern Pangaea (Figure 6.21). This ocean lasted most of the Mesozoic era and separated Laurasia from Gondwana. All the southern glaciers were gone. Cycad and conifer forests flourished (e.g., the **Petrified Forest** of Arizona). The **Andes** began in the Triassic and continued to develop throughout the Jurassic period.

During the end-Triassic extinction event about 30% of all marine life and about half of the land species went extinct. A hot climate and low atmospheric oxygen levels, and massive volcanic eruptions, have all been cited as reasons for this extinction, but no consensus exists.

While large marine reptiles such as **ichthyosaurs** and **plesiosaurs** ruled the seas, modern **bony fish** began to spread and diversify during the **Jurassic** (206–144 Ma). **Ammonites** became important marine **index fossils**, that is, fossils used to identify

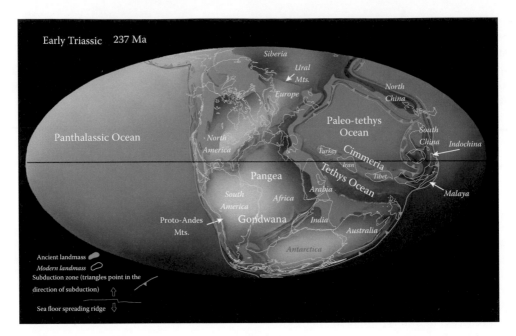

FIGURE 6.20
Early Triassic plate tectonics. The components of Pangaea are shown. (From Scotese, C. R. 2001. *Atlas of Earth History, Volume 1, Paleogeography*, PALEOMAP Project, Arlington, Texas, 52 pp., with permission; http://www .scotese.com/earth.htm.)

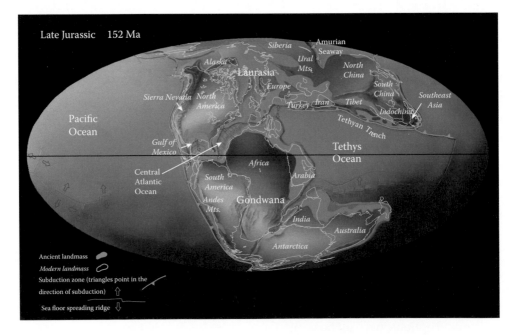

FIGURE 6.21
Late Jurassic geography. Pangaea is breaking up and the Atlantic Ocean is opening. (From Scotese, C. R. 2001. *Atlas of Earth History, Volume 1, Paleogeography*, PALEOMAP Project, Arlington, Texas, 52 pp., with permission; http://www.scotese.com/earth.htm.)

TABLE 6.1

Some Key Index Fossils

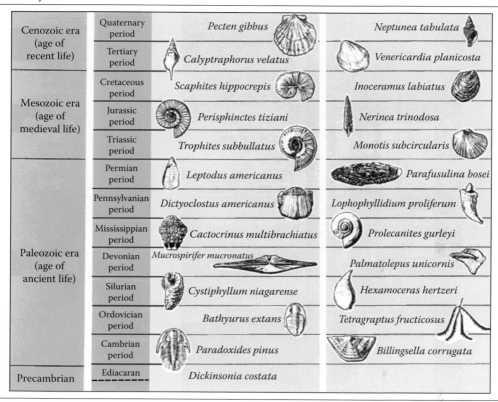

Cenozoic era (age of recent life)	Quaternary period	*Pecten gibbus*	*Neptunea tabulata*
	Tertiary period	*Calyptraphorus velatus*	*Venericardia planicosta*
Mesozoic era (age of medieval life)	Cretaceous period	*Scaphites hippocrepis*	*Inoceramus labiatus*
	Jurassic period	*Perisphinctes tiziani*	*Nerinea trinodosa*
	Triassic period	*Trophites subbullatus*	*Monotis subcircularis*
Paleozoic era (age of ancient life)	Permian period	*Leptodus americanus*	*Parafusulina bosei*
	Pennsylvanian period	*Dictyoclostus americanus*	*Lophophyllidium proliferum*
	Mississippian period	*Cactocrinus multibrachiatus*	*Prolecanites gurleyi*
	Devonian period	*Mucrospirifer mucronatus*	*Palmatolepus unicornis*
	Silurian period	*Cystiphyllum niagarense*	*Hexamoceras hertzeri*
	Ordovician period	*Bathyurus extans*	*Tetragraptus fructicosus*
	Cambrian period	*Paradoxides pinus*	*Billingsella corrugata*
Precambrian	Ediacaran	*Dickinsonia costata*	

Source: US Geological Survey (https://en.wikipedia.org/wiki/Index_fossil).

specific time intervals (Table 6.1). **Pterosaurs**, flying reptiles, roamed the skies. This was the peak of **Ornithischian dinosaurs** like the armored Stegosaurus, and Diplodocidae like Brontosaurus, Apatosaurus, and Diplodocus. The first **birds** appear during the Jurassic. Forests containing cycads, conifers, ferns, and ginkgoes covered the land.

The **Nevadan Orogeny** raised up the ancestral Sierra Nevada through the emplacement of granite batholiths between about 180 and 87 Ma. Convergence of the Pacific and North American tectonic plates caused the **Sevier orogeny**, building much of the Rocky Mountains from Nevada to Alaska.

Flowering Plants

The **Cretaceous** period lasted from 144 to 65 million years ago. The first **flowering plants** appeared at this time. This is the time of the largest horned dinosaurs (Ceratopsians like Triceratops), carnivorous Theropods like Tyrannosaurus, and the duck-billed Hadrosaur.

The Brooks Range of northern Alaska formed in the early Cretaceous about 126 Ma. The **Laramide orogeny** in late Cretaceous time uplifted the eastern Rocky Mountains from New Mexico in the United States to the Northwest Territories in Canada. A shallow inland sea extended from the Gulf of Mexico to the Arctic Sea, dividing North America into two

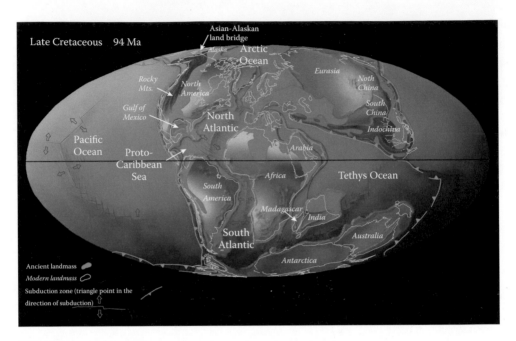

FIGURE 6.22
Late Cretaceous geography. The Atlantic now exists, a shallow sea cuts North America in half, and India is moving north toward Asia. (From Scotese, C. R. 2001. *Atlas of Earth History, Volume 1, Paleogeography*, PALEOMAP Project, Arlington, Texas, 52 pp., with permission; http://www.scotese.com/earth.htm.)

land areas (Figure 6.22). The Atlantic continued to widen. India split from Africa, Australia is almost separated from Antarctica, and New Zealand rifted away from Australia. The Mediterranean formed as Africa moved north into Europe and closed off the last of the Tethys Ocean. The Andes formed by convergence of the Pacific and South American plates. The Pyrenees formed in the early Cretaceous when the block containing Spain plowed into Europe. The **Alpine orogeny** is a result of the collision of the African and Eurasian tectonic plates starting in the Cretaceous. During this time, the Tethys, formerly between these continents, disappeared. Mesozoic and early Cenozoic strata of the Tethys basin were pushed against the stable Eurasian landmass by the northward-moving African landmass, mainly during the Oligocene and Miocene (6–35 Ma). This orogeny formed the Alps, Atlas, the Rif, the Apennine Mountains, the Hellenides, the Carpathians, the Balkan Mountains, the Taurus, the Caucasus, the Zagros, the Hindu Kush, Karakoram, and Himalayas. Much of this convergence and uplift continues today.

End of the Dinosaurs

The end of the Cretaceous period 65 million years ago is marked by another great extinction event known at the **Cretaceous-Paleogene (K-Pg) extinction** or **Cretaceous-Tertiary (KT) extinction**. About 75% of all animal and plant species disappeared, including all dinosaurs except birds. The event is marked in the geologic record by a worldwide thin layer with high levels of iridium, an element found mainly in asteroids (Figure 6.23). It is believed that a large asteroid hit Earth at the **Chicxulub** crater in Yucatan, Mexico, at this time. Everything near the impact was instantly vaporized. Beyond that, there were massive fires. The impact filled the atmosphere with so much dust that it caused an "**impact**

FIGURE 6.23
K–T boundary iridium layer near Starkville, Colorado, showing the light gray boundary claystone and overlying thin coal bed (https://commons.wikimedia.org/wiki/File:K-T_boundary_at_Starkville_South.jpg).

winter" that stopped photosynthesis for many years. This may have been exacerbated by the eruption of massive flood basalts, and associated carbon dioxide and sulfur dioxide, of the Deccan Traps in India. The ecological niches made available by this extinction led to an explosion of new species known as the **Paleocene radiation**. This diversification was especially pronounced in mammals and birds.

Key Developments of the Cenozoic Era

The **Cenozoic era** covers the past 65 million years. It is the time when mammals and **angiosperms** (flowering plants) dominated the land, mammals became the largest marine creatures, and birds diversified and ruled the skies. Angiosperms became the most diverse land plants: they are characterized by flowers and seed-containing fruits. The continents we know today are moving into place (Figure 6.24).

Mammals

Following the breakup of Pangaea, mammals were separated and evolved along different paths. Placental mammals took over in Europe–Asia–Africa–North America while

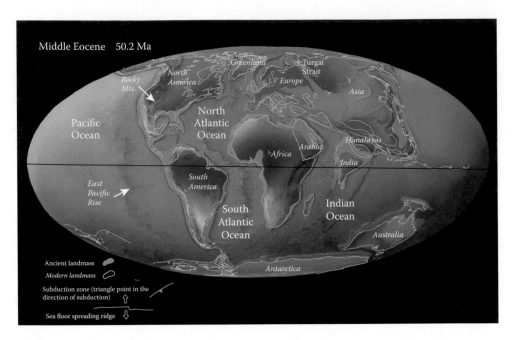

FIGURE 6.24
Middle Eocene geography. (From Scotese, C. R. 2001. *Atlas of Earth History, Volume 1, Paleogeography*, PALEOMAP Project, Arlington, Texas, 52 pp., with permission; http://www.scotese.com/earth.htm.)

marsupials filled the ecological niches in Australia, and placentals and marsupials co-existed in South America. At this time, South America was an island continent similar to Australia. The Americas linked up about 3 million years ago when volcanic erup-tions in Central America built a land bridge between the continents. This led to the **great American interchange**, where South American animals such as armadillos, opossums, and porcupines migrated north and North American animals such as deer, bear, and horses migrated south. The exchange presumably led to the extinction of animals that were less competitive.

Separation of the continents provided a wonderful example of **evolutionary conver-gence**. Placentals and marsupials occupying the same niche evolved similar traits. Examples include the European placental saber-toothed cat and the South American marsupial saber-toothed tiger, the European wolf and Tasmanian wolf, and marsupial and placental moles.

Disappearance of the dinosaurs and subsequent availability of ecological niches led directly to the diversification and radiation of mammals. **Megafauna**, unusually large mammals, existed during the Paleogene (65–23 Ma). At the same time, mammals returned to the sea where they diversified and became the largest marine creatures. The earliest whale ancestors (*Pakicetus*) were carnivorous, dog-sized land animals that lived in and around estuaries about 50 million years ago (Figure 6.25). As more of their lives were spent in the water (e.g., *Ambulocetus*, 50–48 Ma), they evolved paddle-like limbs, nostrils farther back on the head, and a longer and stronger tail. Eventually, they lost their external limbs altogether. Marine mammals also include seals, dolphins, porpoises, narwhals, manatees, sea otters, walrus, and polar bears.

The human lineage (*Homo*) is thought to have diverged from a common ances-tor with chimpanzees during the Miocene epoch, between 5 and 13 Ma (see Ongoing Evolution of Humans, Chapter 8). The large land animals associated with the last ice age

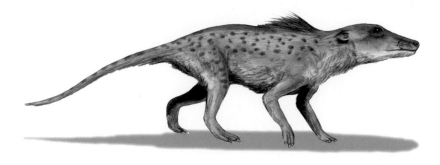

(a)

(b)

(c)

FIGURE 6.25
Evolution of the whale from a dog-sized land animal to what we know today. (a) Artist drawing of an Early Eocene Pakicetus. This land-dwelling ancestor of the whale lived 48–49 million years ago and was 1 to 2 m long. (b) The crocodile-like Rhodocetus lived in shallow marine waters around 48 million years ago. (c) Late Eocene Dorudon lived in a fully marine setting around 35 million years ago. (Courtesy of Nobu Tamura; https://en.wikipedia.org/wiki/Evolution_of_cetaceans.) *(Continued)*

(75,000–12,000 years ago) such as mammoths, mastodons, Megatherium (giant sloth), Paraceratherium (giant rhinoceros), Bison latifrons, Megaloceros giganteus (giant Irish Elk), Arctotherium (a giant short-faced bear), *Proborhyaenidae* (a hyena-like marsupial carnivore), giant flightless birds such as the Moa, all became extinct between 50,000 and

(d)

FIGURE 6.25 (CONTINUED)
Evolution of the whale from a dog-sized land animal to what we know today. (d) Cetotherium lived around 18 million years ago during the Miocene. (Courtesy of Nobu Tamura; https://en.wikipedia.org/wiki /Evolution_of_cetaceans.)

10,000 years ago. One theory for their demise is that they were hunted to extinction by Stone Age humans. Environmental changes associated with a warming climate likely also played a role.

Ice Ages

Most of the Cenozoic was a time of moderate to cool worldwide temperatures. However, there was an episode around 55.5 Ma, the **Paleocene–Eocene Thermal Maximum (PETM)**, when Earth's temperatures were 5°C–8°C higher than they are today. The cause for this 200,000-year temperature spike is unknown. During the PETM, Earth was essentially ice-free and sea levels would have been about 70 m (230 ft) higher than today.

The first effects of **global cooling** would likely have been felt in Antarctica. During the Cretaceous, 100 million years ago, Antarctica was covered by lush rain forests similar to those seen in New Zealand today. Between 37 and 34 Ma (at the end of Eocene), it was covered by woodland and tundra dominated by conifers and beech. After separating from Africa–Australia–South America between 30 and 35 million years ago, the Circumpolar Current cut Antarctica off from warmer waters, cooling it significantly. Fossils of Gondwanan seed ferns (Figure 6.26) and extensive coal deposits indicate that the climate had been warmer and that Antarctica had the same flora and fauna as Gondwana (Africa–India–Australia–South America). By 15 Ma, Antarctica was frozen under a thick ice sheet. All vertebrate life had been wiped out except for penguins and seals. The flora now consists only of lichens, mosses, and liverworts.

The most recent ice age, the **Quaternary glaciation** or **Pleistocene glaciation**, began about 2.6 million years ago and lasted until 11,500 years ago (Figure 6.27). There have been a number of cold periods (**glacials**) and warm periods (**interglacials**) during this time. The primary driver for this cold snap is controversial but probably has something to do with variations in solar irradiance, the amount of energy Earth gets from the sun. See Chapter 17, Climate Change, for a detailed discussion.

Ice sheets covered much of Europe and the northern parts of North America and Asia (Figure 6.28). Antarctica and Greenland were buried beneath glaciers. Grassland savanna replaced much of the tropical rain forest of central Africa, a change that may

FIGURE 6.26
Frond of *Dicroidium dubium* from the Middle Triassic of New South Wales, Australia. (Courtesy of Timotheus K.T. Wolterbeek, Utrecht University.)

have spurred the evolution of humans. A wet, temperate climate in the Middle East between 9000 and 12,000 years ago was instrumental in the initiation and development of early agriculture.

The Earth We Know

Many of the mountain belts seen around the world developed during the Cenozoic. The **Atlas Mountains** of Morocco and Algeria were uplifted when Africa collided with Europe 66 million years ago. The Himalayas grew when the Indian subcontinent collided with Asia around 50 million years ago. The **Indian Plate** is still moving north about 65 mm/year and continues to push up the **Himalayas**, **Karakoram Range**, **Hindu Kush**, **Arakan Yoma**, and ranges as far inland as the **Tien Shan** in northern China.

Rifting is another consequence of Cenozoic tectonics. The **Red Sea–Dead Sea rift** extends from Syria south to the Afar Depression in Ethiopia. The rift separates the African and Arabian tectonic plates, which are spreading apart about 1 cm/year. Rifting began between 40 and 55 million years ago and, after a 30-million-year lull, started again 5 million years ago. The rifting has been accompanied by extensive basaltic volcanism.

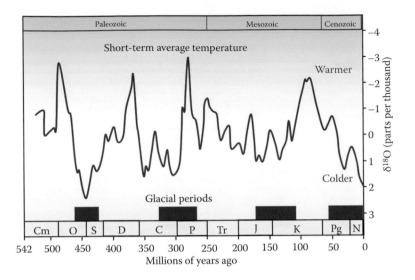

FIGURE 6.27
Glacial periods over the history of the Earth. The most recent ice age is at the right end of graph. The oxygen isotope ratio $^{18}O/^{16}O$, known as $\delta^{18}O$ ("delta O-18"), measured in ice cores is used to determine the temperature of precipitation through time. $\delta^{18}O$ is a proxy for temperature. (Courtesy of Global Warming Art; https://en.wikipedia.org/wiki/Timeline_of_glaciation.)

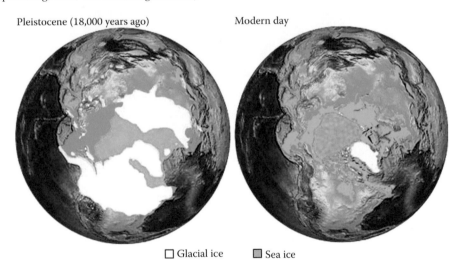

FIGURE 6.28
Maximum extent of Northern Hemisphere glaciers 18,000 years ago and at present. North polar view. (Courtesy of NASA Earth Observatory; http://earthobservatory.nasa.gov/Features/BorealMigration/boreal_migration2.php.)

The **Southern Alps** of New Zealand lie along the Alpine Fault, which marks the boundary between the **Australian Plate** and **Pacific Plate**. Convergence of the plates along this boundary over the past 45 million years has uplifted these spectacular mountains, most rapidly in the past 5 million years (Figure 6.29).

Convergence of the **Farallon Plate** and **North American Plate** led to subduction and development of a volcanic arc, the **Cascade Ranges**, starting about 37 million years ago.

FIGURE 6.29
Mt. Cook, Southern Alps, South Island, New Zealand. (Courtesy of G. Prost.)

This andesitic volcanic arc extends from British Columbia to northern California and is still active: Mt. Lassen last erupted in 1914–1917, and Mt. St. Helens erupted in 1980. The **Cascadia Fault** offshore Oregon and Washington is the subduction zone plate boundary. It is of concern because it is capable of generating large tsunamis.

The **Great Basin** and **Rio Grande rift** of the western United States and northern Mexico represent crustal extension that began about 35 Ma and continues today. East–west extension in this region, the **Basin-and-Range geologic province**, is accommodated by normal faulting and tilting of the down-dropped basins and uplifted ranges (Figure 6.30). About

FIGURE 6.30
Galiuros Mountains of Arizona are tilted fault-blocks in the Basin-and-Range province. The West Range lies in the middle distance, while the East Range forms the background. The tilt directions of the two are quite different. (Courtesy of Dr. Dick Henderson, Geology Walks; http://www.saguaro-juniper.com/i_and_i/geology /geology_walk/3_basin_range.html.)

20 million years ago, as part of this extension, there was extensive volcanism in the Sierra Nevada of California. About 10 million years ago, the extension caused block faulting along the east side of the Sierra Nevada and westward tilting of the entire range.

The **San Andreas Fault system** marks a boundary between the Pacific Plate and North American Plate in California. This fracture zone developed 28–30 million years ago. It has many strands in addition to the San Andreas Fault, such as the Calaveras and Hayward faults east of San Francisco. The oldest strands of this system are farther west, offshore, and the most recent movements appear to be in eastern California. The Salton Trough, at the south end of the San Andreas system, is a result of crustal stretching and sinking caused by both the fault and spreading of the East Pacific Rise as it comes onshore. Once onshore, the two sides of the fault slip sideways past each other. The San Andreas system is part of continuous, alternating zones of lateral movement and subduction that run from the Gulf of California to the Aleutians in Alaska. North of the San Andreas Fault is the **Cascadia subduction zone** offshore Oregon and Washington. The **Queen Charlotte–Fairweather fault system** is a plate margin similar to the San Andreas Fault, but offshore British Columbia and Alaska. It has been active for the past 42 million years. This system links northward into the Aleutians subduction zone.

The **Zagros Mountains** of Iran and the **Caucasus Mountains** of Turkey are both the result of collision of the **Arabian plate** with Eurasia starting about 28 Ma. This is related to the general northward movement of both India and Africa and could be considered part of the Himalayan and Alpine mountain building events.

The sierras that extend in Central America from Mexico to Colombia are largely a Cenozoic volcanic arc developed east of a subduction zone where the **Cocos plate** (Pacific side) plunges beneath the **Caribbean plate**. This **Central American Volcanic Arc** is part of the Pacific Rim "**Ring of Fire**," active volcanoes related to subduction around the margins of the Pacific. For the most part, these Central American subduction-related volcanoes are less than 30 million years old, and many have erupted in modern times.

Between 25 and 22 Ma, the **East African rift** began to split Africa in two. This developing plate boundary is spreading between 6 and 7 mm/year and, if it continues at this rate, a new ocean will form here in the next 10 million years.

The **Alaska Range** and Mt. Denali (highest peak in North America) began to rise about 6 Ma. The range is a result of compression and crumpling of the crust related to subduction of the Pacific plate beneath Alaska.

The Colorado River carved the **Grand Canyon** of Arizona through a high plateau that exposes rocks almost 2 billion years old. The age and origin of the canyon has been debated for close to 150 years. Several western tributary canyons are up to 70 million years old. Some geologists think the eastern half of the canyon was eroded only 15 to 25 million years ago by an early Colorado River that flowed northeast off the Kaibab Plateau (Figure 6.31). One theory holds that about 5–6 million years ago, a river draining into the Gulf of California eroded upstream by **headward erosion** into the plateau and captured the ancestral Colorado River by **stream piracy**, reversing its flow. In the last 1.8 Ma, at least 13 lava flows have dammed the canyon, forming lakes up to 600 m (2000 ft) deep. These eroded lava dams now form rapids in the canyon.

The Mediterranean Sea dried out completely between 5 and 6 million years ago. Africa, moving slowly north toward Europe, closed off the Strait of Gibraltar. Sediment samples from the ocean floor show soil zones, plant fossils, and thick salt deposits. It is estimated that it took about 1000 years for the sea to dry out completely, with only a few Dead Sea–like lakes 3 to 5 km (2 to 3 mi) below sea level. This drying out is called the **Messinian salinity crisis** (Figure 6.32).

FIGURE 6.31
Google Earth view to the northeast over Las Vegas and the Grand Canyon. One hypothesis for the development of the Grand Canyon has the Ancestral Colorado River flowing toward the northeast from the Kaibab Plateau. Another river flowing into the Gulf of California eroded upstream until it captured the Ancestral Colorado and reversed its flow from northeast to southwest. The age of the combined rivers flowing toward Las Vegas is taken from the earliest Grand Canyon gravels found in the Las Vegas area. Red arrows show original river flow direction; blue arrows show current flow direction. (Courtesy of Google Earth.)

The Strait of Gibraltar was breached around 5.3 Ma in what is called the **Zanclean flood**. It is estimated that the ocean poured into the dry basin at a rate about 1000 times the volume of the Amazon, and that it took about 2 years to fill the Mediterranean basin.

The **Altai Mountains** at the junction of Russia, Mongolia, Kazakhstan, and China are relatively young mountains. A result of the collision of India and Asia, they have been raised up in only the past 2.6 Ma. Ongoing earthquake activity indicates that these mountains are still rising.

The largest volcanic eruption in the past 3 million years was from **Mount Toba**, a **supervolcano** on the island of Sumatra, Indonesia. About 75,000 years ago, the eruption put 2900 km³ (700 mi³) of ash into the atmosphere and left a caldera 100 by 30 km (60 × 18 mi) in size. A layer of ash 14 cm (6 in) thick was deposited over all of south Asia. The eruption of Krakatoa (Indonesia) in 1883 released a mere 12.5 km³ of material. Geneticists have determined, based on analysis of male (Y-chromosomal) DNA, that, around 70,000 years ago, some kind of catastrophe affected humanity, causing the worldwide population to drop to between 1000 and 10,000 breeding pairs. All humans are descended from these few thousand survivors. This **genetic bottleneck** may have been the result of dramatic environmental changes, such as a volcanic winter following the eruption of Mount Toba, when global surface temperatures dropped 3°C–5°C for decades or centuries (Dawkins 2004; Witze and Kanipe 2014).

The **Black Sea deluge** hypothesis refers to a sudden rise in the level of the Black Sea around 7600 years ago. The Black Sea basin, which extends as much as 2200 m (7250 ft)

FIGURE 6.32
Mediterranean Sea at maximum drawdown during the Messinian salinity crisis. The breach at the Strait of Gibraltar is shown. (Courtesy of Roger Pibernat under supervision of Daniel Garcia-Castellanos; https://commons .wikimedia.org/wiki/File:Messinian_Mediterranean_and_Gibraltar_-_reconstructed_landscape.jpg.)

below sea level, contained a lake whose surface was almost 150 m (500 ft) lower than the present surface (Figure 6.33). Early agricultural settlements would have been located around the margins of the lake. Worldwide sea levels were rising as a result of the end of the ice age and melting of glaciers. The rising level of the Mediterranean eventually breached the Bosphorus and began to pour into the Black Sea basin. The rising waters

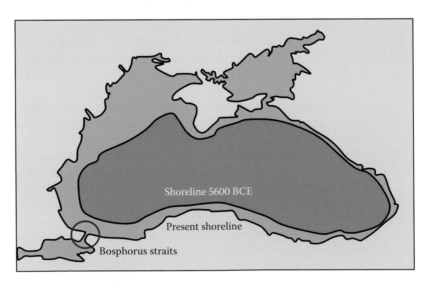

FIGURE 6.33
Black Sea outline today (light blue) and in 5600 BCE (dark blue). (Drawing by GFDL; https://en.wikipedia.org /wiki/Black_Sea_deluge_hypothesis.)

flooded low-lying coastal communities, forcing their inhabitants to move elsewhere. Some speculate that the oral histories of their hurried evacuation evolved into the **Epic of Gilgamesh**, the Biblical story of **Noah's flood**, and similar tales told throughout the region (Ryan and Pitman 2000).

What the Future Holds

If current tectonics continue, we can make a few predictions, although we won't be around to see whether they actually happen. The Yellowstone volcanic center will eventually become dormant and new volcanism will erupt farther east as the North American tectonic plate moves westward over a mantle plume. A new island will form southeast of the big island of Hawaii as the Pacific Plate continues to move northwest over the Hawaiian hotspot. If movement on the San Andreas Fault continues as it is today, then the Gulf of California will one day extend into the Salton Sea and eventually the area west of the fault (Baja California to San Francisco) will become an island off the west coast of North America. Likewise, with continued movement on the Red Sea Transform, the Red Sea will connect to the Mediterranean and Africa will become an island continent. Until it crashes into Europe. The Gulf of Aqaba will one day extend into the Dead Sea. The Rio Grande Rift will extend to the Gulf of Mexico and central Colorado will have ocean-front property. The East African Rift will split eastern Africa from the main body of Africa, forming a new continent in the Indian Ocean. Continued movement of Africa north into Europe will close the Mediterranean basin and its sediments will be raised in a Himalaya-like mountain range. Australia will continue moving north and will grow by accreting the Indonesian–Philippian archipelago. Eventually, it will collide with Asia in much the same way the Indian plate merged with Asia (Figure 6.34).

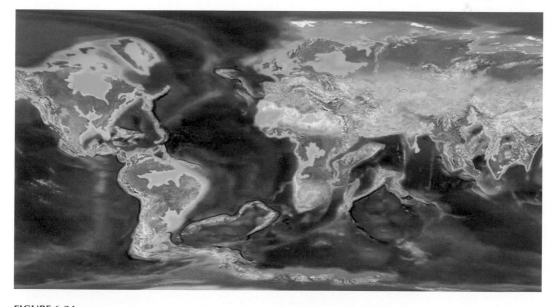

FIGURE 6.34
One version of what the continents might look like in 100 million years. (Courtesy of SpaceRip, YouTube; https://www.youtube.com/watch?v=uGcDed4xVD4&ebc=ANyPxKp5MB2yl2lGY8s-Swm8qSP3jfOJuNfA6jFX ZJCiF3nhYI2qO0pWWHfnRjLJk23M7Kp_u-OpJCMwg0tM1gpMGLAzQ8yuEw.)

The collision and breakup of tectonic plates has been going on since the beginning of plate tectonics 4 billion years ago, and we expect it to continue well into the future.

References and General Reading

Benton, M. J. 2008. *When Life Nearly Died—The Greatest Mass Extinction of All Time*. Thames and Hudson, New York, 336 pp.

Dawkins, R. 2004. *The Ancestor's Tale—A Pilgrimage to the Dawn of Evolution*. Mariner Books, Boston, 405 pp.

Dunbar, C. O., and K. M. Waage. 1969. *Historical Geology*, 3rd ed. John Wiley & Sons, New York, 556 pp.

Hazen, R. M. 2012. *The Story of Earth*. Penguin Books, New York, 306 pp.

Powell, J. L. 2005. *Grand Canyon—Solving Earth's Grandest Puzzle*. Pi Press, New York, 308 pp.

Press, F., and R. Siever. 1982. *Earth*, 3rd ed. W. H. Freeman and Co., San Francisco, 613 pp.

Ranney, W. 2005. *Carving Grand Canyon*. Grand Canyon Association, Grand Canyon, 160 pp.

Ryan, W., and W. Pitman. 2000. *Noah's Flood*. Simon and Schuster, New York, 319 pp.

Scotese, C. R. 2001. *Atlas of Earth History, Volume 1, Paleogeography*, PALEOMAP Project, Arlington, Texas, 52 pp.

Walker, G. 2004. *Snowball Earth*. Bloomsbury Publishing, London, 288 pp.

Ward, P., and J. Kirschvink. 2015. *A New History of Life*. Bloomsbury Press, New York, 400 pp.

Witze, A., and J. Kanipe. 2014. *Island on Fire—The Extraordinary Story of a Forgotten Volcano That Changed the World*. Pegasus Books, New York, pp. 66–67.

7

Fossils

And Xenophanes believes that once the earth was mingled with the sea, but in the course of time it became freed from moisture; and his proofs are such as these: that shells are found in the midst of the land and among the mountains, that in the quarries of Syracuse the imprints of a fish and of seals had been found, and in Paros the imprint of an anchovy at some depth in the stone, and in Melite shallow impressions of all sorts of sea products. He says that these imprints were made when everything long ago was covered with mud, and then the imprint dried in the mud.

Attributed to Xenophanes of Kolopbon (born about 580 BCE) by Hippolytus

It isn't easy to become a fossil.... Only about one bone in a billion, it is thought, becomes fossilized. If that is so, it means that the complete fossil legacy of all the Americans alive today—that's 270 million people with 206 bones each—will only be about 50 bones, one-quarter of a complete skeleton. That's not to say, of course, that any of these bones will ever actually be found.

Bill Bryson
Humorist, in *A Short History of Almost Everything*

All the fossils that we have ever found have always been found in the appropriate place in the time sequence. There are no fossils in the wrong place.

Richard Dawkins
Evolutionary biologist

Paleontology is the study of past life by means of fossils. It involves determining the relationship between fossils, the relationship of fossils to present-day living things, and deriving from fossils and rocks an idea of their ancient environment. By studying changes in fossil body shape (**morphology**), we can understand how they interacted with their environment. Establishing a succession of fossils allows us to determine the relative age of the strata they are found in. Fossils are the original basis for the geologic time scale. And they are fun to find.

Seashells on Mountain Tops—The Development of Paleontology

How do you explain seashells on mountain tops? In the sixth century BCE, the Greek Xenophanes concluded that fossil shells are the remains of sea creatures. In renaissance Europe, the conventional wisdom was that stone objects resembling living things could grow spontaneously in the earth. This view came to be challenged by those who worked in the earth. Leonardo da Vinci was employed by the Duke of Milan as an engineer building roads and tunnels in the years 1482 to 1519. His notes, compiled in the Codex Leicester between 1504 and 1510, show him to be a keen observer (Figure 7.1). He deduced that fossil sea shells were the remains of sea creatures from a time when the sea had covered that part of Italy: "*...it is no marvel if, in our day, no records exist of these seas having covered so many countries.... But sufficient for us is the testimony of things created in the salt waters, and found again in high mountains far from the seas.*" He anticipated stratigraphic correlation when he wrote "*...the shells in Lombardy are at four levels, and thus it is everywhere, having been made at various times.*"

In 1665, the English scientist Robert Hooke published *Micrographia*, a collection of his observations made with a microscope. One observation was a comparison of petrified wood and ordinary wood. He decided that the petrified wood was just ordinary wood that had been "impregnated with stoney and earthy particles." It looked to him like fossil sea shells had gone through the same process.

The Danish doctor/scientist Niels Stensen, better known by his Latin name Nicholas Steno, spent much of his time studying anatomy. In 1667, while working in Italy, he described his dissection of a shark head and concluded that the common fossils known as "tongue stones" looked like shark teeth because they *were* shark teeth (Figure 7.2).

This inspired him to look into the problem of fossils, or as he called it, "a solid within a solid." He determined that objects such as rock crystals grew within the rock, whereas fossils were the remains of creatures formed outside of the rock that had been deposited in and buried by sediment that later became rock.

In *Epochs of Nature* (1778), the French naturalist Buffon used elephant and rhinoceros fossils found in northern Europe to deduce that the climate there had once been tropical.

(a)

(b)

FIGURE 7.1
(a) Page of da Vinci's *Paris Manuscript I* with sketches of marine fossils including possibly Paleodictyon (arrow). (From BIBLIOTECA LEONARDIANA.) (b) Oligo-Miocene Paleodictyon from Algeciras, Spain. (Courtesy of Falconaumanni; https://commons.wikimedia.org/wiki/File:Paleodictyon_P.San_Garc%C3%ADa_Algeciras _I03.JPG.)

FIGURE 7.2
Shark head with tooth and fossil tooth. (From Steno's 1667 paper; https://en.wikipedia.org/wiki/Nicolas_Steno.)

The French naturalist Léopold Chrétien Frédéric Dagobert Cuvier (he preferred being called Georges Cuvier) in 1796 published a paper describing the skeletons of elephants, mammoths, and mastodons. He determined that these were all different species and, furthermore, that mammoths and mastodons were extinct. This contradicted common thinking at the time, which was that extinction implied imperfection. Since God had made Earth and all its creatures in perfect form, there was no reason to have removed species. In 1808, Cuvier identified a fossil found in Holland as a marine reptile. He identified a fossil found in Bavaria as a flying reptile, and reasoned that an age of reptiles had come before the age of mammals. Cuvier recognized that such an epoch required a large-scale extinction that had wiped out the giant reptiles. When he realized that large mammals like mammoths and mastodons were also extinct, he proposed that there had been multiple extinctions.

While working as a surveyor and mining engineer in the 1780s and 1790s, William Smith carefully noted the fossils in each layer of rock he encountered. He found that he could identify the same strata at isolated locations across England using unique groups of fossils he saw in each layer (Figure 7.3). Each sedimentary layer had a specific fossil assemblage that falls within a predictable sequence, now known as the "principle of faunal succession." He used these fossil sequences to identify equivalent strata across England.

The English paleontologist Mary Anning (1799–1847) collected many fossil marine reptiles, thus supporting Cuvier's claim of an earlier age of reptiles. Her discoveries, in Jurassic strata near Lyme Regis on the English Channel, include the first **plesiosaur** and **ichthyosaur** skeletons ever found (Figure 7.4).

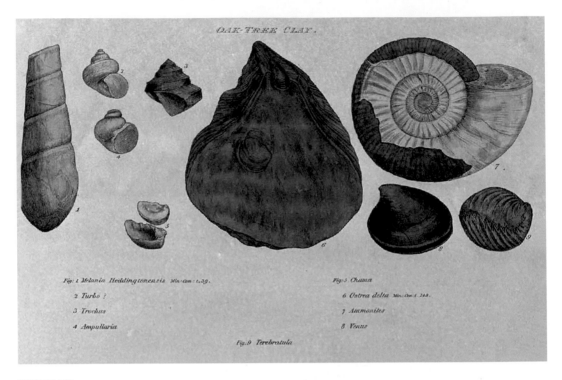

FIGURE 7.3
Figure from William Smith's 1815 publication *Strata by Organized Fossils* (https://en.wikipedia.org/wiki/File:Smith_fossils3.jpg).

FIGURE 7.4
Cast of an early Plesiosaur skeleton, *Thalassiodracon hawkinsi* skeleton at Bristol City Museum, Bristol, England. Found in the county of Somerset, England. (Courtesy of Adrian Pingstone; https://commons.wikimedia.org/wiki/File:Plesiosaur_cast_arp.jpg.)

Evidence continued to mount during the first half of the 1800s that there had been an **age of reptiles** when giant creatures that no longer exist roamed the earth. In 1822, Gideon Mantell, an English obstetrician and geologist, found teeth near Tilgate that looked like they belonged to a giant iguana. He called the creature *Iguanadon*. In 1832, he found the bones of an armored reptile he called *Hylaeosaurus*. In 1824, William Buckland, an English theologian and geologist, found bones of a giant carnivorous land reptile near Stonesfield he called *Megalosaurus*. To account for these discoveries, the English anatomist Richard Owen in 1842 defined a new order of reptiles called **Dinosauria**.

At the same time, botanists were also classifying plants. The French botanist Adolphe Brongniart, considered the "father of paleobotany," published his *Histoire des Vegetaux Fossiles* (History of Fossil Plants) in 1828 (Figure 7.5). In it, he grouped plants into four periods. The oldest period was marked by the appearance of **cryptogams**, spore-producers

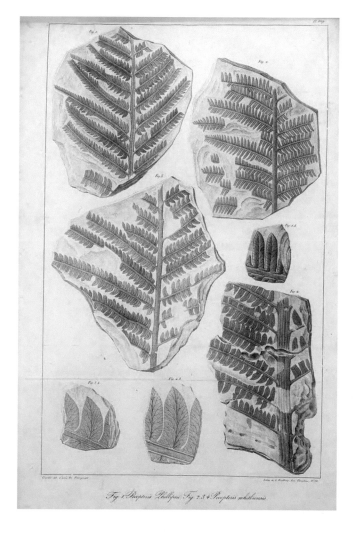

FIGURE 7.5
Page from Brongniart's *Histoire des Végétaux Fossiles* (1828–1837) (https://commons.wikimedia.org/wiki /File:Histoire_des_v%C3%A9g%C3%A9taux_fossiles,_ou,_Recherches_botaniques_et_g%C3%A9ologiques _sur_les_v%C3%A9g%C3%A9taux_renferm%C3%A9s_dans_les_diverses_couches_du_globe_(Pl._109) _BHL40155201.jpg).

that included algae, lichens, mosses, fungi, bacteria, and ferns. The second group to emerge comprised **conifers**, cone-bearing, seed-producing woody plants that include cedars, firs, pines, redwoods, and spruce trees. The third period was characterized by **cycads**, woody cone-bearing seed plants where the leaves grow directly from the trunk. The Sago Palm (not a real palm) is a cycad. The most recent period is distinguished by the development of **angiosperms** (flowering plants) characterized by flowers, fruits, and seeds. Brongniart noticed that these periods were separated by abrupt changes in the fossil record.

In the mid-1800s, the English geologists Adam Sedgwick and Roderick Murchison described fossils associated with periods they named Cambrian, Silurian, Devonian, and Permian. They recognized that there was a time before the age of reptiles where marine invertebrates had been the dominant life forms.

The publication of Darwin's *Origin of Species* in 1859 led paleontologists to search for transitional forms that would help support the concept of evolution. Key to this was the discovery in 1861 of **Archaeopteryx**, a creature with wings and feathers like a bird but teeth and tail like a reptile (Figure 7.6). Working in Kansas in 1872, Othniel Marsh discovered

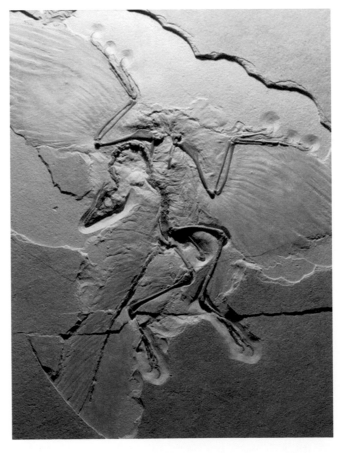

FIGURE 7.6
Archaeopteryx, a transitional form between reptiles and birds. (Courtesy of Natural History Museum, Vienna, Wolfgang Sauber; https://commons.wikimedia.org/wiki/File:NHM_-_Archaeopteryx_lithographica_Fossil.jpg.)

more toothed birds as well as a series of fossils that tracked the evolution of the horse from the small five-toed *Hyracotherium* to the modern single-toed *Equus*. Marsh and another American paleontologist, Edward Drinker Cope, were fiercely competitive in finding and describing fossils across North America in the late 1800s. Cope is credited with discovering up to 1000 fossil species, and between them they described more than 120 new species of dinosaur alone.

In 1909, Charles Doolittle Walcott of the US Geological Survey described a number of early Cambrian fossils from the Burgess Shale in British Columbia, Canada (Figure 7.7). The sudden appearance of plentiful fossils in the early Cambrian above rocks that contained no apparent life has been called the "Cambrian Explosion." These fossils were featured in the book *Wonderful Life* (1989) by the American paleontologist Stephen Jay Gould.

Soft-bodied fossils are notoriously difficult to find. One place they have been found is the Ediacara Hills of South Australia's Flinders Range. In 1946, the Australian geologist Reginald Claude Sprigg discovered some fossils in this group and named them after the location. This **Ediacara fauna** dates from 600 to 540 Ma, predating and slightly overlapping the Cambrian period when life was originally thought to have begun (Figure 7.8).

In addition to the continuing search for new species, recent efforts in paleontology include evaluating carbon inclusions in minerals as evidence of life, attempting to extract DNA from fossils, finding ever-earlier evidence of feathered dinosaurs, and filling in the gaps between *Homo sapiens* and their earliest ancestors.

FIGURE 7.7
Fossil specimen of Wiwaxia from the Burgess Shale on display at the Smithsonian Museum in Washington, D.C. (Courtesy of Jstuby at English Wikipedia; https://en.wikipedia.org/wiki/Fossils_of_the_Burgess_Shale.)

FIGURE 7.8
Ediacara fauna *Dickinsonia costata* from Australia. (Courtesy of Verisimilus; https://en.wikipedia.org/wiki/Dickinsonia.)

How Fossils Are Formed and Preserved

Fossils can be imprints, carbon smudges, shell or bone replaced by minerals, and even mummies. How do they form?

Soft tissue preservation is rare. Freeze-dried mammoths found in permafrost or insects trapped in amber are examples (Figure 7.9). Some original material is preserved in tar, as in the La Brea Tar Pits in Los Angeles, or in peat bogs, as with the 8000-year-old Koelbjerg woman from Denmark.

Organic material that is buried and slowly cooked by Earth's heat turns to carbon. Coal is mostly the carbonized remains of plants.

Many fossils are formed by replacing the original shell, bone, or wood one molecule at a time. Silica, calcite, pyrite, or other minerals dissolved in groundwater precipitate in the pores of the original material. This is how petrified wood, pyrite sand dollars, and mineralized ammonites are formed (Figure 7.10).

Most evidence of soft tissues come from imprints in originally soft sediment. **Trace fossils** include worm tracks, burrows, and footprints (Figure 7.11). Impressions include skin and feather imprints (Figure 7.12). **Molds** are the casts or impressions of shells or other fossils that have since been dissolved away by groundwater (Figure 7.13).

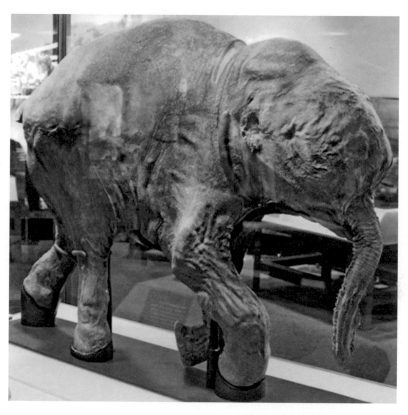

FIGURE 7.9
Lyuba, a baby mammoth frozen 42,000 years ago, found in Siberia in 2007 (https://en.wikipedia.org/wiki
/Lyuba).

The Fossil's Tale

Paleontologists study fossils because of what these traces of former life can tell us. First
and foremost, a fossil tells what it was: plant or animal; its size, age, and shape; how it
moved; whether it was a carnivore or herbivore. Second, it gives clues as to how it died,
whether hunted and killed (bite marks), trapped in a flood or landslide (encased unharmed
in mud), or buried in an ash fall (the people in Pompeii).

Some of the most important information gleaned is how an organism is related to other
lineages. Slight changes in the shape of bones, teeth, or leaves tell much about evolution of
groups over time.

Fossils provide clues to the ancient environment. Was it moist or dry, marine or freshwater, warm or cool, near the equator or at high latitude? This, along with information
from the rocks themselves, helps reconstruct the former world (Figure 7.14).

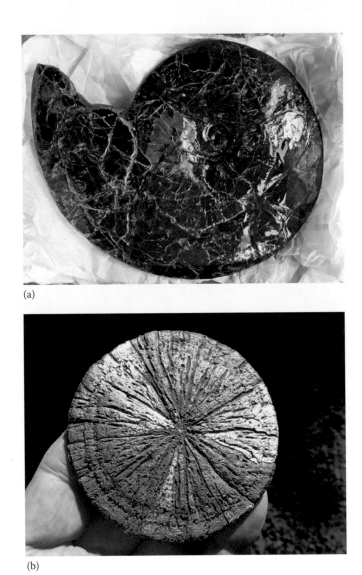

(a)

(b)

FIGURE 7.10
(a) Ammonite replaced by opaline silica known as Ammolite. (Courtesy of Mike Peel; https://commons.wikimedia
.org/wiki/File:BLW_Ammolite_ammonite_(2).jpg.) (b) Echinoderm (sand dollar) replaced by pyrite. (Courtesy of
Dan Edwards, Spirit Rock Shop; http://www.spiritrockshop.com/Pyrite_Suns.html.)

Finally, fossils tell us the relative age of the rocks and what other rock layers are the
same age. Identification of **Olenellus** (a trilobite) tells you that the strata is Cambrian, no
matter where you are. Likewise, **crinoids** are typically found in the Carboniferous,
Inoceramus shells are always Jurassic–Cretaceous, **baculites** fossils are always Cretaceous,
Vesperopsis is always middle Cretaceous, and **nummulites** are generally Eocene to
Miocene (Figures 7.15 through 7.18).

FIGURE 7.11
Permian lizard tracks in Coconino Sandstone, northern Arizona. (Courtesy of G. Prost.)

FIGURE 7.12
Impressions of the feathered dinosaur Anchiornis. (Courtesy of Kumiko; https://en.wikipedia.org/wiki/Theropoda#/media/File:Anchiornis_feathers.jpg.)

Major Categories of Life

Early attempts to classify life were pretty simple. Aristotle (384–322 BCE) classified life into plants and animals, and he subdivided animals into **vertebrates** (with a backbone) and **invertebrates** (without a backbone). The Roman Pliny the Elder (23–79 CE) listed all the plants and animals he was aware of in his multivolume encyclopedia *Naturalis Historiae* (Natural History). It described animals and mentioned uses for various plants.

FIGURE 7.13
Mold of a Paleozoic bivalve mollusk. (Courtesy of Mark Wilson (Wilson44691); https://en.wikipedia.org/wiki
/Fossil#Casts_and_molds.)

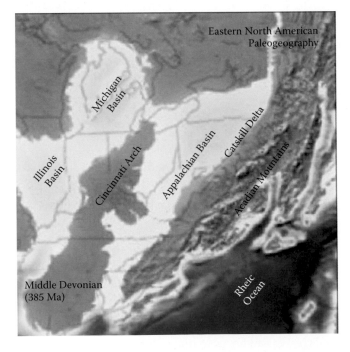

FIGURE 7.14
Paleogeography of eastern North America in the middle Devonian period 385 million years ago, with key
geographic features labeled. (Courtesy of Ron Blakey; http://jan.ucc.nau.edu/~rcb7/; https://en.wikipedia.org
/wiki/Palaeogeography.)

(a)

(b)

FIGURE 7.15
Index fossils date the strata they are found in. (a) *Olenellus*, a Cambrian trilobite from Pioche Shale, Nevada. (Courtesy of SNP; https://en.wikipedia.org/wiki/Olenellus.) (b) The Cretaceous bivalve mollusc *Inoceramus*. (Courtesy of Kozuch, National Museum, Prague; https://en.wikipedia.org/wiki/Inoceramus.)

Taxonomy is the science concerned with the classification of organisms. Modern taxonomy began with the Swedish physician, botanist, and zoologist Carl Linnaeus (1707–1778) and his *Systema Naturae* (The System of Nature, 1735). **Linnaean taxonomy** grouped animals and plants into Kingdom–phylum–class–order–family–genus–species on the basis of shared physical characteristics. More than 4400 species of animals and 7700 species of plants were classified. He popularized and consistently used a two-part naming system for plants and animals wherein each is identified by genus and species. He established three Kingdoms: Animal, Vegetable, and Mineral. His classification of animals, although much modified, is still used. It is based on observable characteristics (form and structure). Animals were divided into six classes: Mammalia, Aves (birds), Amphibia, Pisces (fish), Insecta, and Vermes (worms

FIGURE 7.16
A key Cretaceous index fossil is the cephalopod mollusk *baculites*. These are from the Pierre Shale of South Dakota. (Courtesy of Children's Museum of Indianapolis, Michelle Pemberton; https://commons.wikimedia .org/wiki/File:The_Childrens_Museum_of_Indianapolis_-_Baculites.jpg.)

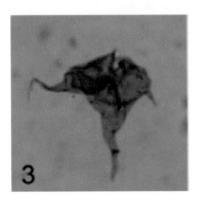
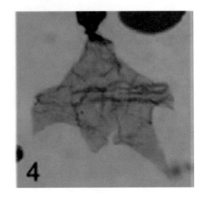

FIGURE 7.17
Microfossils can only be identified under the microscope. They are excellent age and environment indicators. These are *Vesperopsis*, the dormant zygote stage of marine plankton. (Courtesy of Graham Dolby; https://www .researchgate.net/profile/Graham_Dolby.)

and other invertebrates). Interestingly, he combined humans and monkeys in the group Anthropomorpha, since they have essentially the same anatomy.

Linnaeus classified plants on the basis of how they reproduced rather than physical characteristics. This system is no longer in use. He classified Minerals into Rocks, Minerals, Fossils, and Ores.

Some of the more common animal fossil groups and their classification are shown in Table 7.1. The table is followed by illustrations of most major fossil groups.

Common plant fossils include the groups shown in Table 7.2.

Plant pollen and spores are important as well. The study of spores, known as **palynology**, is used extensively by the energy industry to determine the relative age of sedimentary

FIGURE 7.18
Eocene numulitic limestone, Gerona-stone, Spain. Nummulites are the rice-shaped fossils. (Courtesy of PePeEfe; https://commons.wikimedia.org/wiki/File:Nummulites-secciones_axiales-Eoceno-Gerona.JPG.)

rocks, the ancient environment, and the amount of heat a rock has been subjected to: the more a pollen has been cooked, the darker it appears under a microscope (Figure 7.62).

Fossil Hunting

Do you know what you are looking for? If it is fossil sea shells, you want to find rocks deposited in a shallow marine setting (sandstone and shale) or on a reef (limestone). If you are looking for leaf fossils, you want to look at muddy lake or swamp deposits or in sandstone or shale near coal beds. What you are looking for determines where and in what rocks to look.

If you have no idea where to go or what to look for, contact a gem and mineral club, your state geological survey or mining department, local museums, rock shops, or a tourist information center. Get a book on local mineral and fossil sites. Do an online search using "fossils" and your area of interest.

Before starting, you should have a map of the search area and permission from landowners to enter or cross their land. It goes without saying that fossil collecting (like rock collecting in general) is not permitted in national parks and monuments. Collecting without permission on private land or reservations can lead to fines or worse.

As with any such adventure, have sturdy shoes, a sun hat, lots of water, and know the limitations of your vehicle with respect to driving off road. Check weather reports before you go and be prepared.

TABLE 7.1

Some of the More Common Animal Fossil Groups

Phylum	Subphylum or Class	Class or Subclass
	Invertebrates	
Protozoans	Radiolaria (Figure 7.19)	
	Foraminifera (Figure 7.20)	
Coelenterates/Cnidaria	Hexacorals (Figure 7.21)	Sea anemones (Figure 7.22), corals
	Octocorals	Corals
	Ceriantharia	
Porifera (Sponges) (Figure 7.23)	Stromatoporoids (Figure 7.24)	
Bryozoans (Figure 7.25)		
Brachiopods (Figure 7.26)		
Annelids (Worms) (Figure 7.27)		
Arthropods	Trilobites (Figure 7.28)	
	Crustaceans	Ostracods (Figure 7.29)
		Branchiopods
		Remipedia
		Cephalocarida (horseshoe shrimp)
		Maxillopoda (barnacles) (Figure 7.30)
		Malacostraca (shrimps, crabs, lobsters, crayfish) (Figure 7.31)
	Chelicerates (scorpions, mites, ticks, arachnids, eurypterids, horseshoe crabs) (Figure 7.32)	
	Myriapods (centipedes, millipedes) (Figure 7.33)	
	Insects (Figure 7.34)	
Echinoderms	Crinoids (Figure 7.35)	
	Asteroids (Figure 7.36)	
	Ophiuroids (Figure 7.37)	
	Holothurioids	
	Echinoids (Figure 7.38)	
	Edrioasteroids (Figure 7.39)	
	Cystoids (Figure 7.40)	
	Blastoids (Figure 7.41)	
Mollusks	Gastropods (snails) (Figure 7.42)	
	Cephalopods (ammonites, nautilus, squid, octopus) (Figure 7.43)	
	Pelecypods (oysters, mussels, clams) (Figure 7.44)	
	Scaphopods (Figure 7.45)	
	Amphineura (Figure 7.46)	
Graptolites (Figure 7.47)		

(Continued)

TABLE 7.1 (CONTINUED)

Some of the More Common Animal Fossil Groups

Phylum	Subphylum or Class	Class or Subclass
	Vertebrates	
Chordata	Agnatha (jawless fish) (Figure 7.48)	
	Chondrichthyes (cartilaginous fish) (Figure 7.49)	
	Osteichthyes (boney fish) (Figure 7.50)	
	Amphibians (Figure 7.51)	
	Reptiles (Figure 7.52)	
	Birds (Figure 7.53)	
	Mammals	

You can practice "catch and release" fossil hunting or you can collect the fossils you find. There are good arguments for each. Photographing the fossil and leaving it in place allows others the excitement of finding it again. If you are going to put the fossil in a collection that will someday get thrown away, consider leaving it in place. If you plan to identify the fossil and preserve it, collecting may be the way to go.

TABLE 7.2

Plant Classification

Informal Group	Division Name	Common Name
Green algae	Chlorophyta	Green algae (Figures 7.54 and 7.55)
	Charophyta	Desmids, stoneworts
Bryophytes	Marchantiophyta	Liverworts
	Anthocerotophyta	Hornworts
	Bryophyta	Mosses
Pteridophytes	Lycopodiophyta	Club mosses (Figure 7.56)
	Pteridophyta	Ferns, whisk ferns, horsetails (Figure 7.57)
Seed plants	Cycadophyta	Cycads (Figure 7.58)
	Ginkgophyta	Ginkgo (Figure 7.59)
	Pinophyta	Conifers (Figure 7.60)
	Gnetophyta	Gnetum, ephedra, welwitschia
	Magnoliophyta	Flowering plants (Figure 7.61)

FIGURE 7.19
Radiolaria, single-celled planktonic animals with a siliceous skeleton. (Courtesy of Ernst Haeckel, 1904, Kunstformen der Natur, plate 91; https://upload.wikimedia.org/wikipedia/commons/0/0c/Haeckel_Spumellaria.jpg.)

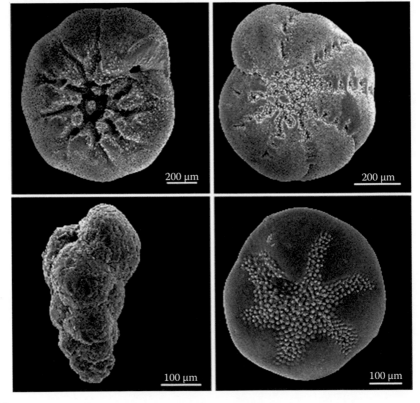

FIGURE 7.20
Foraminifera, single-celled planktonic and sea-bottom-dwelling animals commonly having a calcium carbonate shell. (Courtesy of US Geological Survey; https://commons.wikimedia.org/wiki/File:Benthic_foraminifera.jpg.)

FIGURE 7.21
Devonian Favosites coral, Alberta, Canada. (Courtesy of G. Prost.)

FIGURE 7.22
Cambrian sea anemone from the Heilinpu Formation, Yunnan Province, China. (Courtesy of The Virtual Fossil Museum; http://www.fossilmuseum.net/Fossil_Sites/Chengjiang/Xianguangia-sinica/Xianguangia.htm.)

FIGURE 7.23
Porifera (sponge) from Rowland's Reef near Gold Point, Nevada. (Courtesy of Killamator; https://commons
.wikimedia.org/wiki/Category:Porifera.)

FIGURE 7.24
Stromatoporoids in Silurian-Devonian Keyser Formation, Blair County, Pennsylvania. (Courtesy of Jstuby;
https://commons.wikimedia.org/wiki/File:Stromatoporoid1_Keyser_Formation.jpg.)

FIGURE 7.25
Devonian Bryozoa (arrows). Alberta, Canada. (Courtesy of G. Prost.)

FIGURE 7.26
Brachiopods (*Cincinnetina meeki*) from the Ordovician Waynesville/Bull Fork Formation of southern Ohio. (Courtesy of Wilson44691; https://en.wikipedia.org/wiki/Brachiopod.)

FIGURE 7.27
Middle Cretaceous annelid worm, Haqel, Lebanon. (Courtesy of The Virtual Fossil Museum; http://www
.fossilmuseum.net/fossils/Worms/Polychaeta/Invert7.htm.)

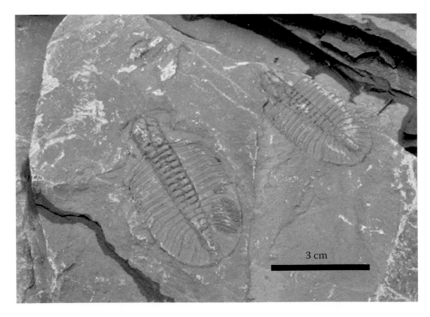

FIGURE 7.28
Cambrian trilobites from Mt. Stephen, British Columbia, Canada. (Courtesy of G. Prost.)

FIGURE 7.29
Herrmannina, a large Silurian ostracod from the Paadla Formation, Saaremaa Island, Estonia. (Courtesy of Wilson44691; https://commons.wikimedia.org/wiki/File:HerrmanninaSilurianEstonia.jpg.)

FIGURE 7.30
Miocene barnacle from Maryland. (Courtesy of Wilson44691; https://commons.wikimedia.org/wiki /File:Chesaconcavus_base_detail.jpg.)

FIGURE 7.31
Cycleryon, a Jurassic crab of the Class Malacostraca. (Courtesy of Ghedoghedo, Teylers Museum, Haarlem, Netherlands; https://commons.wikimedia.org/wiki/File:Cycleryon_propinquus_56.jpg.)

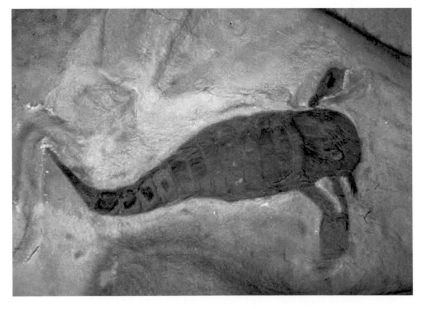

FIGURE 7.32
Chelicerate (eurypterid), Silurian Fiddler Green Fm., Herkimer Co., New York. (Copyright Alan Goldstein.)

FIGURE 7.33
Late Carboniferous *Euphoberia* (millipede). (Courtesy of Ghedoghedo; https://commons.wikimedia.org/wiki
/File:Euphoberia_sp_43.JPG.)

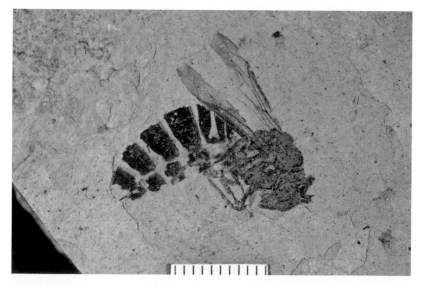

FIGURE 7.34
Eocene–Oligocene insect (the wasp *Palaeovespa* sp.), Florissant Fossil Beds, Colorado. (Courtesy of National Park
Service, US Department of the Interior; http://nature.nps.gov/geology/nationalfossilday/cenozoic_flfo.cfm.)

FIGURE 7.35
Carboniferous crinoid from Indiana. (Courtesy of Vassil; https://commons.wikimedia.org/wiki/File:Agaricocrinus
_americanus_Carboniferous_Indiana.jpg.)

FIGURE 7.36
Asteroidea (star fish) from Boulogne-sur-Mer, Pas-de-Calais, France. (Courtesy of Didier Descouens; https://
commons.wikimedia.org/wiki/File:Astropecten_lorioli.jpg.)

FIGURE 7.37
A Middle Jurassic ophiuroid from the Lias Formation, Dorset, England. Specimen measures 9.5 cm across (https://en.wikipedia.org/wiki/Paleocoma).

FIGURE 7.38
Miocene echinoid (sand dollar). Topanga Canyon near Los Angeles, California. (Courtesy of G. Prost.)

FIGURE 7.39
Edrioasteroid (13 mm across) from the Bellevue Formation (Upper Ordovician), Kentucky, USA. (Courtesy of James St. John; https://commons.wikimedia.org/wiki/File:Streptaster_vorticellatus_(13_mm_across)_from_the _Bellevue_Formation_(Upper_Ordovician)_at_the_Maysville_West_roadcut_of_northern_Kentucky,_USA.jpg.)

FIGURE 7.40
A Middle Ordovician Cystoid, 17 mm, from the Bromide Formation, Oklahoma, USA. (Courtesy of Dwergenpaartje; https://commons.wikimedia.org/wiki/File:Oklahomacystis_tribrachiatus.jpg.)

FIGURE 7.41
A Lower Carboniferous blastoid from Illinois. (Courtesy of Wilson44691; https://commons.wikimedia.org /wiki/File:Pentremites_Glen_Dean_Fm_KY.jpg.)

FIGURE 7.42
Eocene gastropod *Turritella* and unidentified snail. Maryland. (Courtesy of Glen Kuban, PaleoScene; http:// paleo.cc/fossils/gastro.htm.)

(a)

(b)

FIGURE 7.43

(a) The Cephalopod *Orthoceras* in limestone from late Silurian-Early Devonian of Erfoud, Morocco. (Courtesy of Antonov; https://commons.wikimedia.org/wiki/File:Orthoceras_Limestone_Erfoud_Maroc.jpg.) (b) *Orthoceras* from Morocco as commonly seen in rock shops around the world. (Courtesy of Rafandalucia; https://commons .wikimedia.org/wiki/File:F%C3%B3sil_de_orthoceras_(II).jpg.)

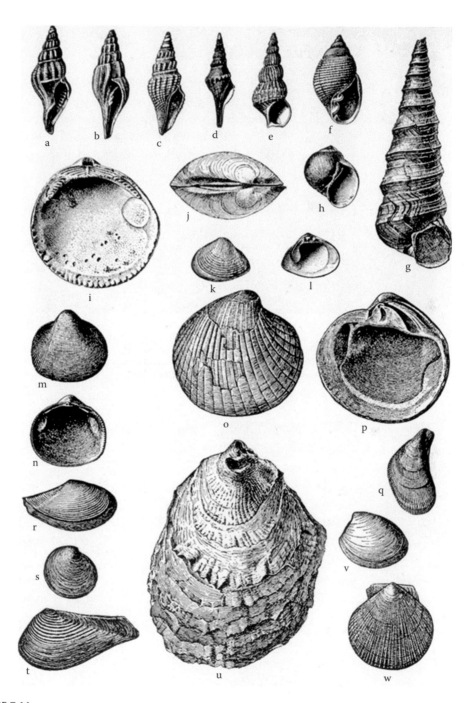

FIGURE 7.44
A sampling of pelecypods. (From Chamberlain and Salisbury, 1907; https://commons.wikimedia.org/wiki
/File:Geology_(1907)_(14796302093).jpg.)

FIGURE 7.45
Late Pleistocene Scaphopod, center. These are also known as "tusk shells." This fossil-rich sandstone is from Sternberg, Mecklenburg-Vorpommern, Germany. (Courtesy of Stadtmuseum Berlin; https://commons.wikimedia .org/wiki/File:StadtmuseumBerlin_GeologischeSammlung_SM-2012-4234.jp.)

(a)   (b)

FIGURE 7.46
Amphineura. (a) Recent from Guadelupe. (Copyright Hans Hillewaert, with permission.) (b) Late Cambrian Amphineura plates from the Notch Peak Limestone, Steamboat Pass, Utah. (Courtesy of Wilson44691; https:// en.wikipedia.org/wiki/Chiton.)

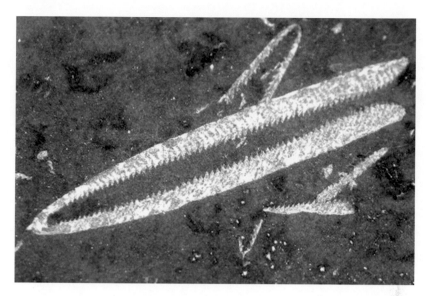

FIGURE 7.47
Ordovician graptolites. (Courtesy of Kevin Walsh; https://simple.wikipedia.org/wiki/Hemichordata.)

Some fossils are lying on the ground just waiting to be picked up. Sometimes it takes a "calibrated eyeball" to see things that are obvious once you know what to look for. It is helpful to have someone along who has found your target fossils before.

Fresh outcrops (cliffs, river cuts), beaches, quarries, coal mines, and road cuts are prime locations to look for fossils (Figure 7.63). Many fossils are inside rocks and need to be released from their stony prison. Most of these are found along bedding planes, so you want to split the rock along the layers to reveal them. For this, you will need a rock hammer, maybe a chisel, and safety glasses. Have sample bags and some way to label where you found each fossil. A Global Positioning System (GPS) or smartphone will provide you with exact coordinates of your location.

Take care when removing excess material around the fossil, as many specimens are ruined or broken by careless chipping away of surrounding rock. Put a dab of white paint on the underside of a specimen and label it with name, location, and age. Wrap carefully and store specimens in a cool, dry environment.

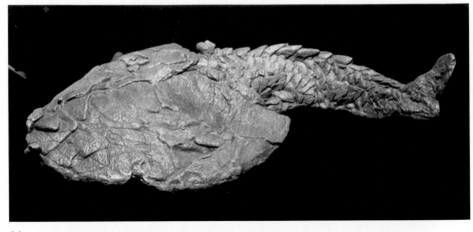

(a)

(b)

FIGURE 7.48

(a) Drepanaspis, an agnatha (jawless fish), National Museum of Nature and Science, Tokyo, Japan. (Courtesy of Momotarou2012; https://commons.wikimedia.org/wiki/File:Drepanaspis_Fossil.jpg.) (b) Life reconstruction of the agnatha Zenaspis pagei. (Courtesy of Nobu Tamura; https://commons.wikimedia.org/wiki/File:Zenaspis _NT_small.jpg.)

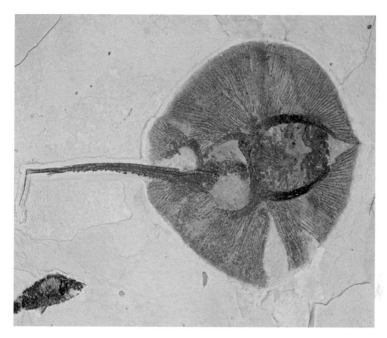

FIGURE 7.49
Eocene Chondrichthyes from Fossil-Lake, Kemmerer, Wyoming. Size: 39.5 × 33.5 × 3 cm. (Courtesy of Didier Descouens; https://commons.wikimedia.org/wiki/File:Heliobatis_radians_Green_River_Formation.jpg.)

FIGURE 7.50
Cretaceous fish from the Mount Lebanon Range, Lebanon. From the Stonerose Interpretive Center Collection, in Republic, Washington. (Courtesy of Kevmin; https://en.wikipedia.org/wiki/List_of_prehistoric_bony_fish _genera#/media/File:Aipichthys_velifer_01.jpg.)

FIGURE 7.51
Amphibian from the lower Permian of Germany. (Courtesy of LZ6387; https://commons.wikimedia.org/wiki
/File:Sclerocephalus.JPG.)

FIGURE 7.52
Early Permian reptiles *Captorhinus aguti* from Dolese Brothers quarry, Comanche County, Oklahoma. (Courtesy
of Didier Descouens; https://en.wikipedia.org/wiki/Reptile.)

FIGURE 7.53
Eocene bird from Green River Fm., Lincoln Co., Wyoming. (Courtesy of Glen Kuban, PaleoScene; http://pw1
.netcom.com/~paleo/ps/kfosscasts.htm.)

FIGURE 7.54
Fossil stromatolites, algal mounds formed by cyanobacteria (blue-green algae). In the old Heeseberg quarry,
Lower Saxony, Germany. (Courtesy of Ginganz-in; https://commons.wikimedia.org/wiki/File:Heeseberg
_SteinBr_2014-06-01_11.47.10.jpg.)

FIGURE 7.55
A shallow marine green algae from the Triassic of Italy. (Courtesy of Hectonichus, Museo Civico di Storia Naturale di Milano; https://commons.wikimedia.org/wiki/File:Diploporaceae_-_Diplopora_annulata.JPG.)

FIGURE 7.56
Upper Carboniferous Pteridophyte lycopodiophyta (club moss) from the Fire Creek Formation, West Virginia. (Copyright FossilMall.com, with permission; http://www.fossilmall.com/Stonerelic/plants/srpf3/srplant fossils3.htm.)

FIGURE 7.57
Carboniferous tree fern *Pecopteris plumosa*, likely from the Carnic Alps, Mt. Corona, Italy. (Courtesy of The Fossil Forum, Plantguy.)

FIGURE 7.58
Fossil Cycad *Zamites feneonis*. (Courtesy of H. Sonntag; https://simple.wikipedia.org/wiki/Cycad.)

FIGURE 7.59
Eocene *Ginkgo biloba* fossil leaf from the Tranquille Shale of MacAbee, British Columbia, Canada. (Courtesy of SNP and tangopaso; https://commons.wikimedia.org/wiki/File:Ginkgo_biloba_MacAbee_BCE.jp.)

FIGURE 7.60
Petrified Eocene redwood, Florissant Fossil Beds National Monument, Colorado. (Courtesy of National Park Service Digital Image Archives; https://commons.wikimedia.org/wiki/File:Florissant_Fossil_Beds_National _Monument_PA272520.jpg.)

FIGURE 7.61
Eocene Sycamore leaf from the Fort Union Formation near Craig, Colorado. (Courtesy of G. Prost.)

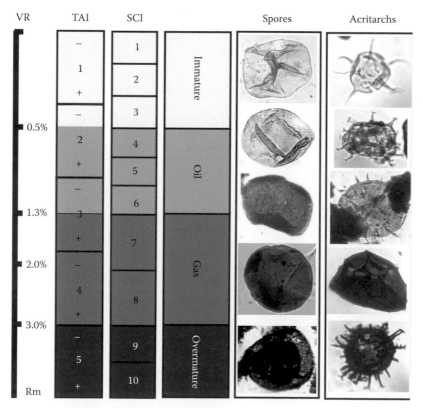

FIGURE 7.62
Pollen spore color chart showing the relationship of color to oil and gas generation and sediment maturity. Acritarchs are marine pollens. (Copyright Hartmut Jäger, GeoResources STC; http://www.georesources.de.)

FIGURE 7.63
Plio-Pleistocene gastropod fossils found in cemented gravel, Cajiloa beach near Punta San Carlos, Baja California, Mexico. (Courtesy of G. Prost.)

References and General Reading

Baucon, A. 2010. Leonardo da Vinci, the founding father of ichnology. *Palaios* v. 25, pp. 361–367.

Bryson, B. 2003. *A Short History of Nearly Everything*. Broadway Books, New York, 560 pp.

Busby III, A. B., R. R. Coenraads, P. Willis, and D. Roots. 1996. *The Nature Company Guide to Rocks and Fossils*. The Nature Company and Time-Life Books, Sydney, NSW, and San Francisco, 288 pp.

Chamberlain, T. C., and R. D. Salisbury. 1907. *Geology*. Vol. 2, Earth History. H. Holt and Co., New York, 692 pp.

Firefly. 2003. *Guide to Fossils*. Firefly Books, Buffalo, NY, 192 pp.

Gould, S. J. 1989. *Wonderful Life—The Burgess Shale and the Nature of History*. W.W. Norton & Company, New York, 347 pp.

Rhodes, F. H. T., H. S. Zim, and P. R. Shaffer. 2001. *Fossils. A Golden Guide*. St. Martin's Press, 160 pp.

8

Evolution

As many more individuals of each species are born than can possibly survive: and as, consequently, there is a frequently recurring struggle for existence, it follows that any being, if it vary however slightly in any manner profitable to itself, under the complex and sometimes varying conditions of life, will have a better chance of surviving, and thus be naturally selected. From the strong principle of inheritance, any selected variety will tend to propagate its new and modified form.

Charles Darwin
Naturalist

For the first half of geological time our ancestors were bacteria. Most creatures still are bacteria, and each one of our trillions of cells is a colony of bacteria.

Richard Dawkins
Evolutionary biologist, from *The Richard Dimbleby Lecture:*
Science, Delusion and the Appetite for Wonder

The capacity to blunder slightly is the real marvel of DNA. Without this special attribute, we would still be anaerobic bacteria and there would be no music.

Lewis Thomas
Physician and author, in *The Medusa and the Snail*

Whether we recognized it or not, we have all seen evidence for evolution. Man-made evolution can be seen in the many breeds of dogs derived from the gray wolf in the past 30,000 years or so. The same applies to all the breeds of cattle, horse, and other domestic animals. These were bred for specific purposes: to herd sheep, to provide more milk, to run faster, or to pull heavier wagons. **Domestication** happens by picking and breeding only the tamest of wild animals. Different **breeds** result from allowing only those animals that are best at a given task to reproduce. The same applies to plants. Wild fruit is small and a little sweet, but by selecting the sweetest each season and planting their seeds, fruit evolved to the large sweet varieties we know today. Corn (maize) is another example. About 10,000 years ago, farmers in Mexico began harvesting the wild grass teosinte. They selected for plants that were larger than the rest, or that had more kernels, or that tasted better. By planting those with the best characteristics over millennia, the cobs became larger, with more kernels, and evolved to become corn as we know it (Figure 8.1).

Now, imagine the gazelle and cheetah, or wolf and caribou. These are prey and predator. If the prey is even a bit faster, it will escape and live to reproduce. If the predator is even a shade more camouflaged, so that it can surprise and eat the prey, it will live long enough to reproduce. The slower gazelle or caribou gets eaten; the slower and less camouflaged cheetah or wolf will starve. This is how evolution works on an individual. Over long periods of

FIGURE 8.1
Teosinte (top), maize-teosinte hybrid (middle), and standard corn, or maize (bottom). (Courtesy of John Doebley; https://en.wikipedia.org/wiki/Zea_(plant).)

time, there is an **evolutionary arms race**, with prey developing ever better ways to hide or escape, and predator developing ever better ways to surprise and catch their victims. The same applies to plants: the tree that grows just a little taller will catch more sunlight and eventually take over a forest; the shrub that can tolerate drought will survive and reproduce where non–drought-tolerant shrubs do not survive.

The Concept of Evolution

The idea that animals could descend from earlier types of animals dates from the Greek philosopher Anaximander of Miletus (c. 610–546 BCE) who proposed that the first animals lived in water during a wet phase of Earth's history. He believed that the first land-dwelling ancestors of mankind must have been born in water, and only spent part of their life on land. In contrast, the Greek Aristotle (384–322 BCE) and his followers believed in the **fixity of species**, that the form of all animals was fixed by divine design and remained unchanged.

The ancient Chinese Taoist philosopher Zhuang Zhou (c. 369–286 BCE) believed in changing rather than fixed species. Taoist philosophers speculated that species developed different traits in response to differing environments. Taoism regards humans, nature, and the heavens as existing in a state of "constant transformation."

The fourth-century bishop Augustine of Hippo (Saint Augustine) wrote that the creation story in Genesis should not be taken too literally. In his book *De Genesi ad Litteram* (Quasten et al. 1982), he stated that, in some cases, new creatures may have come about through the "decomposition" of earlier forms of life. "Plant, fowl and animal life are not perfect… but created in a state of potentiality," unlike the theologically perfect forms of angels and the human soul. Augustine believed that forms of life had transformed slowly over time.

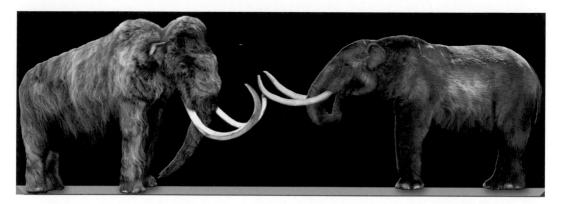

FIGURE 8.2
A Eurasian Woolly mammoth (left) and an American mastodon (right) showing the physical differences between the two animals and the similarity to modern elephants. (Courtesy of Dantheman9758 at en.wikipedia; https://commons.wikimedia.org/wiki/File:MammothVsMastodon.jpg.)

In the *Akhlaq-i-Nasri*, Arab thinker Nasīr al-Dīn al-Tūsī (1201–1274) proposed that the original universe consisted of equal and similar elements. Gradually internal contradictions appeared and, as a result, some substances began developing faster and differently from other substances. Elements evolved into minerals, then plants, then animals, then humans. Al-Tūsī explains why variability is important for the evolution of living things: "Organisms that can gain the new features faster ... gain advantages over other creatures. ... The bodies are changing as a result of the internal and external interactions."

Aristotle's ideas on the fixity of species had a powerful influence on Renaissance thinkers. But they took it a step farther: they believed that ideal species are arranged hierarchically in a **ladder of creation**, with lower life forms on the bottom and higher forms on the top. Large bones found weathering out of rocks were thought to belong to an extinct race of giants, perhaps drowned by the Biblical flood. Shells found on mountains were carried there by The Flood, or else were placed there by the devil to confound humans.

Gradually, naturalists began looking for a logical explanation for why fossils appeared to change over time, why they sometimes disappear altogether at a certain point in the rock record, and why they often to appear similar to present-day life forms (Figure 8.2).

Lamarckian Genetics

In the late 1700s, some European scientists began to suggest that life forms are not fixed. The French naturalist George Louis Leclerc, Comte de Buffon believed that species could change over generations, and proposed this was the result of a changing environment or perhaps pure chance. He also suggested that humans and apes are related. At the time, most European naturalists and clerics believed that God had made animals in a perfect and unalterable condition. It could be dangerous to propose otherwise: at the very least, going against the conventional wisdom invited intense criticism. Buffon avoided broad public criticism by hiding his views in *Histoire Naturelle* (1749–1804) (Buffon 2017), a limited edition book on natural history.

One of the first to publish an explanation for biological change was another French naturalist, Jean-Baptiste Chevalier de Lamarck. An authority on invertebrates, in 1801, he classified them in *Système des animaux sans vertèbres* (Lamarck 1801). Lamarck proposed that animals evolved according to natural laws. His *Philosophie Zoologique* (Lamarck 1809), published in

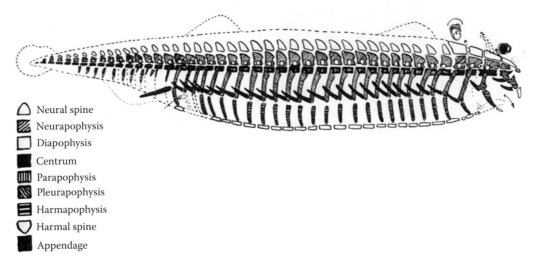

Neural spine
Neurapophysis
Diapophysis
Centrum
Parapophysis
Pleurapophysis
Harmapophysis
Harmal spine
Appendage

FIGURE 8.3
This 1848 Diagram by Richard Owen shows his conceptual archetype for all vertebrates (https://en.wikipedia
.org/wiki/History_of_evolutionary_thought).

1809, explained changes in animal forms as a result of the **inheritance of acquired characteristics**. This theory stated that evolution occurred when an animal used a body part for
a specific task, altering that part or organ during its lifetime. This change was then inherited by its offspring. For example, giraffes evolved their long necks because each generation
stretched farther to reach leaves in trees. This change in body shape was inherited.

The theory of inheritance of acquired characteristics was easy to disprove. If correct, then
dogs with bobbed tails would have pups with stubby tails, the children of musicians would
be born musicians, the children of deaf people would be deaf, the children of people who
had their tonsils removed would be born without tonsils. Inheritance doesn't work that way.

Another theory was that parental characteristics are blended with each generation. The
blending theory, too, was easily discredited since the eventual outcome would be that
everyone and everything would look more and more alike. The children of a tall and short
person would be average height, and the child of a dark haired and blond person would
always be brunette. That didn't work either.

The German biologist-poet Goethe argued that complex body shapes obscured basic
patterns, or **archetypes**, that applied to all life forms. Such patterns included bilateral symmetry (one side is the mirror image of the other side), spinal cords, and basic bone arrange-
ments. In the 1830s, Etienne Geoffroy Saint-Hilaire, a French naturalist, and Richard Owen,
a British anatomist, set out to find the archetype of all vertebrates (Figure 8.3). Saint-Hilaire
argued that every bone in every vertebrate is merely a variation on an archetypal pattern
that has been modified over time (Figure 8.4). He suggested that changes in the environment might disturb an animal's embryonic development causing freaks to be born, thus
leading to new species.

Darwin and Wallace

After his voyage on the Beagle (1831–1836) Charles Darwin sent his Patagonian fossils to
Richard Owen for identification. Owen said that they were giant varieties of animals that

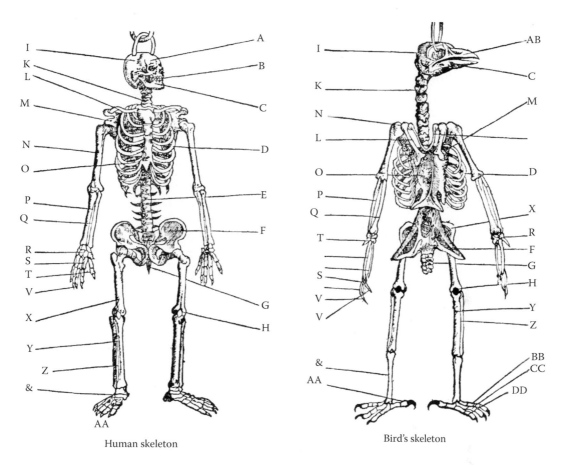

FIGURE 8.4

Comparison of the skeleton of man and bird. (From Belon, *L'Histoire de la nature des oyseaux* [Book of Birds], 1555; https://commons.wikimedia.org/wiki/File:BelonBirdSkel.jpg.)

still exist in South America. Darwin wondered if there was any continuity between the extinct animals and those still living. Could the living animals be descended from the fossil animals?

In the Galapagos Islands, Darwin had tentatively identified finches, wrens, and blackbirds on various islands. He sent his bird collection to the ornithologist James Gould for confirmation. Gould said they were all finches with beaks modified to eat different foods. Darwin wondered if they had evolved from an original pair of finches that colonized each island. The finches would have adapted to the particular island environment by evolving slightly different body forms. He speculated that the same process may have happened to the mammals of Latin America: the giant forms evolved to the smaller forms seen today. After visiting the zoo and seeing similar expressions and behaviors in apes as in humans, Darwin pondered if humans had evolved from other, related forms. He began looking into the breeding of dogs and pigeons and recognized that animals could change form. But he could not figure out what natural process could cause such changes.

He found an answer in the 1798 book by Thomas Malthus, *An Essay on the Principle of Population* (Malthus 1798). Malthus pointed out that human population, if unchecked by famine, disease, or war, would increase rapidly until there was not enough food. The result would be famine and death on a grand scale. Malthus pointed out that famine, disease, and predation applied to all animals. Darwin realized this was the mechanism he needed to explain changes in animals. The purpose of life is to reproduce. Animals with a slight advantage that allows them to live long enough to reproduce will pass on these traits to their offspring. Given enough time, a new species will result. He compared this to farmers breeding plants by using only the seeds from the best plants for the next year's crop.

Robert Chambers, a Scot, in 1845, published *Vestiges of the Natural History of Creation* (Chambers 1845). In it, Chambers proposed that animals evolve by the success of certain birth defects. Whereas most defects cause the death of the newborn animal, some are advantageous in that they help in survival and are passed to offspring. Chambers went out on a limb and suggested that humans are descended from fish. The book introduced the concept of evolution to a broad audience but was roundly denounced by most scientists. The violence of the attacks caused Darwin to pause his work on evolution for 10 years and study barnacles instead.

By 1854, Darwin again began compiling the results of his work on evolution. Then, in 1858, he received a letter from the British naturalist Alfred Russel Wallace. Wallace had traveled broadly and had read the same books by Chambers and Malthus. He and Darwin had been corresponding for years. Wallace asked Darwin to review his ideas on the evolution of species over time. The two men's ideas were nearly identical. Darwin arranged that both of their work be presented to the Linnaean Society on June 30, 1858. There was little negative reaction. Encouraged, in 1859, Darwin published a summary of his ideas called *On the Origin of Species by Means of Natural Selection* (Darwin 1859). In it, he admitted that no one understood how heredity worked, but it was clear that offspring were similar to parents. He claimed that the traits that helped offspring survive would be passed on to descendants while qualities that made animals less fit would be removed by the **natural selection** of famine, disease, and predation. New species would arise over time as a result of this selection. "Natural selection is daily and hourly scrutinizing, through the world, every variation, even the slightest. We see nothing of these slow changes in progress, until the hand of time has marked the long lapses of ages." Competition exists between species for food. Competition between similar species is the most intense, leading eventually to one going extinct. This explained the fossils of animals similar to those seen today.

Darwin described all animals as branches on a universal **tree of life** (Figure 8.5). Regarding the dearth of intermediate forms, Darwin explained that the process of preserving a fossil is haphazard. Few animals are buried and preserved. Once fossilized, they are not all exposed at the surface at a time and place where they will be found; most are eroded and washed away. Gaps are to be expected in the fossil record. It is surprising that we have found as many transitional forms (missing links; Figure 8.6) as we have!

To explain complex bodies, he pointed to flying squirrels as an analog to how bats might have originated: a small tree-climbing mammal with loose skin between its body and limbs jumps or glides ever farther as the skin gradually modifies to wing-like appendages. Likewise for the eye: flatworms have nothing more than nerve endings with light-sensitive pigments. Some animals have light-sensitive cells in a little cup that gives some sense of direction for incoming light. By gradually deepening the cup until it is close to a sphere, a lensless pinhole camera is formed. The marine mollusk Nautilus has such eyes. Some crustaceans have eyes that are a layer of light-sensitive pigment coated by a membrane. Over time, the membrane could separate from the pigment and act as a crude lens. Small

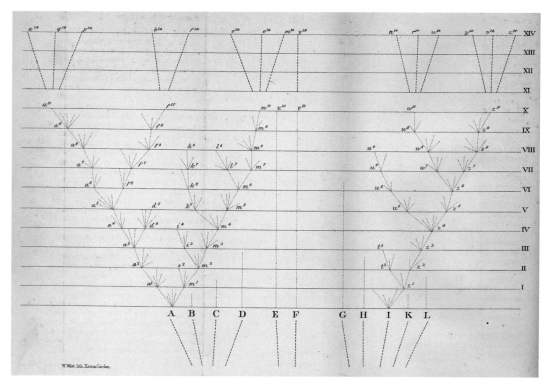

FIGURE 8.5
Tree of life as drawn by Charles Darwin in *Origin of Species* (https://commons.wikimedia.org/wiki/File:On_the_Origin_of_Species_diagram.PNG).

alterations such as these gradually improve the eye until it is the complex organ we know. Because a little sight has more survival value than no sight, each advancement is rewarded by natural selection. "Thus, from the war of nature, from famine and death, the most exalted object which we are capable of conceiving, namely the production of the higher animals, directly follows. There is a grandeur in this view of life ... from so simple a beginning endless forms most beautiful and most wonderful have been, and are being, evolved." In fact, the complex image-forming eye has evolved independently at least 50 times.

Many reviewers believed Darwin's book to be blasphemous, contradicting the revelation of God, as well as "an abuse of science." Yet, by the 1870s, almost all European scientists had accepted the concept of evolution driven by natural selection. **Sexual selection**, a special case of natural selection, is when one sex of a species, usually the males, compete for the ability to mate with the opposite sex. This can lead to extravagant displays, as in peacocks' tail feathers, mating dances among fruit flies, and deep, attractive croaks in bullfrogs.

Mendel and Inheritance of Traits

For thousands of years, farmers and herders have been selectively breeding plants and animals to produce more useful varieties. It was hit or miss because no one knew how heredity worked. Darwin admitted as much in *Origin of Species*. In 1866, Gregor Mendel, a Moravian friar, published the results of crossbreeding pea plants that showed certain

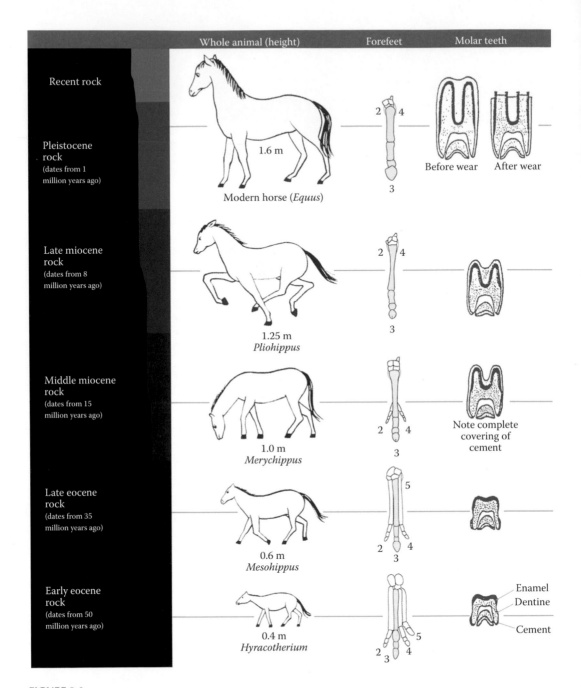

FIGURE 8.6
Transitional forms in the evolution of the horse illustrating changing size, foot, and teeth. Originally described in 1876 by American paleontologist Othniel C. Marsh. (Courtesy of Mcy jerry at English Wikipedia; https://commons.wikimedia.org/wiki/File:Horseevolution.png.)

traits (flower color, stem length, seed shape, seed color) belonging to parents show up in offspring without blending of parental features. He came to three main conclusions:

1. **Inheritance of a trait** is determined by factors passed to offspring intact and unchanged (these "factors" are now called **genes**)

2. An individual inherits one such factor from each parent for each trait (such as eye color)

3. A trait may not show up in an offspring, but can still be carried and passed on to the next generation (like the blood disorder hemophilia).

The Tree of Life

The first evolutionary biologists drew the tree of life as a mighty oak (Figure 8.7). Living things with similar body shapes and parts were close together. Simple bacteria were near the trunk, and mankind was at the summit of evolution. The current tree, based on DNA, has three main branches: Eukaryotes, Bacteria, and Archaea. Our branch, the **Eukaryotes**, includes plants, fungi, and animals, as well as single-celled protozoa such as the amoeba.

Eukaryotes have compartmented cells where proteins are created and energy is generated. Most of the DNA is in the cell nucleus. **Mitochondria** in animal cells use oxygen and sugars to generate energy. **Chloroplasts** are that part of a plant cell that converts energy from sunlight into sugars. Evidence suggests that mitochondria are descended from free-living bacteria that were engulfed by another cell and ended up as a permanent resident in a process called **endosymbiosis** (Figure 8.8). Chloroplasts in plant cells were also once free-living bacteria. The evidence for this is that both mitochondria and chloroplasts have their own DNA that is separate from that of the host cell. They are surrounded by a double membrane within which they duplicate their own DNA and produce their own proteins and enzymes. How did this happen?

About one-and-a-half billion years ago, two kinds of early bacteria entered a mutually beneficial relationship. One type of bacteria used oxygen to generate energy. Its larger neighbor provided a safe, nutrient-rich environment. Over time, the bond deepened and they couldn't live without each other. Their relationship moved to the next stage: one cell enveloped the other. The host cell benefited from the energy the mitochondrion produced, and the mitochondrion gained a protected, nutrient-rich environment.

Prokaryotes are cells where the DNA floats loose inside the cell rather than being confined to the nucleus (Figure 8.9). Prokaryotes include the second and third branches of the modern tree of life, **Bacteria** and **Archaea**. Archaea have some genes and metabolic paths closely related to eukaryotes, which suggest that Eukaryotes and Archaea could have evolved from a common ancestor (Y in Figure 8.10). The Archaea may be the oldest existing form of life on Earth.

Since all living things have certain traits in common (carry genetic information; create proteins), there must have been an even older common ancestor (X in Figure 8.10). That common ancestor is probably extinct due to predation or environmental changes.

The first evidence of **multicellular life** is from cyanobacteria that lived about 2.5 billion years ago. An example of different cells living together in a mutually beneficial relationship can be seen in lichens. **Lichens** are a symbiotic association between algae and fungi

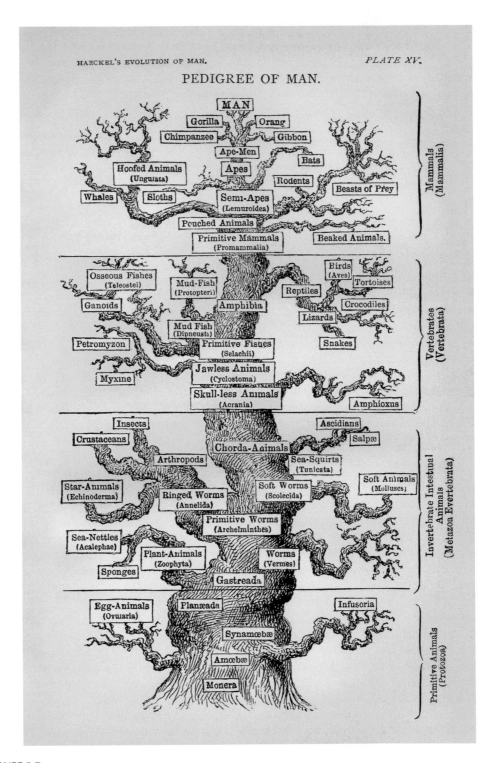

FIGURE 8.7
Tree of Life as drawn by Ernst Haeckel in *The Evolution of Man* (1879). This view of evolution leads progressively towards humans, thought to be at the pinnacle of creation (https://en.wikipedia.org/wiki/History_of_evolutionary_thought#/media/File:Tree_of_life_by_Haeckel.jpg).

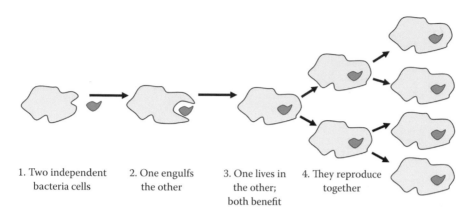

FIGURE 8.8
How mitochondria and chloroplasts are thought to have evolved from separate bacteria.

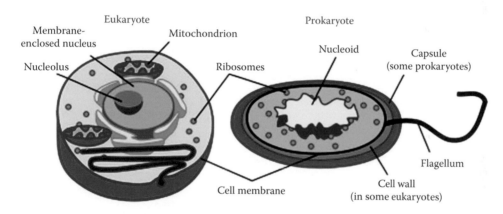

FIGURE 8.9
Differences between prokaryote and eukaryote cells. (Courtesy of Science Primer, National Center for Biotechnology; https://commons.wikimedia.org/wiki/File:Celltypes.svg.)

where the combined entity is different from its constituent organisms. The algae produce food using photosynthesis, and the fungi gather moisture and nutrients and provide a protected environment. By 1.9 billion years ago, Earth's atmosphere had enough oxygen to allow the development of energetic cells that clumped together in colonies that began to resemble primitive animals. Eventually, slight differences in cells provided benefits to the colony: the different cells helped each other and came to depend on one another.

Over time, some colonial cells specialized in pulling nutrients from the surrounding environment, others specialized in digestion, while others circulated nutrients or sensed changes in the environment. Working together, these cells became the first multicellular organisms. Multicellularity happened independently many times: at least once in animals, three times in fungi, six times in algae, and numerous times in bacteria. These organisms had advantages over their noncolonial counterparts and so they multiplied and diversified: their soft-bodied remains are first seen in the 570 million-year-old Ediacara fauna of South Australia, in the 525 million-year-old Chengjiang fauna in Yunnan, China, and in the 510 million-year-old Burgess Shale fauna of British Columbia, Canada.

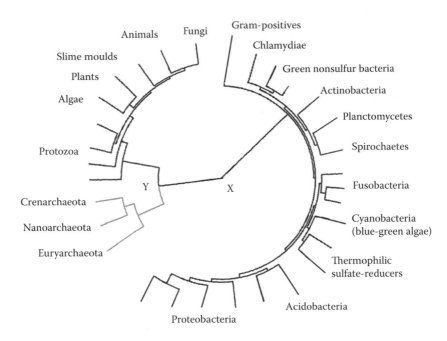

FIGURE 8.10

A DNA-based tree of life showing the diversity of bacteria, compared to other organisms. Eukaryotes are colored red, archaea green, and bacteria blue. X is the common ancestor of all living things; Y is the common ancestor of Archaea and Eukaryotes. (Adapted from Ciccarelli et al. (2006). Toward automatic reconstruction of a highly resolved tree of life. *Science* 311 (5765), 1283–1287; https://en.wikipedia.org/wiki/File:Collapsed_tree _labels_simplified.png.)

The Large Size of Some Animals

Time after time, animals evolved from small to large. The first giants (megafauna) were insects. Between 260 and 330 Ma, some dragonflies had wingspans up to 70 cm (28 in), and mayflies had 45 cm (18 in) wingspans. There were spiders with 43 cm (17 in) legs, and 2 m (6.6 ft) long millipedes and scorpions. During this time, atmospheric oxygen levels, around 35%, were the highest ever measured. High levels of oxygen provided these animals with the energy needed by their metabolisms.

In the late Permian, a **carnivorous** (meat-eating) reptile, Dimetrodon (the one with a large sail on its back), reached 5 m (16.4 ft) long and weighed as much as 250 kg (551 lbs); the **herbivore** (plant-eating) Moschops reached 2.7 m (9 ft); the Gorgonopsians were rhinoceros-sized Permian predators. These were the largest land animals before dinosaurs.

Jurassic to Cretaceous dinosaurs are the best-known giants. The herbivorous *Argentinosaurus huinculensis* was the largest dinosaur, reaching 96.4 metric tons (106 US tons) and 39.7 m (130 ft) long. *Spinosaurus* was the largest predator, reaching 21 metric tons (23 US tons) and 18 m (59 ft) head-to-tail.

Mammals became large after the end-Cretaceous extinction. The largest land mammal, an extinct elephant called *Palaeoloxodon namadicus*, had a maximum shoulder height of 5.2 m (17 ft) and weighed about 22 metric tons (24 US tons). The largest land carnivore, the bear *Arctotherium angustidens*, weighed up to 1600 kg (3527 lbs) and stood 3.3 m (11 ft) on its hindlimbs. The blue whale (*Balaenoptera musculus*), an existing marine mammal, can reach 30 m (98 ft) in length and 180 metric tons (198 US tons). It is the largest animal known to exist.

What is required, and what are the advantages of **gigantism** in animals? The animals have to be fast-growing so that they can achieve large size within a normal lifetime of a few decades. They need efficient respiratory and circulatory systems that allow oxygen to provide energy throughout their body. And they benefit from an oxygen-rich atmosphere (over 20%) that makes breathing easier. Advantages of large size include a slower metabolism, minimizing the energy needed per unit of weight. An elephant (slow metabolism) eats maybe 5% of its weight per day, whereas a shrew (fast metabolism) eats its own weight in food each day. Large size provides insulation, so animals are less affected by extreme temperatures. Large size allows long digestive tracts so that tougher plants can be digested. Great height allows plant eaters access to taller plants. Bulk also provides protection from smaller predators. Adult hippos, for example, are not concerned about lions or crocodiles.

Ongoing Evolution of Humans

According to Will Cuppy, "All Modern Men are descended from a Wormlike creature, but it shows more on some people" (Cuppy 1944). The human lineage has continued to evolve since humans and chimpanzees diverged from their most recent common ancestor between 5 and 13 million years ago. *Homo sapiens* are thought to have diverged from Neanderthals and Denisovans between 350,000 and 800,000 years ago. There is evidence of these groups interbreeding with non-African humans: Europeans and Asians, for example, share between 1% and 4% of their DNA with Neanderthals. The earliest evidence for anatomically modern *H. sapiens* is dated at around 200,000 years ago in East Africa.

Certain traits, like the ability to speak, large brains, and use of hands to make tools, are considered uniquely human (Lynch and Granger 2008). These all evolved in the past 3 to 5 Ma. Other traits are much more recent developments. For example, in the 10,000 years since humans first domesticated sheep, goats, and cattle, certain groups of people have acquired enzymes that allow adults to digest lactose and get nourishment from milk (Cochran and Harpending 2009; Zuk 2013). Presumably, those with the enzyme were able to survive famines and reproduce whereas those without the enzyme did not fare as well.

The same goes for the evolution of resistance to disease. Some human groups, notably sub-Saharan West Africans, have developed the sickle cell genetic mutation that allows them to tolerate malaria. The sickle cell is caused by an abnormality in the hemoglobin protein found in red blood cells. Sickle cell disease causes severe infections, severe pain, stroke, and an increased risk of death. However, a person with only one abnormal hemoglobin gene does not get the disease. The disease requires two copies of the abnormal gene, one from each parent. Thus, the chance of overall survival until reproductive age is improved.

People living in the Andean and Tibetan highlands have adapted to living at high altitudes in one of two ways: by increasing oxygen saturation in hemoglobin, or increasing hemoglobin production. Hemoglobin is the protein in red blood cells that carries oxygen. In both cases, this development occurred in the past 3000 years.

Interestingly, the genes for blond hair and blue eyes only became common in northern Europe in the past 3000 years. The survival value of these particular traits is uncertain.

The point is, all evidence indicates that humans continue to evolve.

Extinction

Extinction marks the end of existence of an organism or group of organisms. It is a normal part of any ecosystem. New groups of animals and plants are gradually evolving all the time. Likewise, individual species are going extinct all the time. The same conditions of environmental stress, changing habitat, and competition that lead to the development of new groups also lead to extinctions. This is a common, continuing, daily process and is not in any way exceptional. It is estimated that more than 5 billion species, or 99% of all species that ever lived, have gone extinct. On average, a species lasts roughly 10 million years, although some, like the coelacanth fish, have gone virtually unchanged for 400 Ma, and ginkgo plants are essentially unchanged for the past 270 Ma.

The ongoing process of evolution and extinction is vastly different from **mass extinction** events. Mass extinctions mark the disappearance of large numbers of species or multiple groups of organisms. The causes tend to be global: oxygenation of the atmosphere 2.3 billion years ago (mainly affected single-celled organisms), global glaciation (such as "Snowball Earth" between 740 and 580 million years ago), massive flood basalts and resulting atmospheric changes (at the end of the Permian), and large asteroid or comet impacts (at the end of Cretaceous time). Nearby supernovas, gamma ray bursts, and other causes have been mentioned.

As Dr. Malcolm points out in the movie Jurassic Park, life breaks free, expands to new territories, finds a way. With each mass extinction, a few well-adapted groups of organisms survive. Diversity is low, and unoccupied environmental niches are plentiful. As organisms move into new environments, they evolve to best fill available niches. Soon, the number of species has increased to compete in all habitable environments. This concept became known as the "**cone of increasing diversity**" (Figure 8.11).

In *Wonderful Life* (1989), Stephen Jay Gould describes the 510 million-year-old fauna of the Burgess Shale. These animals appeared well adapted to their environment and yet most of them went extinct by the end of Cambrian time. He argued that chance events and characteristics that would prove beneficial under future, changed conditions determined

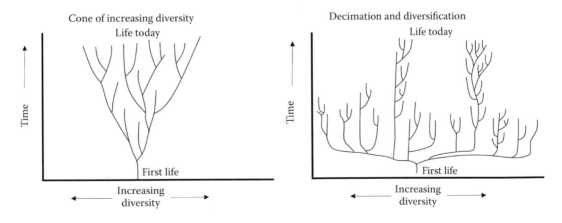

FIGURE 8.11
Concept of the "cone of increasing diversity": phyla, classes, genera, and species all increase over time. Gould's concept of "decimation and diversification": phyla and classes decrease over time, but genera and species increase.

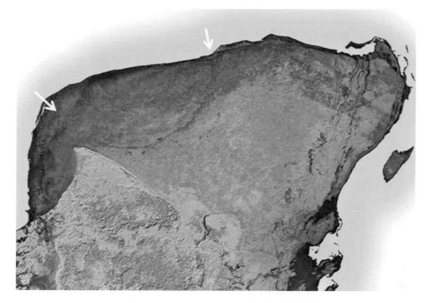

FIGURE 8.12
Scientists think the dinosaurs were wiped out when a large asteroid hit the Yucatan Peninsula, Mexico. NASA's Shuttle Radar Topography Mission reveals the onshore southern half of the 180-km-diameter Chicxulub Crater as outlined by sinkholes (arrows) (https://en.wikipedia.org/wiki/Chicxulub_crater).

which would survive. Gould concluded that one result of extinction events over time was that the number of major categories of life (phyla and classes) decreased, but overall diversity (the number of genera and species) remained the same or even increased. He called this "**decimation and diversification**." Another consequence of mass extinctions is that they make way for entirely new groups of animals to become dominant. Reptiles became the prevailing creatures on Earth after the end-Permian extinction. Had it not been for the impact of a large asteroid at Chicxulub in Yucatan, Mexico, at the end of Cretaceous time, they might still rule the earth (Figure 8.12).

Precambrian mass extinctions are difficult to document because they applied to single-celled organisms or soft-bodied creatures. There are five generally recognized mass extinctions in the past 600 million years, with several additional minor extinction events (Figure 8.13). The extinction of large mammals and birds (mammoths, giant sloths, the dodo bird, carrier pigeons, the "marsupial wolf" thylacine, the moa) over the past 50,000 years is thought by many to be a result of human predation and human-caused habitat change.

Origin of Life

Life is thought to have started on Earth between 3.8 and 4.1 billion years ago. But what is "life"?

There are many definitions, but most agree that **life** has three essential functions that distinguish it from inorganic processes: (1) the ability to reproduce; (2) heredity, the ability to pass parental traits to offspring; and (3) **metabolism**, the process that uses energy to convert chemicals to cellular components. Other features of living things include organization

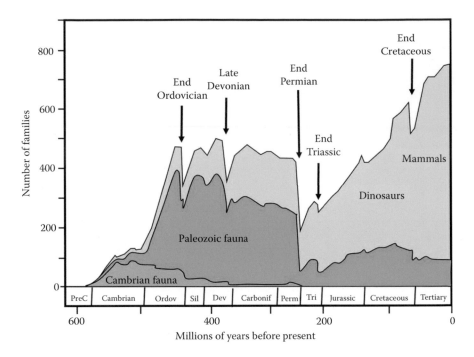

FIGURE 8.13
Five major Phanerozoic mass extinction events. (Modified after ZME Science; http://www.zmescience.com
/ecology/animals-ecology/mass-extinction-humans-14122014/.)

(of components in cells), growth, response to stimulation, and regulation of the environment within the cell or body.

In 1952, Stanley Miller and Harold Urey at the University of Chicago showed that conditions on early Earth could generate complex organic compounds from simpler inorganic molecules. There is evidence that volcanic eruptions 4 billion years ago would have provided a primitive atmosphere consisting of carbon dioxide (CO_2), nitrogen (N_2), hydrogen sulfide (H_2S), and sulfur dioxide (SO_2). The **Miller–Urey experiment** (Figure 8.14) demonstrated that most amino acids can be synthesized from inorganic compounds in seawater and the primitive atmosphere. More than 20 different **amino acids**, chemicals used to build proteins, were produced in their experiments.

A **catalyst** is anything that speeds up a chemical reaction without being consumed by the reaction. The first life-like molecules on Earth may have been formed by metal-based catalysts on the surface of mineral crystals. Metabolism could have existed on these surfaces long before the first cells. The **"metabolism first" theory** proposes that life requires molecules with the ability to catalyze reactions that lead to the production of more molecules like themselves. The simplest self-replicating systems contain two original molecules and a product molecule. The product molecule is a catalyst that provides a template that brings together the original (ancestor) molecules, which, in turn, produce more product molecules.

Autocatalytic sets are groups of molecules that generate self-sustaining chemical reactions in which the product of one reaction is the feedstock for the next reaction. The result is a self-contained cycle of chemical creation. In 1990, Julius Rebek, Jr. and his colleagues at

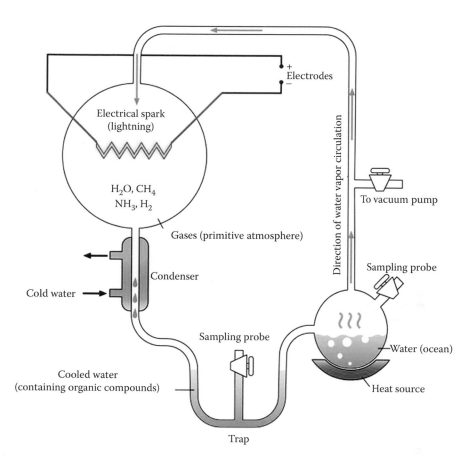

FIGURE 8.14

The Miller–Urey experiment. (Courtesy of GYassineMrabetTalk; https://commons.wikimedia.org/w/index .php?curid=3500500.)

Scripps Research Institute (California) combined amino adenosine and pentafluorophenyl esters with the autocatalyst amino adenosine triacid ester (AATE). One product was a variant of AATE, which catalyzed the synthesis of itself. This demonstrated that autocatalysts could compete within a molecular population showing heredity, which would have been a primitive type of natural selection.

Nucleic acids are large molecules necessary for life. They include **DNA** (deoxyribonucleic acid) and **RNA** (ribonucleic acid). Both consist of **nucleotides** that contain a sugar, a phosphate group, and a nitrogen-containing base (Figure 8.15). If the sugar is deoxyribose, the molecule is DNA. If the sugar is ribose, the molecule is RNA. Nucleic acids transmit genetic information through the arrangement of nucleotides.

It is generally accepted that DNA-based life descended from RNA-based life ("**RNA world**"), although RNA-based life was probably not the first life.

The first molecules with both catalytic activity and information storage ability may have been molecules that resemble RNA but are chemically simpler. The transition from the pre-RNA world to the RNA world would have occurred through creation of RNA using a simpler compound as both template and catalyst. Pre-RNA molecules would have

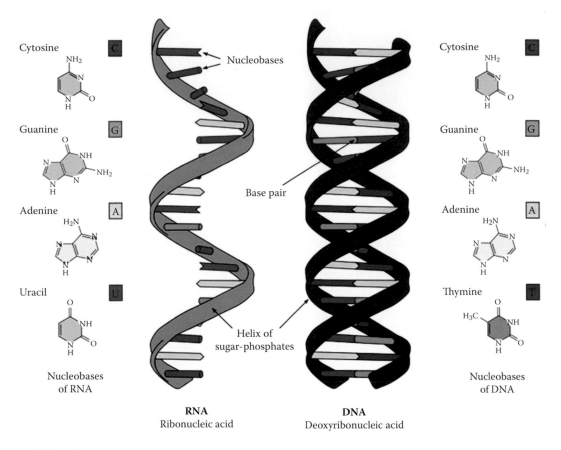

FIGURE 8.15
A comparison of RNA (left) with DNA (right), showing the helices and nucleobases/nucleotides each employs. (Courtesy of Difference_DNA_RNA-DE.svg: Sponk (talk); https://commons.wikimedia.org/wiki /File:Difference_DNA_RNA-EN.svg.)

catalyzed formation of ribonucleic acids from these simpler molecules. Once the first RNA molecules had been produced, they could have slowly evolved to take over functions originally performed by pre-RNA molecules.

An RNA world is based on the unique ability of RNA molecules to act both as an information carrier and as a catalyst. RNA stored genetic information and catalyzed chemical reactions in primitive cells. Only later did DNA take over as the genetic material and proteins become the major catalyst. DNA is thought to have taken over the role of data storage because of its increased stability, while proteins, through a greater variety of amino acids, replaced RNA's role as catalyst. The fact that RNA still catalyzes some fundamental reactions in modern cells makes it a molecular fossil. The transmission of hereditary information today proceeds from DNA to protein through an RNA intermediary.

Complex organic molecules have been found in the solar system and in interstellar space, and some biologists think these molecules provided the starting material for life on Earth. The **panspermia hypothesis** suggests that microscopic life was distributed by meteoroids, asteroids, and comets and that life may exist throughout the universe. It is speculated that life may have begun as early as 13.8 billion years ago, shortly after the big bang. This doesn't answer the question of how life started; it merely shifts the beginning of life to another place and time.

Where were the compounds that first interacted to form life? Darwin speculated that life may have begun as a primordial soup of organic molecules in a sun-warmed scummy pond. The Miller–Urey experiment supported the **primordial soup theory** by creating amino acids out of seawater, a primitive atmosphere, and energy from lightning. However, in 1992, the Australian-born astronomer Thomas Gold proposed instead that life began under the high temperatures and pressures found deep in the earth. All rocks have cracks and pores, usually filled with water. Deep drilling confirms that this space today contains organisms (**hyperthermophile** bacteria and Archaea) that thrive in temperatures up to 120°C and high pressures that prevent water from boiling. Such conditions exist at depths from 5 to 10 km. The temperature and chemistry of this environment are close to that expected near the surface of the early Earth, and thus lends support to the **hot deep rock theory** of first life. Gold thought there may be a greater mass of such organisms than all surface, sun-based life combined. In 2011, the journal *Nature* reported the discovery of complex multicellular life, nematode worms, living at depths of 1.3 km (4250 ft) in water flowing into South African gold mines, and recovered DNA from an unknown species at 3.6 km and 48°C (11,800 ft and 118°F; Borgonie et al. 2011).

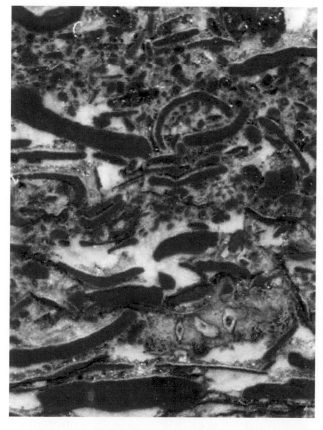

FIGURE 8.16
Limestone rock in Australia contained fossils (blue) that may represent Earth's earliest animal life. (Courtesy of Adam Maloof Lab/Situ Studio, Princeton University; http://www.livescience.com/6885-fossils-earliest-animal -life-possibly-discovered.html.)

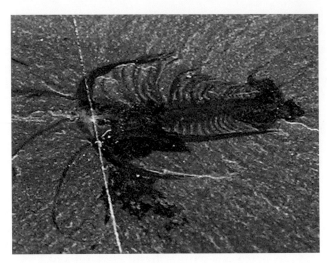

FIGURE 8.17
Five-hundred-five-million-year-old soft-bodied arthropod *Marrella splendens* from the Burgess Shale, British Colombia, Canada. (Courtesy of Verisimilus; https://en.wikipedia.org/wiki/Fossils_of_the_Burgess_Shale.)

The earliest evidence for life on Earth is carbon in the form of graphite inclusions in zircon. The graphite, in 4.1-billion-year-old rocks in the Jack Hills of Western Australia, is thought to be remains of life. This suggests that life generated relatively soon after Earth's crust solidified. Isotopically light carbon (the ratio of C^{12} to C^{13} is higher than that expected for inorganic carbon) has been found as inclusions in apatite in 3.85 billion-year-old metasedimentary rocks on Akilia Island, southern West Greenland. Cyanobacteria fossils have been found in 3.48 billion-year-old layered rocks called **stromatolites** in Western Australia.

In 2010, researchers from Princeton reported the discovery of the oldest hard-bodied fossils, sponge-like marine animals that lived in stromatolite reefs about 650 million years ago (Figure 8.16). The shelly fossils were found in the Trezona Formation of the Flinders Ranges, South Australia, beneath a 635 million-year-old glacial deposit. They are the earliest evidence of animal body forms currently in the fossil record.

The oldest soft-bodied fossils date to around 500 Ma and include the Fezouata Formation of southeastern Morocco and Burgess Shale of British Colombia, Canada (Figure 8.17).

References and General Reading

Borgonie, G., A. Garcia-Moyano, D. Litthauer, W. Bert, A. Bester, E. van Heerden, C. Moller, M. Erasmus, and T. C. Onstott. 2011. Nematoda from the terrestrial deep subsurface of South Africa. *Nature* 474, 79–82 (02 June 2011), doi:10.1038/nature09974, published online 01 June 2011.

Buffon, G. L. L. 2017. *Histoire Naturelle*. Forgotten Books Classic Reprint, French edition, 28 volumes.

Chambers, R. 1845. *Vestiges of the Natural History of Creation*. Open Library, https://www.archive.org/stream/vestigesnatural00cheegoog?ref=ol#page/n6/mode/2up.

Ciccarelli, F. D., T. Doerks, C. von Mering, C. J. Creevey, B. Snel, and P. Bork. 2006. Toward automatic reconstruction of a highly resolved tree of life. *Science* 311 (5765), 1283–1287. Bibcode:2006Sci...311.1283C.doi:10.1126/science.1123061.PMID 16513982.

Cochran, G., and H. Harpending. 2009. *The 10,000 Year Explosion—How Civilization Accelerated Human Evolution*. Basic Books, New York. 288 pp.

Cuppy, W. 1944. *The Great Bustard and Other People*. Murray Hill Books, New York. p. 30.

Darwin, C. 1859. *On the Origin of Species by Means of Natural Selection*. Open Library, https://www.archive.org/stream/6edoriginspecies00darwuoft?ref=ol#page/n3/mode/2up.

Dawkins, R. 1996. *The Richard Dimbleby Lecture: Science, Delusion and the Appetite for Wonder*. BBC1.

Dawkins, R. 2004. *The Ancestor's Tale—A Pilgrimage to the Dawn of Evolution*. Mariner Books, Boston. 673 pp.

Dawkins, R. 2009. *The Greatest Show on Earth—The Evidence for Evolution*. Free Press, New York. 470 pp.

Diamond, J. 1997. *Guns, Germs, and Steel: The Fates of Human Societies*. W.W. Norton and Company, New York. 480 pp.

Gould, S. J. 1983. *Hen's Teeth and Horse's Toes*. W.W. Norton & Company, New York. 413 pp.

Gould, S. J. 1989. *Wonderful Life—The Burgess Shale and the Nature of History*. W.W. Norton & Company, New York. 347 pp.

Lamarck, J. B. 1801. *Système des animaux sans vertèbres*. Muséum d'Histoire Naturelle, Paris. Open Library, https://www.archive.org/stream/systmedesanim00lama?ref=ol#page/n7/mode/2up.

Lamarck, J. B. 1809. *Philosophie zoologique*. Open Library, https://openlibrary.org/works/OL1107237W/Philosophie_zoologique.

Lynch, G., and R. Granger. 2008. *Big Brain—The Origins and Future of Human Intelligence*. Palgrave MacMillan, New York. 259 pp.

Malthus, T. 1798. *An Essay on the Principle of Population*. J. Johnson in St. Paul's Church-Yard, http://www.esp.org/books/malthus/population/malthus.pdf.

Quasten, J., W. J. Burghardt, and T. C. Lawler. 1982. *Ancient Christian Writers*, Books 1–6. Paulist Press, Mahwah, NJ.

Thomas, L. 1979. *The Medusa and the Snail: More Notes of a Biology Watcher*. Viking Press, New York. 175 pp.

Zimmer, C. 2001. *Evolution—The Triumph of an Idea*. Harper Perennial, New York. 487 pp.

Zuk, M. 2013. *Paleofantasy—What Evolution Really Tells Us about Sex, Diet, and How We Live*. W.W. Norton & Company, New York. 328 pp.

9

The Changing Face of the Earth—Structure of the Earth, Plate Tectonics, and Mountain Building

Geology gives us a key to the patience of God.

Josiah Gilbert Holland
Author and poet

Mountains are a source of inspiration, places for recreation, and locations where the Earth's interior is exposed as it rarely is elsewhere. What are mountains, and how are they formed? What do they tell us about Earth processes?

Mountains form slowly by the accumulation of small deformations over long periods of time. During our lifetime, the earth appears to be the one constant, never-changing aspect of our existence. Or is it? During our lives, we experience or hear about earthquakes, landslides, floods, eruptions, tsunamis, and other geological events. These sudden events may be large or small and usually don't affect *us*. There are also gradual changes we don't notice: creep along faults without earthquakes, folding of rocks far in the subsurface, the gradual downhill movement of soil, or the movement of continents. These do not appear remarkable over our lifetimes, but they have dramatic consequences over large spans of time.

Structure of the Earth

In the Shawshank Redemption Morgan Freeman's character states that geology is the study of pressure over time; that all it takes to change the earth is pressure and time. The surface of the earth is subject to stress, mainly from below. Earth's internal heat causes forces that push up and sideways. Rock responds to **stress** (force applied over an area) by deforming. Deformation (**strain**) results in **geologic structures** like folds and faults. These structures provide clues about the stress environment that formed them.

Stress is measured in pascals (Pa) or pounds per square inch (psi), where 1,000,000 pascals (1 megapascal, MPa) is the same as 145 psi. **Lithostatic stress** is stress attributed to the thickness and density of rock above a given depth. It increases at roughly 24 MPa/km (1 psi/ft) as one goes into the earth. Hydrostatic stress is the pressure attributed to groundwater and increases at the rate of 9.79 kPa/m (0.43 psi/ft). This is just like water pressure increasing on your ears when you dive to the bottom of a swimming pool (Figure 9.1). It doesn't take long before there is a lot of stress on a rock. At 6100 m (20,000 ft), the depth of many oil wells, there is 146 MPa (20,000 psi) pushing down on the rock.

Heat also increases the deeper one goes in the earth. The source of heat in the earth today is mainly from decay of the radioactive elements uranium, thorium, and potassium found in minerals. The **geothermal gradient** ranges from about 10°C/km (29°F/mi) in stable

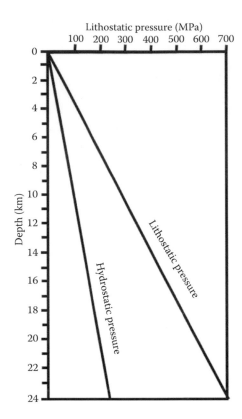

FIGURE 9.1

Typical near-surface lithostatic and hydrostatic pressure gradients. The pressure due to overlying rock increases at 24 MPa/km (1 psi/ft).

continental areas to 30°C/km in areas of active mountain building (Figure 9.2). The deepest mine on Earth, the Tau Tona gold mine in South Africa, extends to 3.9 km depth (2.3 mi) and the rock face is 60°C (140°F). The deepest well on Earth, the Kola Superdeep Borehole on the Kola Peninsula in Russia, reached depths of 12.26 km (40,200 ft) and encountered temperatures of 180°C (356°F).

Earth's structure consists of an inner core, outer core, mantle, and crust. These divisions are defined based on composition and behavior (Figure 9.3). Changing composition with depth is a result of gravity separating elements in the early Earth based on density: heavier elements sank deeper and lighter elements floated toward the surface.

The distance from the center of the earth to the surface of the earth is about 6300 km (4000 mi). The **inner core** lies at depths below 5150 km (3200 mi). It is composed of solid iron and nickel. Although the temperature is around 6000°C (10,800°F), it is solid because of the extreme high pressure.

The **outer core** lies at depths below 2900 km (1800 mi) and is thought to be liquid because it does not transmit seismic shear waves. The temperature ranges from 4000°C (7300°F) near the mantle to 5700°C (10,300°F) near the inner core. The high heat and slightly lower pressure explains why the iron–nickel outer core is molten. The red hot iron and nickel flows around the solid inner core because of the spin of the earth, thereby producing Earth's magnetic field.

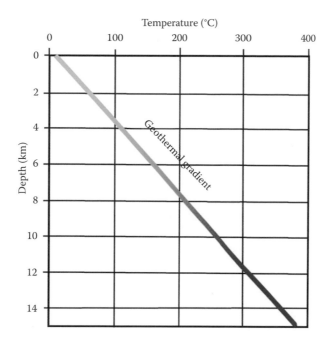

FIGURE 9.2
The typical near-surface geothermal gradient in continental interiors is about 25°C/km (1°F/70 ft).

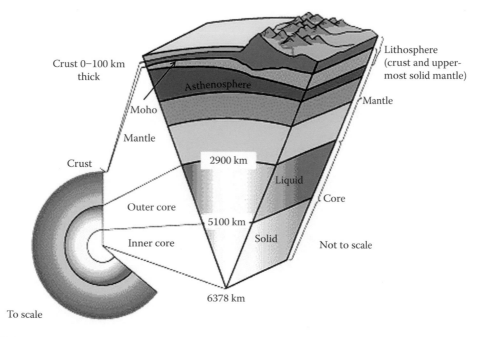

FIGURE 9.3
Structure of the Earth. (Courtesy of USGS; http://pubs.usgs.gov/gip/dynamic/inside.html.)

The **mantle** extends to 2900 km; the upper boundary with the crust ranges from 5 to 50 km depth. The mantle is composed of silicate minerals, typically peridotite in the upper mantle (Figure 9.4). The mantle behaves like a viscous fluid, with convection currents transferring heat from depth to the near surface. The moving mantle breaks and moves the rigid crust floating above. This is the driver for plate tectonics and mountain building. The mantle approaches the surface at **mid-ocean ridges**. Occasionally, as a result of tectonic uplift or squeezing, the mantle is exposed at places like Gros Morne National Park, Newfoundland, Canada (Figure 9.5) and the Troodos massif in Cyprus. The top of the mantle is defined by a sudden increase in seismic velocity (the Mohorovičić discontinuity, or "**Moho**"). The temperature of the mantle–crust boundary is in the range 200°C–400°C (400°F–750°F).

The **crust** is thicker on continents (30–50 km) and thinner in oceans (5–10 km). Continental crust is characterized by rock like granite, rich in silicate and aluminum minerals ("**sial**"). Sial has a lower density and thus floats higher on the mantle than oceanic crust. The denser oceanic crust is characterized by basaltic rock rich in magnesium silicate minerals ("**sima**"). Continents don't end at the coast: in reality, they extend well offshore before ending at a steep drop to the deep ocean floor (Figure 9.6).

Geologists use two methods to classify the outer parts of the earth: one is based on mineral composition, whereas the other is based on variations in rock strength. The distinction between mantle and crust is based on composition. This same region of Earth's interior has also been divided into zones based on strength properties that explain plate tectonics.

Lithosphere is rigid and brittle and forms tectonic plates (like a cracked egg shell). The base of the lithosphere is defined by a temperature of 1300°C. Below this, at depths between 80 and 200 km, is the ductile **asthenosphere**. It is movement in the asthenosphere that drags lithospheric plates across the earth's surface.

FIGURE 9.4

Peridotite from Finero, Ivrea zone, Italian Alps. Peridotite is composed mainly of olivine (green) and pyroxene (greenish black). Collection of the Université de Neuchâtel. Swiss franc (diameter, 23.2 mm) for scale. (Courtesy of Woudloper; https://commons.wikimedia.org/wiki/File:Phlogopite_peridotite.jpg.)

FIGURE 9.5
Ophiolite, an igneous rock formed at a mid-ocean ridge and containing mantle material, is exposed in Gros Morne National Park, Newfoundland (https://en.wikipedia.org/wiki/Ophiolite).

FIGURE 9.6
Continents extend offshore into waters a few hundred meters in depth. The continental shelf (light blue) and deep ocean floor (dark blue) are separated by the continental slope. (Courtesy of National Oceanic and Atmospheric Administration; http://www.virginiaplaces.org/boundaries/ocs.html.)

The crust is mostly rigid. The upper mantle has upper parts that are rigid and lower parts that are ductile. Thus, the transition between lithosphere and asthenosphere is usually in the upper mantle (Figure 9.7).

Continents float higher on the mantle than oceanic crust does because continents are made of less dense rock. One might say continents are the frothy scum of the earth. **Isostasy**, like buoyancy, is the process whereby the crust floats at an elevation that depends on its thickness and density, much like icebergs float on water (Figure 9.8). Within continents,

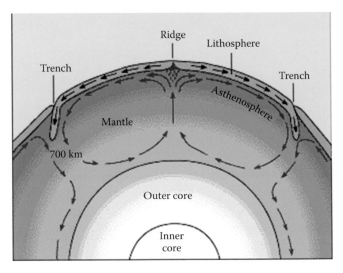

FIGURE 9.7
Structure of the Earth, mantle convection, and plate movement. (Courtesy of US Geological Survey; http://www.geologycafe.com/class/chapter6.html.)

FIGURE 9.8
An iceberg floats at a level that depends on its thickness and its density with respect to sea water. Most of it is under water. (Courtesy of National Oceanic and Atmospheric Administration [NOAA]; http://oceanservice.noaa.gov/gallery/image.php?siteName=nosimages&cat=Iceberg.)

mountains not only are higher than surrounding areas, but the crust under mountains also extends deeper into the mantle. Thus, if continental crust is normally 30–50 km thick, under mountain ranges, it is typically 50–70 km thick (Figure 9.9).

Orogens are extensive areas of deformation, usually linear deformed belts near the margins of Earth's crustal blocks. **Orogenic zones** are characterized by horizontal shortening of Earth's crust and usually contain igneous intrusions, volcanism, and folded and faulted rocks. Mountain building events are called **orogenies**. The large-scale processes that affect the structure of Earth's crust are known as **tectonics**. Tectonic forces are pressures within the earth that uplift or subside, fold, and fault the surface. Mountains are formed by forces that uplift the surface vertically or that squeeze Earth's crust from the side to buckle the surface. Mountains can also be formed by extension, as some faulted crustal blocks drop and others tilt. In contrast to orogenic zones, **cratons** are the stable parts of continents.

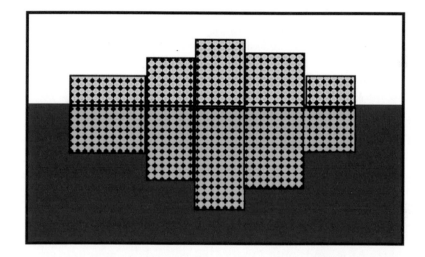

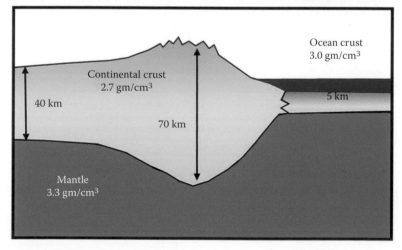

FIGURE 9.9
Isostasy is the state of equilibrium of the Earth's crust as it floats on the underlying mantle. Like icebergs, the wood blocks (above) float to different heights and depths in equilibrium with water. Crustal blocks (below) float to different heights in equilibrium with the upper mantle.

Plate Tectonics and Continental Drift

Mapmakers have long been intrigued by the jigsaw puzzle fit of the coastlines of Africa and South America. As early as 1596, the Dutch mapmaker Abraham Ortelius noticed the fit of the coastlines and suggested that the Americas had been "torn away from Europe and Africa … by earthquakes and floods."

In the late 1800s, the accepted wisdom was that continents were fixed in place on the surface of the earth and grew from a central core, or craton, by accreting mountain ranges around their margins. The American geologists James Hall and James Dana had studied the Appalachian Mountains in the mid-1800s and found that a thick pile of sedimentary rocks had been crumpled and uplifted. In 1857, Hall presented his ideas that mountain belts are the result of thick sediment deposits along the margins of continents that are later compressed and uplifted. He published this **geosynclinal theory** in 1882. Dana, in his *Manual of Geology* (1895), defined a **geosyncline** as a belt of thick, deformed sediments, metamorphic rock, and granitic intrusions along the margin of a continent. A great thickness of sediments must have loaded and depressed the crust until the sediments at the bottom got extremely hot. At that point, the hot, deformed sediments and associated metamorphic and igneous rocks rose up buoyantly to form mountain belts (Figure 9.10).

The geosynclinal theory of mountain chains developed along the margins of continents seemed to work in eastern North America, but it didn't explain why some cratons don't have folded sedimentary mountain ranges along their margins. Nor did it explain why some folded mountain ranges appear within cratons. Clearly, this theory had unanswered questions.

Another suggested mechanism for mountain building was the **shrinking Earth hypothesis** that stated that Earth has been cooling since an original molten state. As it cooled, it contracted, first forming large linear depressions along continental margins that filled with sediment and then compressing the sediment-filled troughs into mountain ranges.

In the late 1800s, the German geographer and naturalist Alexander von Humboldt traveled extensively in Africa and South America. He noticed that mountain ranges containing

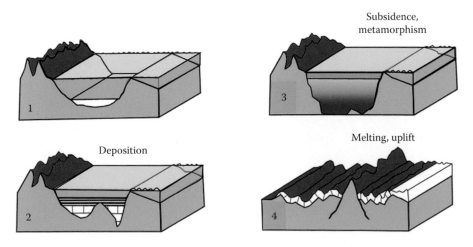

FIGURE 9.10
Development of a geosyncline. (Modified after Benjamín Núñez González; https://commons.wikimedia.org /wiki/File:Geosinclinal,_2016.svg.)

rocks of similar composition and ages exist in both Brazil and Ghana. He noted striking similarities in both fossils and present-day animals on the two continents. To explain these observations, the Austrian geologist Eduard Suess around 1880 proposed that all the continents had originally been joined in a supercontinent he called **Gondwanaland** and had somehow broken apart.

The American geologist Frank Taylor suggested (1908) that mountain ranges are the result of continents drifting across the face of the Earth. Moving continents depress the crust ahead of them, forming a trench that accumulates sediment. Continuing movement, as well as collision of continents, then crumpled the sediment-filled trough into mountains.

In 1909, Howard Baker, another American geologist, began work on a theory that an original supercontinent split about 5 Ma to form the Arctic and Atlantic oceans. The split, he suggested, was caused by a close approach of Venus, which also pulled a huge chunk of Earth out of the present-day Pacific Ocean to form the moon.

Alfred Wegener was a German meteorologist and arctic explorer. He was aware of von Humboldt's observations on the fit of South America and Africa, similar fossils on both continents, and similar rock types and mountain belts across the Atlantic. Wegener found that both continents had indications that glaciers had once been near the equator. As early as 1912, he proposed that the continents had drifted apart as a result of Earth's spin. Wegener's ideas were largely ignored until they were published in English as *The Origin of Continents and Oceans* (1924). Then, as with many novel ideas, they were viciously attacked. The main objection was that there was no known force strong enough to push continents through the rigid ocean floor.

Braving criticism, the South African geologist Alexander du Toit (1937) presented further evidence supporting the idea that South America and Africa had once been joined. Not only did he describe sedimentary formations, glacial deposits, and plant and animal fossils that were the same on both sides of the Atlantic, he proposed that there was an original northern supercontinent he called **Laurasia** and a southern supercontinent, **Gondwana**. A force that could move continents was proposed in 1944 by the British geologist Arthur Holmes. He suggested that mantle convection could do the job.

Most geologists, however, continued to believe that the theory of **continental drift** was supported only by eccentric, fringe geologists. Gradually, however, seemingly unrelated lines of evidence that supported the theory began to emerge.

The study of earthquakes led to an important clue. The American seismologist Hugo Benioff in 1954 interpreted deep earthquake zones around the Pacific as regional-scale inclined faults in the crust. These zones were often associated with volcanic island arcs. Today, these earthquake zones, called **Benioff zones**, are attributed to cold, brittle lithosphere moving downward into warm, ductile asthenosphere in a process called **subduction**. Earthquakes occur as the descending plate sticks and then suddenly becomes unstuck. As the cool lithosphere heats up and melts, the resulting magma rises buoyantly to erupt along volcanic arcs. It turns out when you plot these earthquakes on a map of the world, they pretty well outline the major crustal plates (Figures 9.11 and 9.12).

In 1962, Robert Coates, a Canadian geologist studying the Aleutian volcanic arc, interpreted the dipping seismic zone beneath the arc as a megathrust where continental material moves over oceanic crust. He suggested that volcanoes near oceanic trenches are the result of heating and melting of oceanic crust and sediments beneath the thrust (Figure 9.13).

During the 1930s and 1940s, the Dutch geophysicist Felix Vening Meinesz used a **gravimeter** to measure slight changes in Earth's gravity over ocean trenches. The low gravity values found over trenches suggested to him that trenches are the site of downward-moving low-density lithosphere.

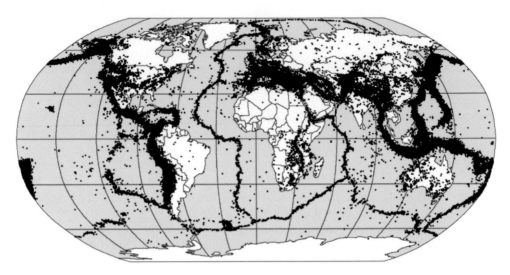

FIGURE 9.11
Worldwide earthquakes (https://en.wikipedia.org/wiki/Plate_reconstruction).

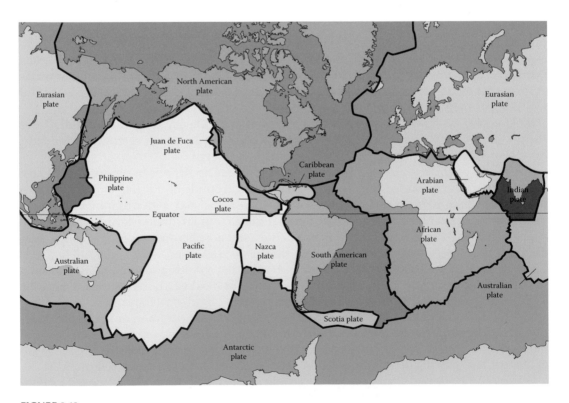

FIGURE 9.12
Major crustal plates of the Earth. (Courtesy of US Geological Survey; https://commons.wikimedia.org/wiki /File:Tectonic_plates.png.)

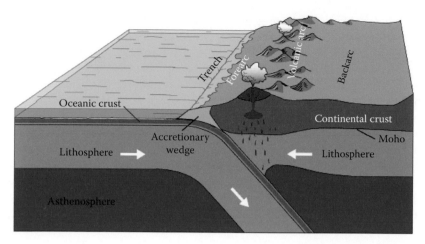

FIGURE 9.13
Components of a volcanic arc. (Courtesy of US Geological Survey Open-File Report 00-365; http://pubs.usgs
.gov/of/2000/ofr-00-0365/report.htm.)

The need to detect submarines during World War I drove the development of **sonar**. Before the invention of sonar, the world's deep oceans were thought to be flat, featureless plains. In the early 1950s, Bruce Heezen and Marie Tharp, American geologists, used sonar to discover and map mid-ocean ridges. They showed that these oceanic mountain ranges are worldwide, are hot, and contain central rifts associated with shallow earthquakes. Their observation that mid-ocean ridges have high heat flow at their crest and get cooler away from the ridge led them to propose that the ridges are spreading and that the earth must therefore be expanding (Figure 9.14).

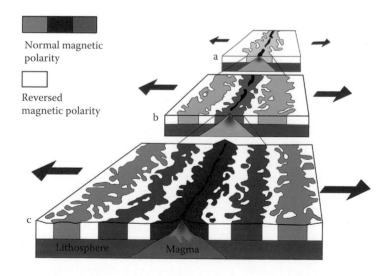

FIGURE 9.14
New basalt added to the ocean floor at spreading centers is magnetized according to Earth's existing magnetic field. (a) Spreading ridge about 5 Ma. (b) About 2 to 3 Ma. (c) Present day. (Courtesy of Chmee2; https://com mons.wikimedia.org/wiki/File:Oceanic.Stripe.Magnetic.Anomalies.Scheme.svg.)

Magnetometers were developed in the 1830s to measure Earth's magnetic field. It was soon discovered that the iron mineral **magnetite** aligns its crystals, like tiny compass needles, in Earth's magnetic field when it is in a magma and, as the volcanic rock solidifies, the magnetite crystals are frozen in place and point to Earth's poles. In 1906, the French geophysicist Bernard Brunhes discovered that different volcanic rocks have been magnetized in opposite directions, as if Earth's north and south poles had switched places. A few decades later, Motonori Matuyama, a Japanese geologist/geophysicist, collected basalt samples in Manchuria and Japan and determined (1929) that a period of **magnetic reversal** had occurred between 2.58 and 0.78 Ma.

"Pole directions" indicate whether magnetite crystals in igneous rocks point to the North Pole or South Pole. In 1957, the British physicist Keith Runcorn compiled magnetic pole directions from volcanic rocks of different ages around the world. He showed that pole directions for young volcanic rocks in North America and Europe point close to the present magnetic pole, but that they diverge as rocks get older. This suggested that the two continents have been moving apart for the past 180 Ma.

The US Coast and Geodetic Survey ran a survey off the west coast of the United States in 1955 that included a magnetometer. To their surprise, they found that the ocean floor contains broad north–south zones of positive and negative magnetism. When the British oceanographer Ronald Mason and American geophysicist Arthur Raff published their findings in 1961, they interpreted the **magnetic stripes** as being caused by multiple reversals of Earth's magnetic poles over time (Figure 9.15).

There was no convincing explanation for why the ocean bottom had magnetic stripes. Then, Allan Cox, Richard Doell, and Brent Dalrymple of the US Geological Survey not only sampled volcanic rocks and measured the magnetic polarity, but in addition they dated the rocks. In 1963, they published their conclusion that one could create a **magnetic time scale** based on Earth's magnetic reversals (Figure 9.16).

Their magnetic time scale hypothesis was tested by the British geologists Frederick Vine and Drummond Matthews. In 1963, Vine and Matthews showed that the age of the seafloor increased as one moved away from the mid-ocean ridges. Heezen and Tharpe had already proposed that these ridges are spreading centers but thought the earth was expanding as a result. Vine and Matthews proposed that, as the plates spread apart, new lava rises from the mantle at the mid-ocean ridge. Magnetite crystals in the cooling lava align themselves to the existing magnetic field. When the magnetic field changes polarity, the alignment of crystals changes polarity as well, causing the stripes of opposite magnetic polarity on the seafloor. The pattern was found to be symmetrical about the ridge, indicating that similar stripes on both sides were formed at the same time. The width and age of the magnetic stripes allowed estimation of the rate of seafloor spreading. Spreading was found to range from 2.5 cm/year on the Mid-Atlantic Ridge and as much as 13 cm/year along the East Pacific Rise.

In 1962, the American geologist Harry Hess was studying sedimentation rates and sediment thickness on the ocean floor. He noticed that there were not enough sediments on the ocean floor considering the age of the earth. He concluded that most of the older sediments must have been recycled, either overridden by continents or somehow welded onto the continents. The issue of what happened to all the sediments was finally resolved in 1968. Leg 3 of the **Deep Sea Drilling Project (DSDP)** sampled a number of sites between Brazil and Senegal and determined that the age of seafloor gets older the farther one goes from the Mid-Atlantic Ridge. Sampling since that time has only confirmed this observation, with new seafloor forming at mid-ocean ridges and the oldest seafloor being only around 200 million years old.

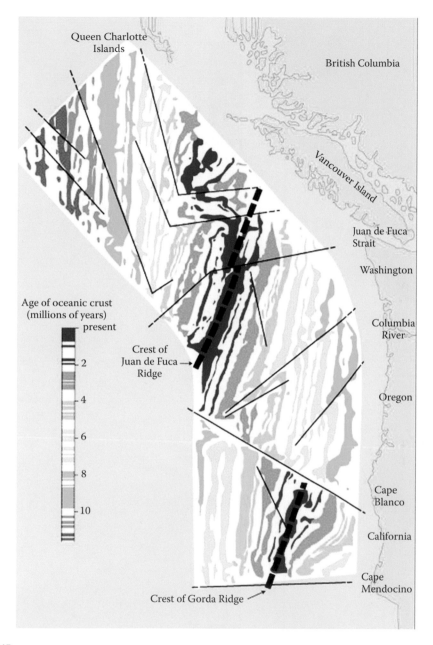

FIGURE 9.15
Magnetic anomalies around the Juan de Fuca and Gorda Ridges, off the west coast of North America, color coded by age. (Courtesy of W. Jacquelyne Kious and Robert I. Tilling, US Geological Survey; http://pubs.usgs .gov/gip/dynamic/graphics/Fig6.gif.)

The Canadian geophysicist John Tuzo Wilson spent his career studying tectonics. His work led him to the concepts of transform faults and hot spots. **Strike-slip** faults are those where opposite sides of the fault move laterally past each other. They have most of their offset in the center of the fault and gradually lose displacement toward their ends. In contrast, **Transform faults**, defined by Wilson in 1965, look like strike-slip faults

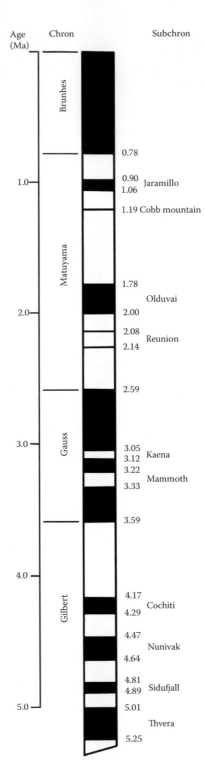

FIGURE 9.16
Magnetic time scale for the past 5 million years. (Courtesy of US Geological Survey; https://pubs.usgs.gov
/of/2003/of03-187/of03-187.pdf.)

except that movement is in the opposite direction one would expect based on offset ridges (Figure 9.17). That is because transform faults only exist between oceanic spreading centers. Along transform faults, the distance between the offset spreading centers remains constant, the faults have an equal amount of offset along their length, and they end at another type of plate boundary such as a fracture zone. Mountains or basins form at bends in these faults, depending on the direction of the bend with respect to the offset (Crowell 1974; Figures 9.18 and 9.19). Wilson made the first plate tectonic map of the world by drawing plate boundaries along seismic zones, transform faults, and spreading centers.

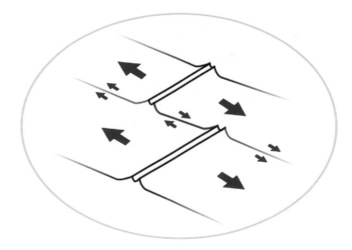

FIGURE 9.17
Transform fault in red is shown offsetting spreading ridges. Arrows indicate movement direction. (Courtesy of Los688; https://commons.wikimedia.org/wiki/File:Transform_fault-1.svg.)

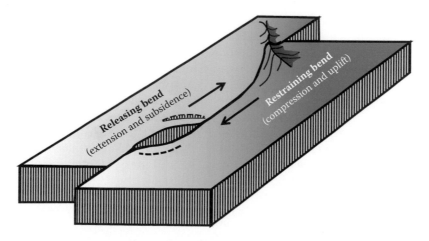

FIGURE 9.18
Mountains and basins related to bends in transform faults (https://en.wikipedia.org/wiki/Transtension #/media/File:Releasing_bend.png; https://en.wikipedia.org/wiki/Transpression).

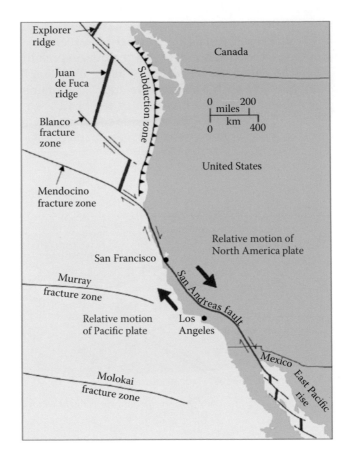

FIGURE 9.19
Transform faults along the west coast of North America. The San Andreas is one of a few transform faults exposed on land. (Courtesy of US Geological Survey; https://pubs.usgs.gov/gip/dynamic/understanding.html.)

   Hot spots, like Hawaii and Yellowstone Park, are areas where a plume of magma rises from the asthenosphere and penetrates the surface. The plumes last many millions or tens of millions of years and are relatively stationary while lithospheric plates move over them.
   Hot spots leave progressively older volcanic rocks in their wake. The easiest explanation is that the hot spot is stationary and the lithosphere is moving over it. The hot spot punches through the overlying plate, erupts lava, is closed off by the moving plate, and punctures the plate again in a new spot. For example, the Hawaiian island chain contains volcanic islands that get older the farther one moves northwest from the present-day big island of Hawaii. In fact, a new island (Lo'ihi) is being born just southeast of Hawaii as the Pacific plate continues moving northwest over the stationary mantle plume (Figure 9.20). Likewise, the Yellowstone Park hot spot is the latest manifestation of a trail of volcanic rocks that began 17.5 Ma in eastern Washington and Oregon, continued erupting between 15 and 1.6 Ma along the Snake River plain in Idaho, and is currently in northwest Wyoming (2.1 million years to 640,000 years ago).

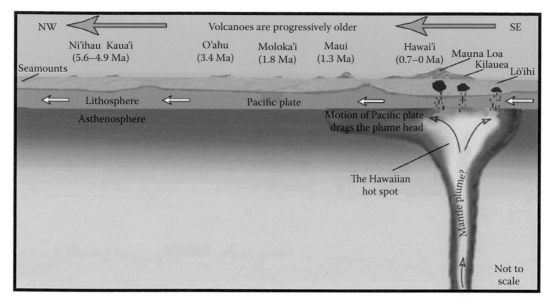

FIGURE 9.20
The Hawaiian hot spot. (Modified from Joel Robinson, USGS, "This Dynamic Planet," Simkin and others, 2006; http://geology.com/usgs/hawaiian-hot-spot/.)

All these lines of evidence came together during the late 1960s. Workers suddenly had a mechanism to move continents and data to support it. Their ideas merged into the theory of plate tectonics.

Plate tectonics is essentially the same concept as that expounded by Wegener and du Toit, but with an explanation for how the continents move. The theory states that Earth is covered by a number of rigid lithospheric plates that move over a ductile asthenosphere. Most of the deformation that builds mountains takes place at or near **plate margins**. These plates are created at and move away from mid-ocean ridges (spreading margins) where mantle convection currents approach the surface. The plates plunge back into the mantle and are consumed at **subduction zones** (convergent margins) and the associated oceanic trenches. Subduction zones are the site of volcanic arcs and deep earthquakes that result from downward-moving and melting plates. The plates slide past each other at **transform margins**, the sites of destructive, near-surface earthquakes (e.g., San Andreas Fault, Queen Charlotte Fault, Alpine Fault, Dead Sea Transform, and North Anatolian Fault). Where continents collide, the two plates crumple and form mountains (e.g., the Himalayas). Less commonly, ocean floor collides with and accretes onto continents (Franciscan mélange and Cascadia terrane in California-Oregon) or is thrust onto continents (Semail ophiolite in Oman, and Coast Range ophiolite in California).

One of the cool things you can do with plate tectonics is recreate the globe at various times in the past by moving plates opposite their present sense of motion. Some examples of past configurations are shown (Figures 9.21 and 9.22).

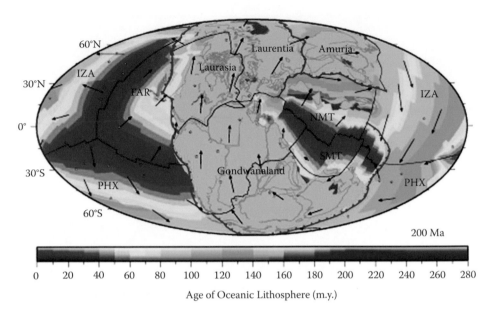

FIGURE 9.21
Plate tectonic reconstruction of the continents and plate movement directions (arrows) 200 million years ago (http://www.earthbyte.org/Resources/global_plate_model_ESR12.html).

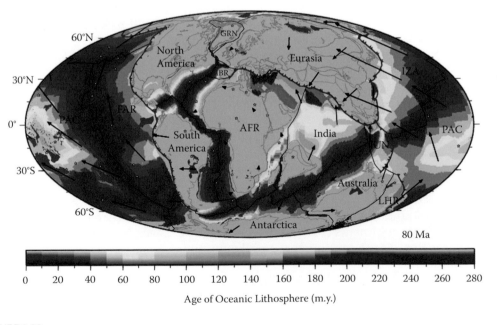

FIGURE 9.22
Plate tectonic reconstruction of the continents as they were 80 million years ago (http://all-geo.org/highlyal lochthonous/2014/05/reconstructing-ocean-spreading-when-half-your-record-is-now-in-the-mantle-or-a-plug -for-my-new-paper/seton_80ma_reconstruction/).

Mountain Building

Deformation styles include compression (shortening), tension (extension), and shear (slip and rotation). Geologic stress also causes translation (moving an object from one place to another) and tilting.

Mountain building, then, is a result of shortening by folding and faulting, of igneous intrusion and eruption, of uplift, and of extension. Shortening of the crust at convergent and transform plate boundaries causes folding, thrust faults, and reverse faults. The Alps and Carpathian Mountains in Europe, the Sierra Madre Oriental of Mexico, the Atlas Mountains of Morocco and Algeria, the Appalachian–Ouachita Mountains in the United States, and the Rocky Mountain thrust belt in the United States and Canada are examples. Convergent plate margins are also the site of igneous intrusion and volcanism. **Volcanic mountain chains** and **volcanic island arcs** are a result of regional melting at depth, usually associated with subduction, and the piling up of erupted material at the surface. Examples include the Aleutian Islands in Alaska, the islands of Japan, the Andes in South America, and the Cascade Range that extends from California to British Columbia.

Epeirogenic uplift is any widespread vertical uplift within the stable craton, usually the interior of a crustal plate. It is a result of regional lithospheric heating. Molten rock is less dense than the same rock when solid, and the density contrast causes the molten rock to rise as a result of buoyancy. When the uplifts occur at plate tectonic spreading centers, they form mid-ocean ridges. When gentle uplift happens over a large, relatively stable part of a continent, it is epeirogenic uplift. Epeirogenic movement has uplifted the Rocky Mountain region of Colorado between 1300 and 2000 m in the past 40 Ma. Epeirogenic uplift can occur with or without faulting and folding. Epeirogenic forces also cause regional subsidence owing to lithospheric cooling or downward movement in the mantle. Examples include many midcontinent basins in the United States (Figure 9.23).

Mountain building by **isostatic uplift** is a result of erosion and unloading of the crust. During erosion, the weight of the crust on the mantle decreases, mantle material flows in from the sides and pushes up the remaining mountain. An example of isostatic uplift is the rebound of Europe and North America after the last glacial ice sheets melted 12,000 years ago. Much as a "memory foam" mattress returns to its original shape after weight is removed, the release of pressure attributed to melting of up to 4 km (2.5 mi) of ice has led to as much as 700 m (2300 ft) of uplift in Scandinavia and as much as 100 m (328 ft) of uplift in North America.

The tilted and fault-bounded mountains of the **Basin-and-Range province** of Arizona–Nevada–Utah are the result of rifting and extension. A special case of uplift associated with extension is the **metamorphic core complex**. These are areas where the crust has extended so much that metamorphic rocks in the lower parts of the crust are exposed at the surface. Core complexes are characterized by low-angle faults (**detachment faults**, also known as **decollements**) that allowed the overlying upper crustal rocks to slide off the uplift (Figure 9.24).

When uplift happens over a shallow, local magma body, it is considered to be of volcanic origin. The Jemez caldera in northern New Mexico, the Absaroka Volcanic field of northwest Wyoming, and the San Juan Mountains in southwest Colorado occur as volcanic highlands (Figure 9.25). The **caldera** itself is a local volcanic depression, or crater, formed by the collapse of the eruptive center following an eruption.

Volcanic mountains are built layer upon layer. Cone-shaped volcanoes of this type are stratovolcanoes, that is, volcanoes built up of many layers of lava and ash. They can form

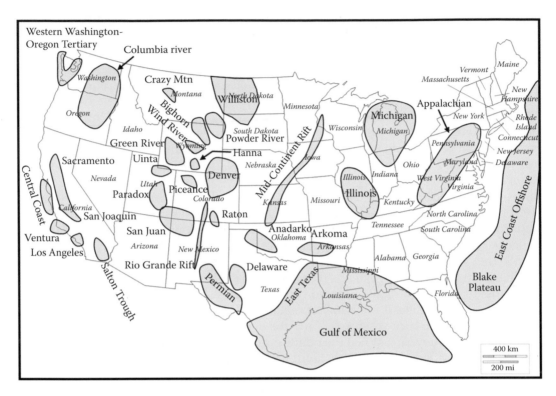

FIGURE 9.23
Major US sedimentary basins. The Williston, Michigan, and Illinois basins are epeirogenic depressions.

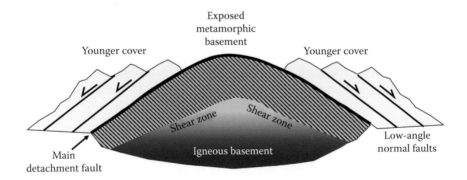

FIGURE 9.24
Schematic of a typical metamorphic core complex, an uplift caused by extension.

entire mountain ranges, notably the Andes of South America and the Cascade Mountains of Oregon and Washington (Figure 9.26). Shield volcanoes, such as those that form the Hawaiian Islands, are built up of many layers of basaltic lava that flows easily and forms a broad, shield-like highland. Although shield volcanoes do not have steep sides, they can be exceedingly high: Mauna Kea on the big island of Hawaii rises 4205 m (13,796 ft) above sea level, but towers 9330 m (30,600 ft) above the ocean floor, making it the tallest mountain in the world (Figure 9.27).

FIGURE 9.25
The San Juan Mountains, an eroded volcanic plateau, as seen from Ridgeway, Colorado. (Courtesy of Murray Foubister; https://commons.wikimedia.org/wiki/File:Ridgeway_Col._Pano.-2.jpg_(8644099251).jpg.)

FIGURE 9.26
Mt. Rainier, Washington, a large stratovolcano in the Cascade Range, looms over Seattle, Washington (https://en.wikipedia.org/wiki/Mount_Rainier).

Rock Deformation

The deformation that forms mountains and basins takes place slowly but continuously over millions of years. Rock deformation is described as brittle, ductile, or some combination of the two. Under relatively low temperatures and pressures (within about 15 km of Earth's surface), rocks undergo mostly brittle deformation. **Brittle** behavior occurs when a material is cool, hard, and rigid and deforms little before it breaks. Rock becomes less brittle and more ductile as temperature and pressure increase, until eventually all deformation is ductile. **Ductile** behavior is the ability to deform extensively without breaking.

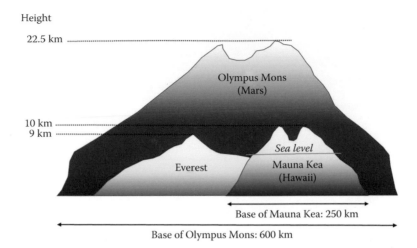

Height

22.5 km ....

10 km ....
9 km ....

Olympus Mons
(Mars)

Sea level

Everest

Mauna Kea
(Hawaii)

Base of Mauna Kea: 250 km

Base of Olympus Mons: 600 km

FIGURE 9.27
Comparison of two shield volcanoes, Mauna Kea and the Martian volcano Olympus Mons, with Mt. Everest.

Rock behavior is also influenced by mineral composition, water content, and strain rate. Some minerals (quartz) are more brittle and others (calcite, mica) are less so. A quartz sandstone is generally more brittle than a limestone. Water in the rock acts to lubricate grains and weaken chemical bonds in minerals such that wet rocks tend to be more ductile than dry rocks. High strain rates promote fracturing, whereas low strain rates cause rocks to fold or flow over time (think of pulling silly putty quickly and slowly).

Increasing heat and pressure in the earth causes a **brittle/ductile transition** at depths between 10 and 20 km (6 to 12 mi).

Brittle Deformation: Faults, Joints, and Earthquakes

Rocks near the surface of the earth are brittle and tend to rupture under stress. Brittle behavior in rocks causes fractures. Some fractures have movement along the fracture surface (**faults**) while others do not (**joints**). An **earthquake** is the result of sudden movement along a fault that releases stress.

Tectonic forces that squeeze Earth's surface layers from the side cause mountains to uplift by buckling, folding, and faulting. Such sideways forces usually occur at converging crustal plate boundaries such as the Pacific–North American or the Indian–Asian boundaries. These mountains are characterized by thrust faulting, where older rocks are pushed laterally up and over younger rocks in great folded and broken sheets of rock.

Truly massive earthquakes, as well as uncounted smaller quakes, created the Himalayan Mountain chain in the 55 Ma since the Indian crustal plate collided with Asia (Figure 9.28). The peak of Mt. Everest consists of limestone deposited on the ocean floor 450 million years ago. How did marine rock get on top of Mt. Everest? The sediments were buried and then uplifted when they were squeezed between India and Asia. Although the Himalayas now stand 9 km (29,000 ft) above sea level, the total uplift is much higher. The combination of roughly 10 mm (0.39 in)/year of uplift and erosion rates of 2 to 12 mm/year have created the mountains we see today. Earthquakes that lift the Himalayas are the result of movement on **thrust faults** (flat to gently inclined surfaces) or **reverse faults** (fault plane greater than 45° inclination; Figures 9.29 through 9.31).

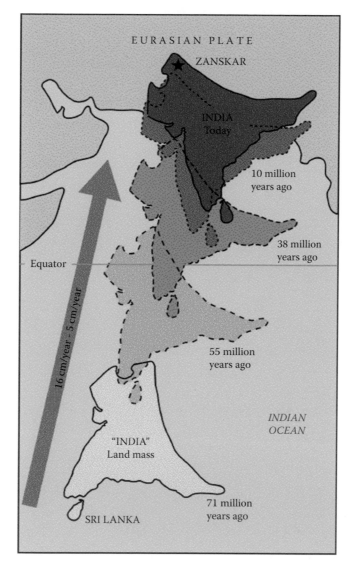

FIGURE 9.28

The northward drift of India from 71 Ma to the present. Collision of the Indian continent with Eurasia occurred starting about 55 Ma. (Courtesy of Moumine, US Geological Survey; www.usgs.org, https://en.wikipedia.org/wiki/Geology_of_the_Himalaya.)

The San Andreas Fault in California is a world-class transform fault. It has mainly lateral offset; that is, the two sides slip past each other without much vertical movement. This fault (actually a zone of many related faults) separates the Pacific crustal plate from the North American plate. The west side of the San Andreas Fault is moving northwest with respect to the east side (Figures 9.32 and 9.33). The largest single earthquake offset measured on this fault, during the Fort Tejon quake of 1857, was 8.8 m (29 ft) in the Carrizo plain. The **scarp** (surface rupture) extended 400 km (250 mi). Most movements along this fault are much smaller, and some sections of the fault deform by ductile movement without earthquakes. The fault has at least 560 km (350 mi) of offset over the past 30 Ma, as shown by matching up equivalent rocks on both sides of the fault.

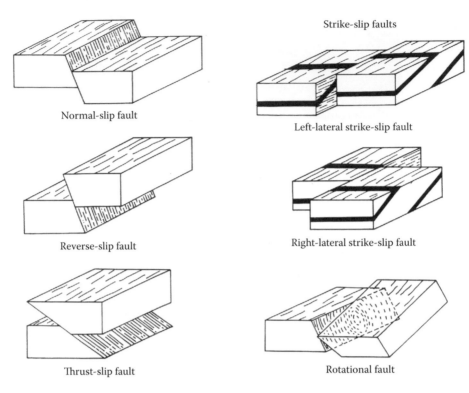

FIGURE 9.29
Fault types. (From Prost, G. L. 2014. *Remote Sensing for Geoscientists*. Taylor & Francis, 702 pp.)

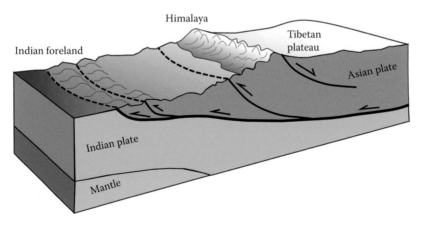

FIGURE 9.30
Thrust faults lift the Himalayas up and over the Indian crustal plate.

Mountains can also form by extension of Earth's crust. When the crust extends, it breaks along near-vertical, roughly parallel fault systems that allow the rock between to rotate or drop down. The resulting fault-bounded mountain range is known as a **horst**, and the valleys between horsts are called **grabens**. They are usually elongated, linear mountain ranges bounded on one or both sides by faults. This type of mountain is common in the Basin-and-Range geologic province.

FIGURE 9.31
View north to the McConnell thrust fault at Mt. Yamnuska, Alberta, Canada. Dashed line is the thrust fault; rocks above the fault moved from west to east (left to right). This iconic view of the Canadian Rockies has been used as a backdrop in many movies. (Courtesy of G. Prost.)

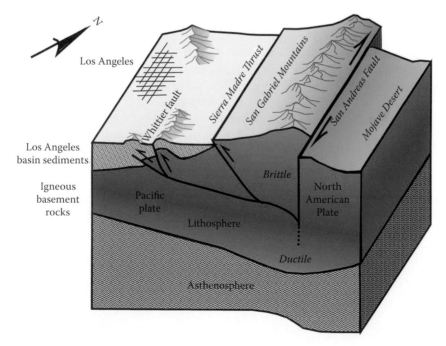

FIGURE 9.32
The San Andreas Fault near Los Angeles.

FIGURE 9.33
Gullies offset across the San Andreas Fault near the Carrizo Plain, California. The lower half of the photo is on the North American tectonic plate and moved to the left; upper half is on the Pacific Plate and moved right (north). (USGS photograph by David K. Lynch, Kenneth W. Hudnut, and David S. P. Dearborn (2009); http://www.sanandreasfault.org/SAFCarrizo24Sep2009.html.)

A spectacular example of recent extensional faulting can be seen in the Sierra Nevada Mountains in east-central California (Figure 9.34). **Normal faults**, typical in areas of extension, have mainly vertical offset. About 10 million years ago, the Sierra Nevada normal fault began to develop along the east side of the range. Granite has dropped at least 4600 m (15,000 ft) vertically from the top of Mt. Whitney to where it is seen in boreholes in the Owens Valley. The Sierra Nevada tilted westward while rocks to the east subsided. The Sierra Nevada fault is the result of many small offsets such as the 1872 Lone Pine quake (magnitude of 7.4 to 7.9). The 140-km-long (85-mi-long) fault scarp had 4.6 to 6.1 m (15 to 20 ft) of vertical offset.

FIGURE 9.34
Sierra Nevada and Owens Valley, view southwest. The Sierra Nevada Fault dropped the Owens Valley at least 4600 m in the past 10 Ma. (Courtesy of G. Thomas; https://en.wikipedia.org/wiki/File:SierraEscarpmentCA.jpg.)

Faults are characterized by broken rock usually separating distinct rock types. **Fault breccia** consist of angular rock fragments in the fault zone. **Fault gouge** is finely ground up rock in the fault zone. **Mylonite** is a fine-grained metamorphic rock that results from shear along a fault at depth. **Slickensides** are scratches caused when the two sides of the fault grind past each other and are a good indicator of the movement direction along the fault (Figure 9.35).

Most faults occur as zones of deformation rather than as a single break (Figure 9.36). Faults with near-equal amounts of vertical and horizontal offset are **oblique-slip faults**.

One type of brittle deformation involves breaking rock without measurable offset. Joints have no measurable offset (Figure 9.37). Joints usually form as a result of expansion, as when a mass of rock is uplifted from deep in the earth to near the surface. As pressure is released, the rock expands and cracks. **Exfoliation joints** in granite are an example (Figure 9.38). Alternatively, as a large mass of rock is uplifted and cools, it shrinks slightly. The change in volume is accomplished through jointing. Cooling and shrinking basalt forms hexagonal joint patterns that create columnar basalts such as those seen at Devils Postpile (California), Devils Tower (Wyoming; Figure 9.39), or the Giant's Causeway (Northern Ireland and Scotland). Joints also form by hydraulic fracturing, as when magma forces its way into overlying rock and forms dikes and sills. Hot, water- and gas-rich fluids derived from magmas also fracture and force their way into the surrounding rock under high pressure and precipitate minerals as they cool. Mineral-filled fractures are **veins**. Ore deposits are frequently found in veins that precipitated from **hydrothermal fluids** (hot, mineral-rich water).

The formation of joints is accompanied by seismic events (the "pop" that goes with fracturing), most of which are too small to notice (less than seismic magnitude 2).

Joints are economically important because they allow fluids such as water, oil, and gas to flow more readily through rocks. They are important to civil engineers because they indicate weak and broken rock that should be avoided or reinforced in road cuts, dams,

FIGURE 9.35
Slickensides in the Carboniferous Mira Formation, Tavira, Portugal. (Courtesy of PePeEfe; https://commons
.wikimedia.org/wiki/File:Striated_fault_plane_in_sedimentary_rock.JPG.)

FIGURE 9.36
Part of the Moab normal fault zone near Moab, Utah. The Moab fault has about 300 m (1000 ft) of total vertical offset, some of which is shown here. (Courtesy of G. Prost.)

FIGURE 9.37
Jointing in the Cretaceous Frontier Formation sandstone at Muddy Gap, Wyoming. (Courtesy of G. Prost.)

FIGURE 9.38
Exfoliation of granite, Enchanted Rock, Texas, USA. (Courtesy of Wing-Chi Poon; https://commons.wikimedia
.org/wiki/File:GeologicalExfoliationOfGraniteRock.jpg.)

FIGURE 9.39
Columnar basalt cooling joints, Devil's Tower, Wyoming. (Courtesy of Colin.faulkingham; https://en.wikipedia
.org/wiki/Devils_Tower.)

home sites, and mines. Joints increase the rate of weathering and erosion by channeling water through rocks.

Ductile Deformation and Folding

Rocks deform by ductile folding under the high temperatures and pressures found at depth: the rocks bend and flow in slow motion. Ductile rock responds to stress by folding rather than faulting. There are two kinds of folding: folding by slip between layers (**flexural slip**—less ductile) and folding by flow of material within the folded layer (**flexural flow**—more ductile; Figure 9.40).

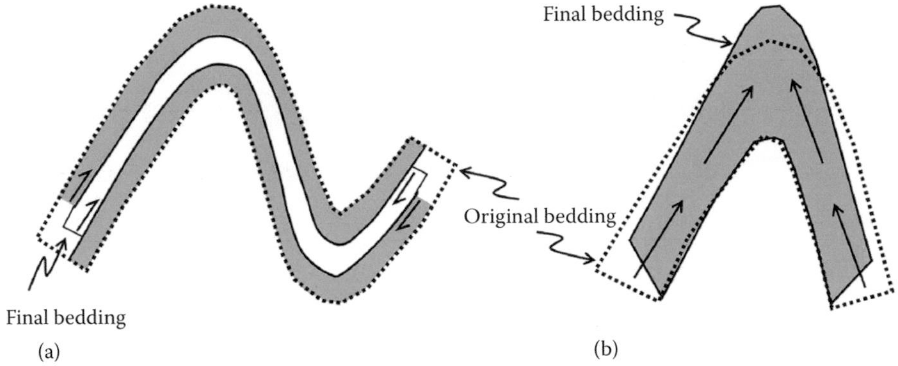

FIGURE 9.40
Styles of folding. (a) Flexural slip: beds allow shear, or slip, on bedding surface. (b) Flexural flow: material flows from the fold limb to the fold axis.

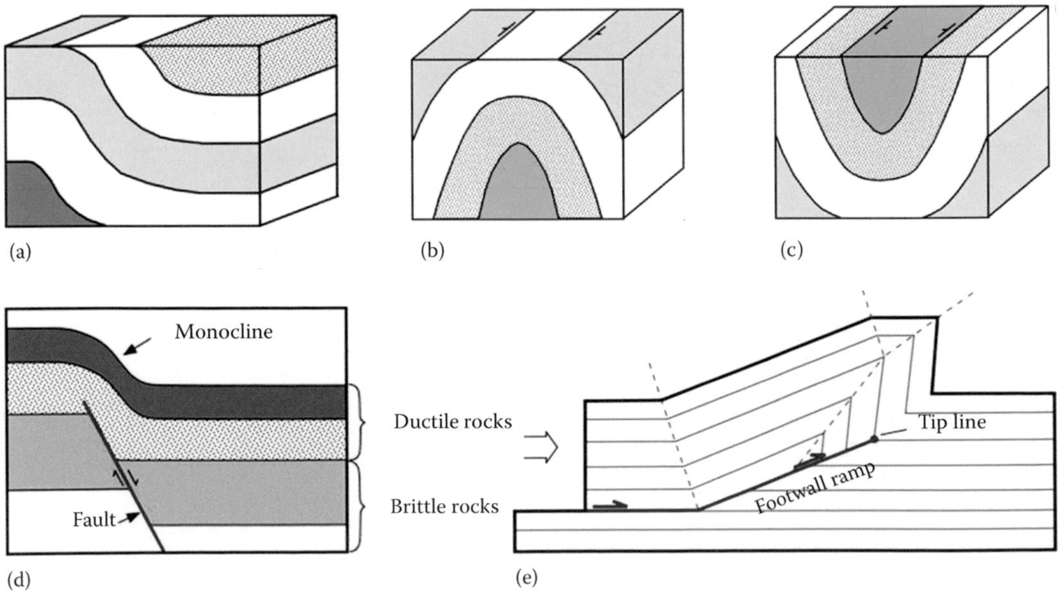

FIGURE 9.41
Fold types. (a) Monocline. (b) Anticline. (c) Syncline. (d) Monocline where beds fold over a fault at depth. (e) Fault-propagation fold where bedding bends over the end of a thrust fault.

Anticlines are folds that are arched up in the center; **synclines** are folds that are dropped down in the center (Figures 9.41 through 9.43). A half-fold, usually developed over a fault, is a **monocline** (Figure 9.44). Folds can be symmetrical or asymmetrical and even overturned. They usually occur in sets, much like a series of waves approaching the shore. In fact, consistently inclined folds indicate where the push came from (Figure 9.45).

FIGURE 9.42
San Blas anticline, Monterrey, Mexico. Dotted line shows trace of bedding. (Courtesy of G. Prost.)

FIGURE 9.43
Lyautey Syncline, Kananaskis Country, Alberta, Canada. Dotted line shows trace of bedding. (Courtesy of Adam Prost.)

FIGURE 9.44
East Kaibab Monocline, Arizona. Solid line is trace of bedding at the surface; dotted line is projection into the subsurface. The monocline is developed over the Palisades Fault at depth. (Courtesy of G. Prost.)

The topographic expression of folding depends on the rock's resistance to erosion. An anticline should be high in the center, but if it has an easily eroded shale in the core, it will form a valley. A syncline should be low in the center, but will form a mountain if it has resistant rock along its axis (Figure 9.46).

FIGURE 9.45
Seismic interpretation of the Port Isabel foldbelt, northwest Gulf of Mexico, showing east-leaning folds. Fault movement is indicated by arrows. The tectonic "push" came from the west. (Modified from interpretation by F. Peel; Courtesy of Mark Rowan, Frank Peel, and Bruno Vendeville; https://commons.wikimedia.org/wiki /Category:Geological_cross_sections#/media/File:Foldbelt.jpg.)

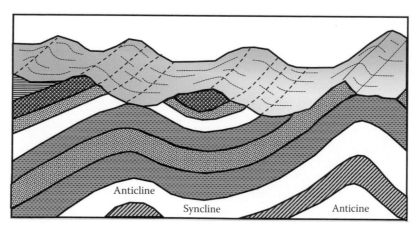

FIGURE 9.46
Topographic expressions of folding.

Erosional Highlands

Plateaus, Mesas, and Buttes are examples of mountains formed by the erosion of uplifted, usually little deformed rock. They are characterized by flat or nearly flat tops and frequently are bounded by faults. Examples include the Colorado Plateau of northern Arizona, Utah, and Colorado; the Edwards Plateau of central Texas; and the Uncompahgre Uplift in western Colorado (Figure 9.47).

Valleys cut into mountain ranges can reveal the internal structure of the mountains, that is, how they formed. A common sight is the same layers of rock on both sides of a valley. This means the layers are continuous across the valley and that a river cut down through the layers and eroded the material that used to be there. A famous example is the Grand

FIGURE 9.47
Mesa and butte landforms developed in the Jurassic Wingate Sandstone on the Colorado Plateau. Colorado River east of Moab, Utah. (Courtesy of G. Prost.)

Canyon that was carved by the Colorado River as it runs through the Kaibab Plateau of northern Arizona (Figure 9.48).

Many valleys expose rocks that are tilted, faulted, or folded (Figure 9.49). These features reveal the internal architecture of the mountain range and provide clues to how the mountains were formed. Near-vertical faults with one side up and the other side down usually indicate vertical uplift or extension. Near-horizontal faults that put older rocks

FIGURE 9.48
The Grand Canyon of Arizona exists because the Colorado River eroded into the Kaibab Plateau. The same layers exist on both sides of the canyon, indicating they were continuous across the area before the canyon formed. (Courtesy of G. Prost.)

FIGURE 9.49
Folded and faulted strata exposed in a canyon wall in Death Valley, California. Recent basalt flows (dark rock) cover some of the units. (Courtesy of G. Prost.)

over younger rocks are frequently associated with inclined folds. Together they indicate compression and shortening.

We have seen that mountain building and rock deformation, whether brittle or ductile, occur in small increments that accumulate over long periods of time. Next, we will examine the process of erosion.

References and General Reading

Baker, H. B. 1912a. The origin of continental forms, II. Michigan Academy of Science Annual Report, 1911–12, pp. 116–141.

Baker, H. B. 1912b. The origin of continental forms, III. Michigan Academy of Science Annual Report, 1911–12, pp. 107–113.

Benioff, H. 1954. Orogenesis and deep crustal structure—Additional evidence from seismology. *Bulletin of the Geological Society of America* 65, 385–400.

Brice, J. 1962. *Workbook in Historical Geology*. William C. Brown Company Publishers, Dubuque, IA, 174 pp.

Collinson, D. W., Creer, K. M., Irving, E., and Runcorn, S. K. 1957. The measurement of the permanent magnetization of rocks. *Philosophical Transactions of the Royal Society A: Mathematical, Physical and Engineering Sciences* 250, 73–82. Bibcode:1957RSPTA.250...73C. doi:10.1098/rsta.1957.0012.

Cox, A. V., G. B. Dalrymple, and R. R. Doell. 1963. Geomagnetic polarity epochs and Pleistocene geochronometry. *Nature* 198, 1049–1051. Bibcode:1963Natur.198.1049C. doi:10.1038/1981049a0.

Crowell, J. C. 1974. Origin of late Cenozoic basins in southern California. In W. R. Dickenson (ed.), *Tectonics and Sedimentation*. SEPM Spec. Pub. 22, 190–204.

Dana, J. D. 1895. *Manual of Geology*. American Book Company, New York, 974 pp.

Dyman, T. S., J. W. Schmoker, and D. H. Root. 1996. Assessment of Deep Conventional and Continuous-type (Unconventional) Natural Gas Plays in the United States. Open-File Report 96-529.

Du Toit, A. L. 1937. *Our Wandering Continents*. Oliver and Boyd, Edinburgh, 366 pp.

Hess, H. H. 1962. The history of ocean basins. In A. E. J. Engel, H. L. James, and B. F. Leonard. (eds.), *Petrologic Studies: A Volume in Honor of A. F. Buddington*. Geological Society of America, Boulder, CO, 599–620.

Holmes, A. 1944. *Principles of Physical Geology*. Thomas Nelson and Son, Edinburgh, 532 pp.

Matyuama, M. 1929. On the direction of magnetization of basalt in Japan, Tyosen and Manchuria. *Proceedings of the Imperial Academy of Japan* 5, 203–205.

Pfiffner, O. A., and L. Gonzalez. 2013. Mesozoic–Cenozoic evolution of the western margin of South America: Case study of the Peruvian Andes. *Geosciences* 2013, 3 (2), 262–310. doi:10.3390/geosciences3020262.

Prost, G. L. 2014. *Remote Sensing for Geoscientists*. Taylor & Francis, Boca Raton, FL, 702 pp.

Raff, A. D., and R. G. Mason. 1961. Magnetic survey off the west coast of North America, 40° N. latitude to 52° N. latitude. *Geological Society of America Bulletin* 72 (8), 1259–1266. doi:10.1130/0016-7606(1961)72[1259:MSOTWC]2.0.CO;2. ISSN 0016-7606.

Seton, M., R. D. Müller, S. Zahirovic, C. Gaina, T. H. Torsvik, G. Shephard, A. Talsma, M. Gurnis, M. Turner, S. Maus, and M. Chandler. 2012. Global continental and ocean basin reconstructions since 200 Ma. *Earth-Science Reviews* 113 (3–4), 212–270. ISSN 0012-8252, 10.1016/j.earscirev.2012.03.002.

Simkin, T., Tilling, R. I., Vogt, P. R., Kirby, S. H., Kimberly, P., and Stewart, D. B. 2006. This dynamic planet: World map of volcanoes, earthquakes, impact craters, and plate tectonics: U.S. Geological Survey Geologic Investigations Series Map I-2800, 1:30,000,000 (http://mineralsciences.si.edu/tdpmap/).

Suess, E. 1885. *Das Antlitz der Erde* (The Face of the Earth), vol. 1. G. Freytag, Leipzig, Germany.

Tarbuck, E. J., and F. K. Lutgens. 2013. *Earth: An Introduction to Physical Geology.* Pearson, London, 912 pp.

Tharp, M., and B. C. Heezen. 1956. Physiographic diagram of the North Atlantic. *Geological Society of America Bulletin* 67, 1704.

Wegener, A. 1924. *La Genèse des Continents et des Océans.* Translated by Manfred Reihel, Collection de monographies scientifiques étrangeres, 6. Paris, Librarie Scientifigque Albert Blanchard.

Wilson, J. T. 1965. Transform faults, oceanic ridges, and magnetic anomalies southwest of Vancouver Island. *Science* 150 (3695), 482–485.

10

The Changing Face of the Earth—
Weathering, Erosion, and Deposition

A rill in a barnyard and the Grand Canyon represent, in the main, stages of valley erosion that began some millions of years apart.

George Gaylord Simpson
Paleontologist, in *Uniformitarianism. An Inquiry into Principle, Theory, and Method*

How many years can a mountain exist, before it is washed to the sea?… The answer, my friend, is blowing in the wind.

Bob Dylan
Singer/songwriter

Weathering, Erosion, and Deposition

Given enough time, all mountains are washed to the sea. **Weathering** is the physical and chemical processes that break down rock and turn it into sediment. **Physical weathering** includes breaking of rock by releasing confining pressure (expansion) or cooling (contraction) during uplift and erosion. Expansion in granite causes pressure-release joints to form (Figure 10.1). Physical weathering also includes the mechanical breakdown of rocks by growing roots, especially tree roots that grow into existing cracks and wedge them apart (Figure 10.2). The same process tilts and cracks sidewalks on city streets. When water freezes, it expands slightly: water seeping into rock sometimes freezes, expands, and cracks the rock. Faults grind up rocks along the fault zone (Figure 10.3). Moving glaciers pluck rocks from beneath the ice and grind them against the underlying bedrock. Rivers, especially when flooding, and flash floods move sand, boulders, and cobbles that grind away at the underlying rock as well as bumping and grinding against each other (Figure 10.4). Rocks falling off cliffs or carried by avalanches break and form talus piles at the foot of a slope (Figure 10.5).

Chemical weathering requires water, oxygen, or biologic activity. Minerals in the rock are chemically changed, such as micas or feldspars in granite turning into clay minerals, thus allowing the rock to decompose to gravel or sand (Figure 10.6). The olivine in basalt and gabbro alters readily to clay in the presence of water. Chemical weathering can involve hot, acid-rich water, altering and breaking down rocks around hot springs. It involves slightly acidic rainfall falling over eons to slowly dissolve rock. Rain falling through the atmosphere picks up carbon dioxide and changes into weak **carbonic acid** which, over time, dissolves limestone, dolomite, and the calcite cement that holds many sandstones together (Figure 10.7). Most **caves**, **caverns** (Figure 10.8), and **sinkholes**

FIGURE 10.1
Expansion joints parallel to the surface are attributed to release of confining pressure. Exfoliating granite, Tokopah Valley, Sequoia National Park, California. (Courtesy of Adam Prost.)

FIGURE 10.2
Bristlecone Pine growing into and breaking up rock, Inyo, California. (Courtesy of daveynin; https://commons .wikimedia.org/wiki/File:Roots_through_rocks_-_Flickr_-_daveynin.jpg.)

FIGURE 10.3
Fractured sandstone in the Moab Fault zone, Utah. (Courtesy of G. Prost.)

FIGURE 10.4
Rocks moved by the Kaweah River during flood stage, Sequoia National Park. (Courtesy of G. Prost.)

FIGURE 10.5
Talus coming off cliffs above Trumbull Lake, California. (Courtesy of G. Prost.)

FIGURE 10.6
Weathering granite and the sand derived from it. Great Basin National Park, Nevada. (Courtesy of Qfl247; https://commons.wikimedia.org/wiki/File:GrusSand.JPG.)

FIGURE 10.7
Weathering Redwall limestone, Grand Canyon. (Courtesy of Adam Prost.)

FIGURE 10.8
Vasey's Paradise, Grand Canyon. Groundwater gushes out of a cavern exposed in the limestone canyon wall. (Courtesy of G. Prost.)

(Figure 10.9) are a result of slightly acid rain and groundwater slowly dissolving limestone. Some cave systems extend for tens or even hundreds of kilometers. Caves also form where rock such as salt or gypsum is dissolved by surface water or groundwater.

Lichen growing on rock surfaces slowly breaks down rock (Figure 10.10). Humic acids produced by plant decay also slowly dissolve rocks.

FIGURE 10.9
Sinkhole at Gaping Gill, North Yorkshire, UK. Fell Beck stream flows into a 105-m-deep hole. (Courtesy of Abcdef123456; https://commons.wikimedia.org/wiki/File:GapingGillSurface.jpg.)

FIGURE 10.10
Several types and colors of lichen growing on granite, Baja California, Mexico. (Courtesy of G. Prost.)

In general, physical weathering occurs in cool, dry climates and results in granular material (sand and gravel) and angular landforms (ridges and cliffs). Chemical weathering occurs in warm, humid areas and creates soil and a more rounded landscape.

Erosion is the process of wearing away the Earth's surface and moving material downslope, whether by flowing water, ice, wind, or waves. This process is driven by gravity and leads to lowered elevation and low relief landscapes.

Deposition is the process where the **agents of erosion** (water, wind, and ice) dump material in low areas. It is the source of all sedimentary rocks. **Talus** is loose, broken rock that collects at the base of cliffs. Rivers deposit boulders, pebbles, sand, silt, and mud (**alluvium**) in the adjacent floodplain during flooding and build deltas at their mouths. Glaciers grind rock and deposit the poorly sorted material (**outwash**, **till**, and **drift**) along their base; rock debris that collects at the front and along the margins of glaciers forms **moraines** (Figure 10.11). Dunes form where there is enough windblown sand, and **loess** (pronounced "luss") deposits contain windblown dust from deserts or glaciers.

Grove Karl Gilbert's *Report on the Geology of the Henry Mountains* (1877) proposed the concept of **dynamic equilibrium** regarding river erosion and deposition. Gilbert said that streams work toward a balance where there is neither erosion of the bed nor deposition of sediment. Where slopes are steep, rivers flow fast, carry lots of sediment, and erode their channels until the slope is shallow. Where slopes are shallow, rivers lose their energy and carrying capacity and deposit sediment such that the slope is sufficient for the river to continue flowing. The space available for sediment to accumulate is called **accommodation space**. Accommodation space is required for deposition to occur.

FIGURE 10.11
Moraine and glacial outwash, Franz Joseph Glacier, South Island, New Zealand. (Courtesy of G. Prost.)

Erosional and Depositional Landscapes

The study of landscapes is called **geomorphology**. **Landscapes** are the result of erosion and deposition. Erosional landscapes include canyons, sea cliffs, mountains, and plateaus. Depositional landscapes include river channels and deltas, floodplains, swamps, lakes, estuaries, and beaches. Whereas all rivers eventually drain to the ocean, some empty into large depressions that are temporarily cut off from the sea. Examples include the Bonneville basin of Utah, the Humboldt sink in Nevada, and the Dead Sea between Israel and Jordan.

Base level is the lowest level that rivers can possibly erode to. The **ultimate base level** is sea level, but **local base level** can be the elevation of a resistant rock layer or a lake. **River channels** are places where rivers deposit their load if there is more sediment than the water can carry. Channels are also where the river cuts into bedrock when flow is too strong to deposit sediments. **Floodplains** are areas along rivers that flood regularly. They are lowlands that flank the river, and they have relatively steep margins where the flooding river has carved into the surrounding countryside (Figure 10.12). **Gullies**, **arroyos**, **valleys**, and **canyons** are the result of slow erosion by rivers or glaciers and by the fast erosion of flash floods. Oftentimes, valleys are located where a fault has broken the bedrock and rendered it more susceptible to weathering and erosion. **Escarpments** are steep slopes caused by fault offset, but can also be any steep, abrupt slope or rock face caused by erosion of a resistant rock. **Terraces** are formed by river erosion, beach erosion by storm waves, or by fault movements (Figures 10.13). **River deltas** are depositional features where the bulk of the sediment carried by a river drops out of suspension owing to decreased velocity of the water. Deltas can be **river-dominated**, such as the Mississippi delta, where the river extends its channel into deep water. They can be **tide-dominated**, such as the Ganges–Brahmaputra, where tides move most of the sediment. Or they can be **wave-dominated**, such as the Nile delta, where waves are the main force shaping the delta (Figures 10.14 through 10.17). River deltas contain floodplains, swamps, and sandbars. **Swamps** are low areas mainly under water because of subsidence. All deltas subside over time owing to compaction and dewatering of the waterlogged sediment. That explains why cities such

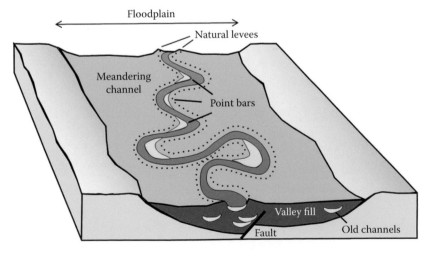

FIGURE 10.12
Features of a floodplain.

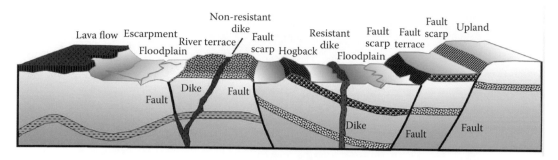

FIGURE 10.13
Topographic expression of river and fault terraces, fault scarps, dikes, and lava flows.

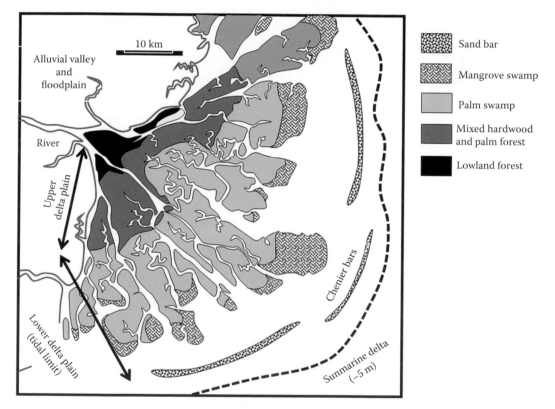

FIGURE 10.14
Components of a typical river delta.

as New Orleans (Mississippi delta), Amsterdam (Rhine delta), and Venice (sandbars just north of the Po delta) have been subsiding. **Sandbars** occur off deltas and beaches and consist of sand that has been winnowed by waves and piled up by storms or currents.

Coastlines are areas where sediment, mostly sand-sized but sometimes cobbles or silt, accumulates to form **beaches** subject to the action of waves and currents. Material is brought to the beach by rivers and then is transported along the beach by **longshore currents** driven by waves hitting the coast at an angle. Offshore sandbars, **barrier islands**, and **spits** are constantly shifting bodies of sand moved by waves (Figures 10.18 and 10.19),

FIGURE 10.15
True color satellite image of the river-dominated Mississippi delta, Louisiana. Sediments (orange brown) are being deposited far out into deep (black) water. (Courtesy of NASA; https://en.wikipedia.org/wiki /Mississippi_River_Delta.)

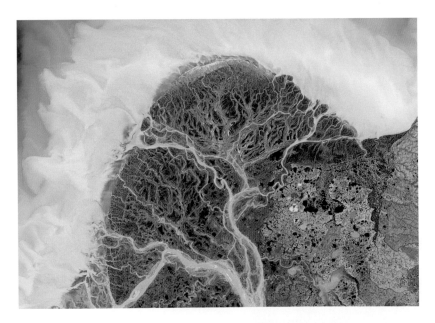

FIGURE 10.16
Color-infrared Landsat 7 satellite image of the wave-dominated Yukon delta, Alaska. The river discharges sediments which are smoothly distributed along the coast by wave action. Vegetation appears red in this image. (Courtesy of NASA; https://en.wikipedia.org/wiki/Yukon–Kuskokwim_Delta.)

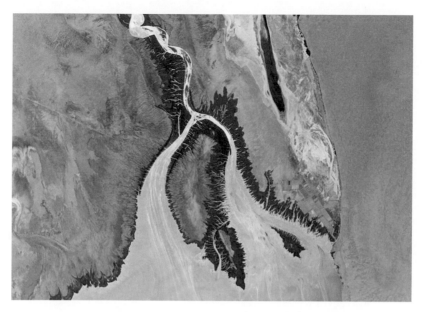

FIGURE 10.17
True color satellite image of the tide-dominated Colorado River delta, Baja California, Mexico. Notice how sandbars are aligned in the direction of the tide movement. (Courtesy of NASA; https://commons.wikimedia.org/wiki/File:ColoradoRiverDelta_ISS009-E-09839.jpg.)

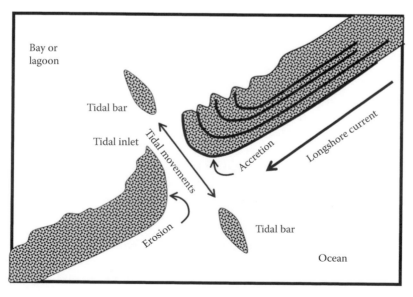

FIGURE 10.18
Shifting of barrier island sandbars by moving sand. The longshore current moves sand along the coast.

FIGURE 10.19
Barrier Island, Outer Banks of North Carolina. (Courtesy of National Oceanic and Atmospheric Administration [NOAA]; https://en.wikipedia.org/wiki/Outer_Banks.)

which is a good reason why you should think twice (or three times) about building or buying a home on a sandbar. Longshore currents eventually move sediment to a submarine valley that funnels it to deeper water. Otherwise, sediments accumulate to the point where a submarine landslide (**turbidity current**) moves the material farther offshore. **Sea cliffs** occur where waves cut into resistant rock or sediments that have been recently uplifted by faulting or isostatic adjustment (called **emergent coastlines**; Figure 10.20). **Bays, estuaries,** and **fjords** are drowned river valleys caused by sea level rising since the last ice age. They will eventually fill with sediment brought by rivers.

Areas with limestone and dolomite are characterized by landforms caused by dissolving of the rock. It is common to see rivers that go underground and reappear, caverns, and **karst** terrain that includes **poljes** (large closed depressions), **karst towers** or **hums** (residual hills), and sinkholes (Figure 10.21).

Natural **lakes** are short-lived depressions filled with water. They are destined to drain or fill with sediment. Lakes are formed by volcanic eruptions (**crater lakes**; Figure 10.22), by glacial scour (**tarns**; Figure 10.23), by sinkholes (Figure 10.24), and by landslides, lava flows, moraines, and fault escarpments that form temporary dams. Eventually, an outflowing river will drain the lake or the lake will fill with sediment washed in from surrounding highlands. When this happens, the lake becomes a meadow. In the case of enclosed, undrained basins like the Dead Sea or Great Salt Lake, the lake evaporates more than it takes in and becomes a **salt lake** or playa lake. **Playas** are shallow lakes that only have

FIGURE 10.20
Sea cliffs at Point Reyes National Seashore, California, indicate an emergent coastline. (Courtesy of Oleg Alexandrov; https://commons.wikimedia.org/wiki/File:Tule_elk_on_Tomales_Point_Trail.JPG.)

FIGURE 10.21
Karst towers, Guilin, China. (Courtesy of Mr. Tickle; https://de.wikipedia.org/wiki/Karst.)

water after a rain or during snowmelt. The rest of the time, the lakebed consists of minerals and salts that precipitate from evaporating lake water.

Dunes and playa lakes are depositional features of desert areas. **Sand dunes** are loose sand blown into large piles by the wind (Figures 10.25 and 10.26). Deserts also have erosional landforms, notably **gullies**, **arroyos**, and **wadis** that are dry streambeds formed during rare flash floods. **Yardangs** and **rock fins** are features caused by wind erosion when storms pick up grains and sandblast everything in their path (Figure 10.27).

FIGURE 10.22
Crater Lake is a natural lake in a volcanic crater on Mt. Mazama, Oregon. (Courtesy of G. Prost.)

FIGURE 10.23
Tarns are lakes formed by glacial scour and are sometimes dammed by moraines. Near Glen Pass, Kings Canyon National Park, California. (Courtesy of G. Prost.)

FIGURE 10.24
Kingsley Lake, Florida, is an almost perfectly round sinkhole lake. (Copyright Google Earth.)

FIGURE 10.25
Dunes, Death Valley National Park, CA. (Courtesy of G. Prost.)

Ridges, **hogbacks**, or **flatirons** occur where an inclined layer of rock has been eroded (Figure 10.28). **Plains**, **prairies**, **steppes**, and **savannahs** are the result of soil accumulating over nearly flat bedrock or sediment and soil accumulating in widespread low areas. **Badlands** are areas where clay-rich sediment is being eroded into a tangle of gullies (Figure 10.29). **Plateaus**, **mesas**, and **buttes** are the erosional remnants of nearly flat but erosion-resistant rock layers (Figure 10.30). Gently rounded hills can be caused by long-term weathering of resistant rock, recent weathering of soft rock, or by deposition of material by glaciers.

FIGURE 10.26
Dune sandstone of the Jurassic Navajo Formation, Zion National Park, Utah. The "cross-bedding" is preserved in the rock and is typical of ancient dunes. (Courtesy of G. Prost.)

FIGURE 10.27
Rock fin caused by wind erosion along joints in the Jurassic Entrada sandstone, Arches National Park, Utah. (Courtesy of G. Prost.)

FIGURE 10.28
Dakota hogback, also known as "the flatirons," near Boulder, Colorado. Cretaceous Dakota sandstone is steeply tilted along the east edge of the Rocky Mountains. (Courtesy of ProTrails; http://www.protrails.com/trail/138 /boulder-denver-golden-fort-collins-lyons-flatiron-1.)

FIGURE 10.29
Badlands at Zabriskie Point, Death Valley, California. (Courtesy of G. Prost.)

FIGURE 10.30
Mesas and buttes, Monument Valley, Arizona. (Courtesy of G. Prost.)

Soil and Agriculture

People depend on good rich soil to grow crops. **Soil** is loose mineral and organic matter that accumulates over bedrock or other sediments. Minerals are derived from weathering of the underlying rock or are brought in by flooding rivers. Some of these minerals contain plant **nutrients** such as nitrogen, phosphorus, potassium, calcium, sulfur, and magnesium: these are the primary components of fertilizers. Other minerals, like quartz, have little nutrient value, but create sandy soils that drain easily, which favors plants such as potatoes, rye, corn, carrots, onions, and blackberries.

The biological component of soil comes from decay of **organic matter**, chiefly other plants, but also from animals and animal manure. Plants can both add nutrients and withdraw nutrients from soil. Legumes such as peas, beans, alfalfa, and clover, for example, have **nitrogen fixing nodules** on their roots that add nitrogen to soil. Most plants deplete nitrogen and other nutrients from soil and eventually leave it barren and unproductive. Some plants, like tobacco, deplete nutrients rapidly. Tobacco-related **soil depletion** was one of the reasons for the westward expansion of plantations of the southern United States in the early 1800s. Nutrient depletion is the main reason for **crop rotation**. Over a 3-year cycle, the farmer rotates crops between nutrient-depleting crops, nutrient-adding crops, and fallow (no crop) conditions.

Soil fertility is maintained naturally by the recycling of nutrients between plants and soil, by adding nutrients from underlying bedrock, and by adding new sediments to the mix. Plant material and animal wastes decompose to add nutrients to soil. Microorganisms like bacteria living in the soil aid this process, which means moisture conditions, soil acidity, salinity, and other factors must be favorable for these organisms as well as for plants. The factors that make a fertile soil include bedrock composition, sediment supply, climate (temperature, moisture), the mix of plants growing on the soil, organisms in the soil, topography (slope steepness and direction), and mixing or churning of the soil.

The French term **terroir** describes the role of geology and soil in producing the best wine, coffee, tea, and other crops. Bedrock is the parent material for soil and provides much of the original nutrients. Minerals such as feldspars, micas, amphiboles, and pyroxenes break down through chemical weathering and release nutrients. Perhaps as important as nutrients is the soil texture, including its **porosity** (spaces between grains) and **permeability** (how easily fluids move through it). The presence of inert minerals such as quartz contributes to the soil's ability to hold moisture or drain well.

Minerals are provided when new sediment is deposited on existing soil by flooding rivers. This was a key reason for the sustainable agriculture of early civilizations of Egypt along the Nile, of Mesopotamia between the Tigris and Euphrates rivers, and of China along the Huang He and Yangtze. Soils can also be replenished by volcanic ash falls. Weathering of volcanic ash releases iron, magnesium, potassium, and aluminum minerals that enrich soils. Volcanic soils tend to be light and fluffy and retain moisture. The area around Naples, Italy, contains rich soils because of volcanic eruptions 35,000 and 12,000 years ago. Hawaiian volcanic soils grow some of the best coffee in the world (Figure 10.31). Likewise, the North Island of New Zealand has rich volcanic soils that, combined with a mild climate and sufficient rain, grow an abundance of crops.

Plants growing on the soil, whether grasses, shrubs, or trees, eventually die, decompose with the help of bacteria and fungi, and get mixed back into the soil with the help of earthworms, moles, and other burrowing creatures.

Climate determines the availability and timing of moisture. Constant moisture such as in tropical rain forests determines not only which plants will grow but how much of the soil nutrients will be leached. Arctic climates inhibit chemical weathering by limiting the activity of microorganisms. Temperate climates have seasonal variations in moisture and plant growth and some of the best developed and most fertile soils.

Topography is crucial to soil development. Steep slopes do not develop thick soils: the soil moves downslope and accumulates in low, flat areas. Flat and gently rolling topography is

FIGURE 10.31
World famous Kona coffee trees thrive in the fertile volcanic soil of Hawaii. (Courtesy of Hawaii Exports International, US Department of Agriculture; https://commons.wikimedia.org/wiki/File:Coffee_farm_-_Flickr_-_USDAgov.jpg.)

where the thickest soils develop. Slope direction (e.g., north-facing) determines how much moisture is retained.

Many naturally rich soils in the Northern Hemisphere are the result of **glacial mixing**. Glaciers over the past 100,000 years have plowed up existing soil and sediment and mixed them with rock flour derived from bedrock beneath the glaciers. This finely ground mineral matter allows plants direct access to nutrients.

The rich diversity of plant life growing on tropical soils obscures the fact that these soils are not particularly rich. Although a warm climate and abundant rainfall causes deep weathering, it also leaches many of the nutrients from the soil. Only a thin topsoil, where plant material decays and is recycled, contains the nutrients needed for plant life. When the forest is burned for **slash-and-burn agriculture**, the cycle is broken, at least temporarily. Crops rapidly deplete the soil of nutrients and rain washes away the soil. Fields are abandoned after a few years and farmers move to new areas to repeat the process (Figure 10.32). It can take decades for plants to reestablish themselves and regenerate topsoil.

Layers within soil are called **soil horizons**. Most soils have three or four horizons (Figure 10.33). The **O Horizon** is dominantly organic matter near the surface, and much of it is fresh. The **A Horizon** (topsoil) contains rich organic matter mixed with mineral material. This is the layer that plants prefer to grow on, and it contains a diversity of soil organisms. The **B Horizon**, or subsoil, contains heavily weathered bedrock and minimal organic matter (mainly deep roots). The **C Horizon** is slightly weathered parent rock. Beneath the C Horizon is unaltered bedrock.

FIGURE 10.32
Slash-and-burn agriculture results in depletion of soil and erosion, as in this example from the Democratic Republic of the Congo. (Courtesy of Max.kit; https://commons.wikimedia.org/wiki/File:Slash_and_burn_in_DRC.jpg.)

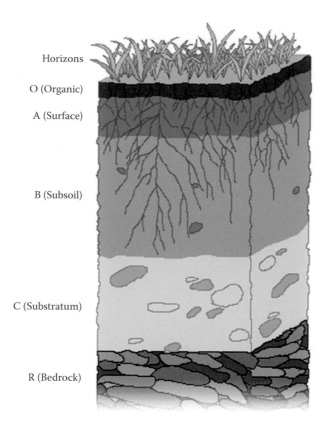

Horizons

O (Organic)

A (Surface)

B (Subsoil)

C (Substratum)

R (Bedrock)

FIGURE 10.33
Soil horizons. (Courtesy of "Horizons" by Wilsonbiggs—derived work from File:SOIL PROFILE.png by Hridith Sudev Nambiar at English Wikipedia; https://commons.wikimedia.org/wiki/File:Horizons.gif#/media/File: Horizons.gif.)

Soil classification by farmers is based mainly on grain size, but also on mineral and organic content, acidity, and ability to retain moisture (Table 10.1). Soil engineers are more interested in soil as a secure base for building homes or for stabilizing roadcuts. Engineers classify soils based on engineering properties such as grain size, clay content, organic content, porosity, moisture content, permeability, **thixotropy** (ability to liquefy when agitated), angle of repose (steepest slope before it slips), strength, **cohesion** (ability to stick together), and **plasticity** (how thin the soil can be rolled before it breaks; Table 10.2).

Soil erosion is a major problem. In most natural settings, soil moves downslope one grain at a time in a process called creep, or gets washed into rivers during heavy rains. When areas have been plowed or overgrazed, however, the soil is primed to be washed or blown away by the next downpour or windstorm. Such was the case in the US and Canadian prairies during the "dustbowl" of the 1930s (Figure 10.34). Man-made soil erosion continues around the world due in large part to slash-and-burn agriculture, logging operations, and overgrazing (Figure 10.35).

TABLE 10.1

USDA Soil Classification System

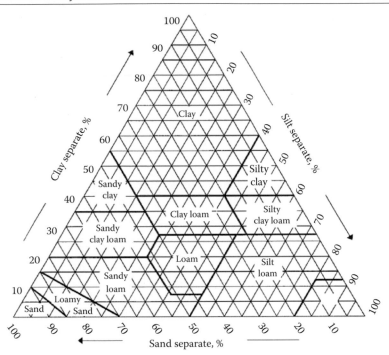

Comparison of particle size scales

USDA	Gravel		Sand					Silt	Clay
			Very coarse	Coarse	Medium	Fine	Very fine		

Unified	Gravel			Sand			Silt or clay	
	Coarse	Fine	Coarse	Medium		Fine		

AASHO	Gravel or stone			Sand		Silt or clay	
	Coarse	Medium	Fine	Coarse	Fine	Silt	Clay

Grain size in millimeters

100   50   10   5   2   1   0.5 0.42 0.25   0.1 0.074 0.05   0.02 0.01 0.005   0.002 0.001

Source: US Department of Agriculture; https://commons.wikimedia.org/wiki/File:SoilTextureTriangle.jpg.

TABLE 10.2

Example of an Engineering Soil Classification Scheme

Letter	Definition
G	Gravel
S	Sand
M	Silt
C	Clay
O	Organic

Letter	Definition
P	Poorly graded (uniform particle sizes)
W	Well-graded (diversified particle sizes)
H	High plasticity
L	Low plasticity

Major divisions			Group symbol	Group name
Coarse grained soils more than 50% retained on or above no. 200 (0.075 mm) sieve	Gravel > 50 of coarse fraction retained on no. 4 (4.75 mm) sieve	Clean gravel <5% smaller than #200 sieve	GW	Well-graded gravel, fine to coarse gravel
			GP	Poorly graded gravel
		Gravel with >12% fines	GM	Silty gravel
			GC	Clayey gravel
	Sand ≥ 50% of coarse fraction passes no. 4 sieve	Clean sand	SW	Well-graded sand, fine to coarse sand
			SP	Poorly graded sand
		Sand with >12% fines	SM	Silty sand
			SC	clayey sand
Fine grained soils 50% or more passing the no. 200 sieve	Silt and clay liquid limit < 50	inorganic	ML	Silt
			CL	Clay of low plasticity, lean clay
		organic	OL	Organic silt, organic clay
	Silt and clay liquid limit < 50	inorganic	MH	Silt of high plasticity, elastic silt
			CH	Clay of high plasticity, fat clay
		organic	OH	Organic clay, organic silt
Highly organic soils			Pt	Peat

Source: https://en.wikipedia.org/wiki/Unified_Soil_Classification_System.

FIGURE 10.34
Dust storm approaching Stratford, Texas, April 18, 1935. (Courtesy of NOAA George E. Marsh Album; https://commons.wikimedia.org/wiki/File:Dust-storm-Texas-1935.png.)

FIGURE 10.35
Soil erosion and downhill creep attributed to cattle overgrazing. Southern slopes of the Serranía del Interior, Venezuela. (Courtesy of Fev; https://commons.wikimedia.org/wiki/File:Reptaci%C3%B3n.JPG.)

References

Gilbert, G. K. 1877. *Report on the Geology of the Henry Mountains*. US Geological Survey, 163 pp.
Simpson, G. 1970. Uniformitarianism. An Inquiry into Principle, Theory, and Method in Geohistory and Biohistory. In *Essays in Evolution and Genetics in Honor of Theodosius Dobzhansky*, M.K. Hecht and W.C. Steere (eds.). Appleton-Century-Crofts, New York. p. 43–96.

11

Maps

The world is the geologist's great puzzle-box; he stands before it like the child to whom the separate pieces of his puzzle remain a mystery till he detects their relation and sees where they fit, and then his fragments grow at once into a connected picture beneath his hand.

Louis Agassiz
Biologist/Geologist

Beneath all the wealth of detail in a geological map lies an elegant, orderly simplicity.

J. Tuzo Wilson
Geophysicist and Geologist, from *In Celebration of Canadian Scientists*

Geologic maps provide information about how rocks are distributed and deformed. The rocks are usually shown as various colors superimposed on a **topographic map**. "Topo maps" show the ground surface by means of contours of equal elevation. Sometimes, geology is posted directly on air photos or satellite imagery instead of topographic maps (Figure 11.1).

Types of Geological Maps

There are general geologic maps as well as specialized geologic maps for many purposes. A few different types of geological maps are described.

Standard Geological Maps

A **standard geologic map** displays the geology in an area (Figure 11.2). In these maps, the different units are indicated by abbreviations that are explained in the **map legend** (Figure 11.3). Symbols help explain the structures seen on the map (Figure 11.4). We describe geologic structures by defining the orientation of planar surfaces such as rock layers, faults, dikes, and veins. **Strike** is the compass bearing (azimuth) of a horizontal line drawn on an inclined surface. **Dip** is the inclination of the surface measured perpendicular to the strike (Figure 11.5).

The fundamental unit of mapping is the **formation**, a distinct body of rock that is mappable on the surface or traceable in the subsurface. Formations can be subdivided into **members**, or combined into **groups** of related units. Boundaries between formations are called **contacts** and are indicated on maps by thin black lines. Where rocks are covered by soil or surface deposits, the distribution of these **unconsolidated** (not cemented) surface units is shown and the bedrock is extrapolated in the subsurface using dashed or dotted lines.

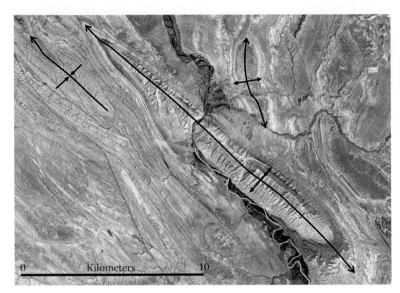

FIGURE 11.1
Landsat satellite image and structure interpretation of Sheep Mountain north of Greybull, Wyoming. Anticline and syncline symbols are shown. The imagery serves as a base for the geologic mapping. (Copyright Google Earth.)

FIGURE 11.2
Part of the geologic map of Oakland, California. See Figure 11.3 for an explanation of some of the unit name abbreviations. (From R. W. Graymer. 2000. Geologic map and map database of the Oakland metropolitan area, Alameda, Contra Costa, and San Francisco Counties, California. US Geological Survey Miscellaneous Field Studies map MF-2342; https://store.usgs.gov/yimages/PDFX/113221_comp.pdf, https://pubs.usgs.gov/mf/2000/2342/mf2342i.pdf.)

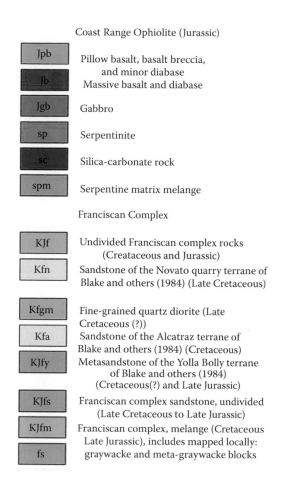

FIGURE 11.3
Part of the legend for the Geologic Map of Oakland, California. The legend gives the unit name and, in this example, who described it and when. It also describes the rock type and provides the age of the unit. (From R. W. Graymer. 2000. Geologic map and map database of the Oakland metropolitan area, Alameda, Contra Costa, and San Francisco Counties, California. US Geological Survey Miscellaneous Field Studies map MF-2342; https://store.usgs.gov/yimages/PDFX/113221_comp.pdf, https://pubs.usgs.gov/mf/2000/2342/mf 2342i.pdf.)

Maps are made by traversing a region on foot or from the air and noting the rock type and age of units at the surface. Symbols show the inclination and orientation of nonhorizontal units. Abrupt changes in rock type and evidence of offset are mapped as unconformities (erosion surfaces) or faults. The offset on the fault (upthrown and downthrown side or horizontal slip direction) is shown. If folding is indicated, then the symbol for anticline or syncline is added to the map.

Well and seismic data that provide subsurface control are incorporated in cross sections. **Cross sections** represent slices through the Earth and show the disposition of rocks in the subsurface (Figure 11.6).

Geologic maps show, for example, where a coal bed outcrops and how thick it is. They can be used to find the best place to look for oil and gas accumulations (the top of anticlines or domes) or mineral deposits (like the contact between igneous rocks and limestones).

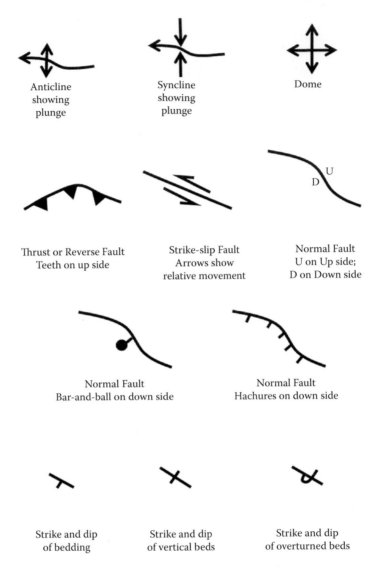

FIGURE 11.4
Some common geologic map symbols.

Collectors can use geologic maps to find minerals (for example in veins, along faults, or in abandoned mines) and fossils (such as trilobites in Cambrian shales or crinoids in Carboniferous limestone). Civil engineers and homeowners use geologic maps to avoid locating a dam, power plant, or housing development on a fault or landslide.

Geological Hazards Maps

Urban planners use geological maps to show where potential hazards (e.g., faults, landslides, and floods) exist. Sometimes, they create a derivative map called a geologic **hazard map** that only shows these features (Figure 11.7).

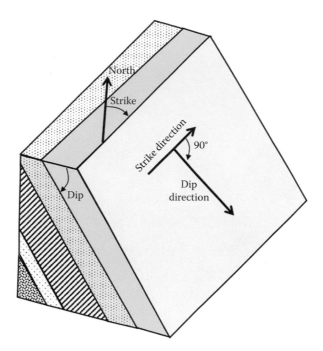

FIGURE 11.5
Graphic explanation of the terms "strike" and "dip."

Seismic Hazard Maps

One specific type of hazard map is the **seismic hazard map**. Earthquake-related landslide maps, ground liquefaction maps, probability of earthquake occurrence maps, and ground shaking intensity maps are generated from geologic maps with additional information from the history of seismic activity in an area (Figures 11.8 through 11.10). They are used by urban planners and insurance companies.

Paleogeographic Maps

Paleogeographic maps show the environment of deposition that existed over an area at a point in the past. Ancient environments are derived from the stratigraphy (such as channel sand, coal beds, and delta muds), fossils, and paleomagnetic surveys that indicate the ancient environment and latitude of the region (Figure 11.11).

Soil Maps

Farmers use **soil maps** (Figure 11.12) to plan where to plant crops based on properties such as sandiness, drainage, moisture, clay content, and acidity. For example, there are some soil types that are ideal for growing wine grapes. The most important factors for grapes are acidity, organic content, and drainage. Grapes cannot tolerate wet soils, extremely acidic or alkaline soils, or those that are severely deficient in nutrients. Wheat prefers a

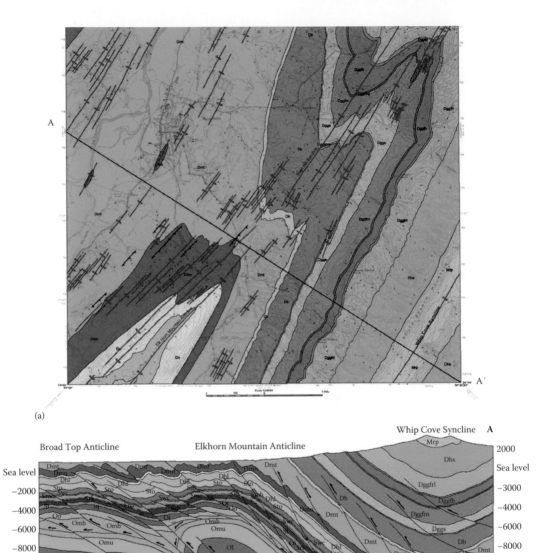

(a)

(b)

FIGURE 11.6

(a) Geologic map of the Moorefield Quadrangle, Hardy County, West Virginia. (b) Cross section is taken along the line A–A′. The cross section is derived from surface dip and strike measurements and information from wells. (Courtesy of West Virginia Geological and Economic Survey; http://www.wvgs.wvnet.edu/www /statemap/483/483_Bed.htm.)

nearly neutral (pH about 6.4), well-drained, **loamy soil**, a mixture of sand, silt, and clay. Blueberries require **acidic soils** (pH between 4.1 and 5.0) and good drainage.

Highway engineers use soil maps to understand engineering properties such as soil strength and cohesiveness. They want to know if a soil slope will fail (slump) if subjected to compaction, vibration, or heavy rain. Environmental firms use soil maps to define wetlands, crop areas, wildlife habitat, and so forth.

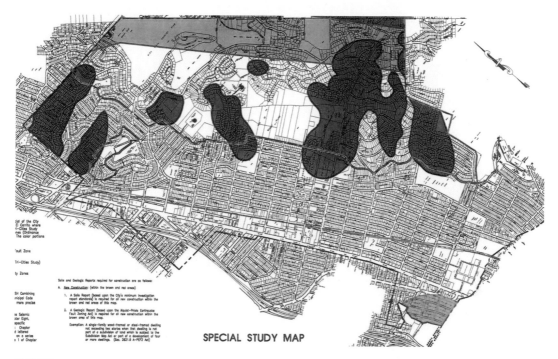

SPECIAL STUDY MAP

FIGURE 11.7

Hazards map of El Cerrito, California. Orange is the Hayward Fault zone; red is potential landslide zone; blue is potential flood zone. (Courtesy of Harris and Associates Inc.; City of El Cerrito; http://www.el-cerrito.org /DocumentCenter/Home/View/463.)

Groundwater or Potentiometric Surface Maps

Groundwater geologists (**hydrologists**) make maps that show the depth to the water table (Figure 11.13). Called a "**potentiometric surface map**," it has contours that show the elevation of the water table in the subsurface, or above the surface in the case of artesian water tables. Water wells are shown, and the direction of water flow in the subsurface may be shown.

Structure Contour Maps

Oil companies and coal companies use a type of geologic map called a "**structure contour map**." These maps show the depth to a specific horizon, usually a contact such as the "top McMurray Formation." When rocks are folded or faulted, these maps show the structure as it exists in the subsurface (Figure 11.14).

Mineral Alteration Maps

Metallic mineral deposits are sometimes revealed at the Earth's surface by altered rock. **Alteration** might be iron stain related to pyrite weathering in the rock, or feldspars that have been converted to clay by the action of hydrothermal fluids. When mining company geologists locate altered rocks, they take samples and have them assayed to test for various

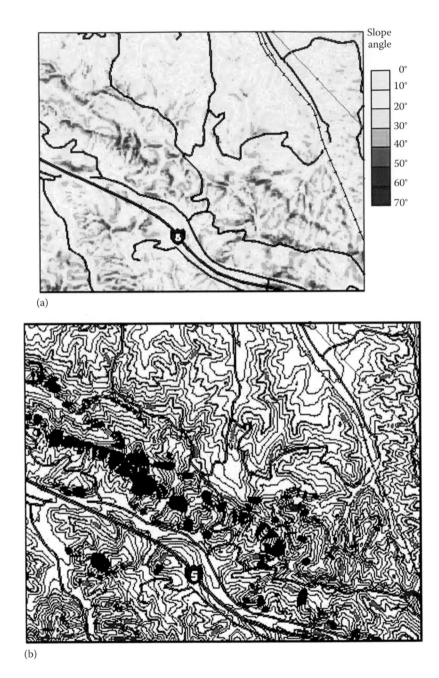

(a)

(b)

FIGURE 11.8
(a) Slope map derived from the Digital Elevation Map of part of the Oat Mountain quadrangle. (b) Map showing landslides (black) triggered by the 1994 Northridge earthquake in the Oat Mountain quadrangle, California. (From Jibson et al. 1998. A Method for Producing Digital Probabilistic Seismic Landslide Hazard Maps: An Example from the Los Angeles, California, Area. US Geological Survey Open-File Report 98-113; http://pubs .usgs.gov/of/1998/ofr-98-113/ofr98-113.html.)

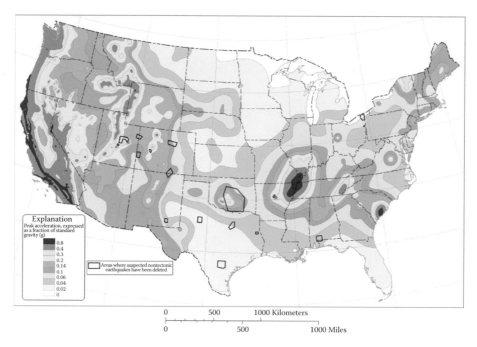

FIGURE 11.9
Map of peak seismic ground acceleration with 2% probability of exceedance in 50 years. (Courtesy of USGS; https://en.wikipedia.org/wiki/Seismic_hazard#/media/File:2014_pga2pct50yrs_(vector).svg.)

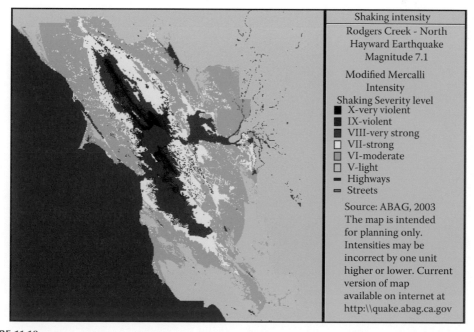

FIGURE 11.10
Surface motion map for a hypothetical earthquake on the northern portion of the Hayward Fault Zone and its northern extension, the Rodgers Creek Fault Zone. (Courtesy of Leonard G; https://en.wikipedia.org/wiki/Seismic_hazard.)

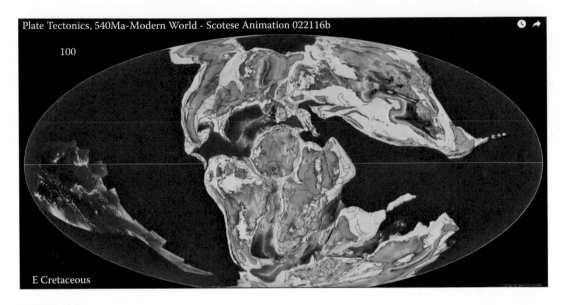

FIGURE 11.11
Early Cretaceous paleogeographic map of the world. (From Scotese, C. R. 2001. *Atlas of Earth History, Volume 1, Paleogeography*, PALEOMAP Project, Arlington, Texas, 52 pp., with permission; http://www.scotese.com/earth .htm, https://www.youtube.com/watch?v=IJO04EV3dYY.)

FIGURE 11.12
Soil map of part of the Napa Valley, California. Soil units in this world-famous winegrowing region are posted directly onto an air photo. Orange lines outline the units. (Courtesy of US Department of Agriculture; http:// lergp.org/2-years-pre-plant/review-soil-maps.)

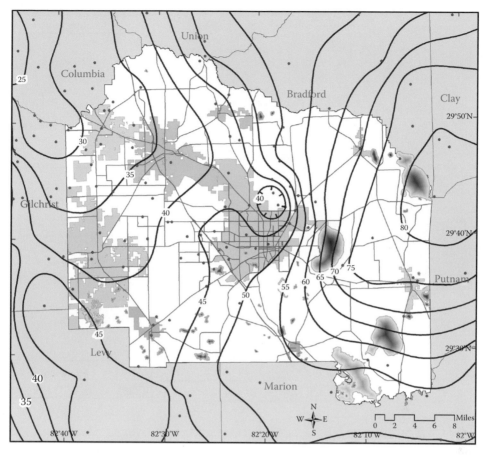

FIGURE 11.13
Potentiometric surface map of Alachua County, Florida. Depth of the water table is given by the contour lines. Well control is shown. (Courtesy of Alachua County, Florida, Environmental Protection Department, Water Resources; http://www.alachuacounty.us/Depts/epd/WaterResources/WaterData/Pages/Maps.aspx.)

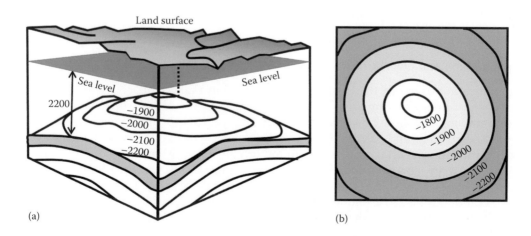

FIGURE 11.14
(a) 3D representation of a geologic structure at depth. (b) Structure contour map of the same feature. Contour values are given as depth below sea level.

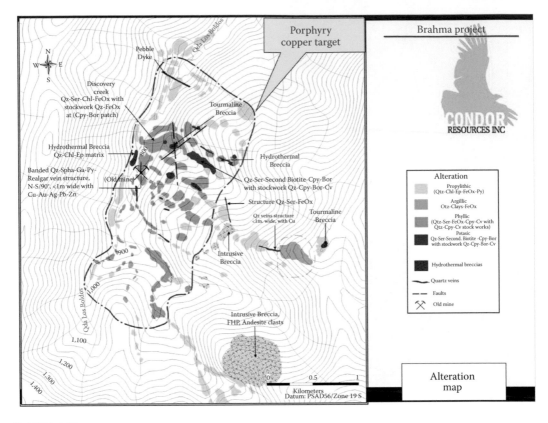

FIGURE 11.15
Alteration map showing the extent of a porphyry copper deposit at Condor Resource's Brahma project, Chile. (Copyright Condor Resources Inc; www.condorresources.com, http://www.condorresources.com/s/photogallery .asp?ReportID=302455&_Title=Brahma-Alteration-Map.)

metals. If the metal concentrations are high enough, they will drill into the subsurface to see if the deposit is rich enough and extensive enough to develop. The first step is to map alteration at the surface (Figure 11.15).

Isopach Maps and Overburden Thickness Maps

Companies interested in the thickness of a formation make **isopach** maps. These maps have contour lines showing areas of equal thickness (Figure 11.16). Isopach maps are used by oil companies to show how thick the hydrocarbon-bearing unit is so that they can plan where to drill wells and calculate reserves. Coal companies do the same when they want to determine how much coal is in the ground or where to start mining.

A similar map shows **overburden thickness** (Figure 11.17). **Overburden** is all the rock and soil lying over the deposit. Mining companies use these to determine where a coal bed or mineral deposit is close enough to the surface to mine using open pit methods, and to calculate the costs of removing the overburden.

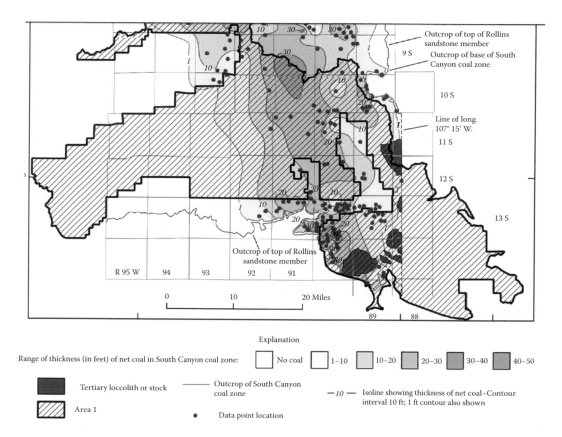

Explanation

Range of thickness (in feet) of net coal in South Canyon coal zone: No coal   1–10   10–20   20–30   30–40   40–50

Tertiary loccolith or stock

Area 1

Outcrop of South Canyon coal zone

Data point location

— *10* — Isoline showing thickness of net coal–Contour interval 10 ft; 1 ft contour also shown

FIGURE 11.16
Coal isopach (thickness) map of South Canyon coal zone, Colorado. (From Hettinger et al. 2008. Coal resources and coal resource potential, in Chapter M of *Resource Potential and Geology of the Grand Mesa, Uncompahgre, and Gunnison National Forests and Vicinity, Colorado*. US Geological Survey Bulletin 2213–M; http://pubs.usgs.gov /bul/2213/M/Chp_M.pdf.)

A Resource Guide—Where to Find Maps and Imagery

Geologic maps are published by the US Geological Survey (USGS) and the Bureau of Land Management (BLM), the Geological Survey of Canada, and other geological surveys. They are published by local governments such as state geological surveys, water bureaus, or departments of mineral resources.

Maps of limited areas (such as the Geological Highway Map series are published by geological organizations such as the Geological Society of America, the American Association of Petroleum Geologists (http://aapg.org), and the Canadian Society of Petroleum Geologists (http://cspg.org).

Maps are also published by national parks and monuments, by universities, by local gem and mineral societies, and by authors interested in the geology of specific areas (such as the *Roadside Geology* books in the United States). Companies interested in maps of out-of-the-way areas will hire companies that specialize in geologic mapping or can make their own maps by analysis of air photos and satellite images and by sending geologists into the field.

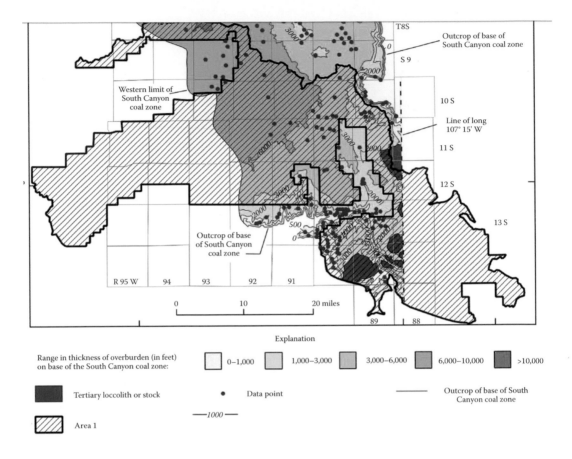

FIGURE 11.17

Overburden thickness map, South Canyon coal zone, Colorado. (From Hettinger et al. 2008. Coal resources and coal resource potential, in Chapter M of *Resource Potential and Geology of the Grand Mesa, Uncompahgre, and Gunnison National Forests and Vicinity, Colorado.* US Geological Survey Bulletin 2213–M; http://pubs.usgs.gov/bul/2213/M/Chp_M.pdf.)

Topographic maps at various scales can be obtained from the USGS and other national agencies. See, for example, http://store.usgs.gov. More regional and less detailed topography, especially outside of North America, can be found on aeronautical charts.

Air photos and satellite images are obtained in the United States from the National Aeronautics and Space Administration (NASA), the USGS, BLM, and state surveys. They are also available from the European Space Agency (ESA), the Indian Space Agency (ISRO), and other national agencies. Go to http://eros.usgs.gov, https://www.blm.gov, or https://earth.esa.int.

An online search using the key words "geologic map" and your area of interest will bring up a list of available maps.

References

Graymer, R. W. 2000. Geologic map and map database of the Oakland metropolitan area, Alameda, Contra Costa, and San Francisco Counties, California. US Geological Survey Miscellaneous Field Studies map MF-2342.

Hettinger, R. D., L. N. R. Roberts, and M. A. Kirschbaum. 2008. Coal resources and coal resource potential, in Chapter M of *Resource Potential and Geology of the Grand Mesa, Uncompahgre, and Gunnison National Forests and Vicinity, Colorado.* US Geological Survey Bulletin 2213–M.

Jibson, R. W., E. L. Harp, and J. A. Michael. 1998. A Method for Producing Digital Probabilistic Seismic Landslide Hazard Maps: An Example from the Los Angeles, California, Area. US Geological Survey Open-File Report 98-113.

Wilson, J. T. 1990. in: Geraldine A. Kenney-Wallace et al. In Celebration of Canadian Scientists: A Decade of Killam Laureates, Charles Babbage Research Centre. p. 285.

12

Acts of God? Earthquakes, Volcanoes, and Other Natural Disasters

Civilization exists by geological consent, subject to change without notice.

Will Durant
Writer, historian, philosopher

Earthquakes don't kill people. Buildings kill people.

Anonymous

I feel the earth move, under my feet, I feel the sky tumbling down, tumbling down…

Carole King
Singer and songwriter, in *I Feel the Earth Move*

We humans have a pretty self-centered view of natural disasters. We are only really concerned about how they affect us, and we tend to see them as rare, divine, almost whimsical acts. Yet, over time, these are the events that shape the earth. Earthquakes raise mountains a few centimeters at a time over millions of years. Flash floods shape desert and mountain landscapes even though they occur once a decade or once a century. Explosive volcanic eruptions happen on a time scale of tens of thousands to hundreds of thousands of years. Large impacts happen once every 50 to 100 million years. These more-or-less random events have always been present, working slowly or suddenly, not only shaping the earth, but altering the course of life over geologic time.

Earthquakes and Tsunamis

In April 2015, a magnitude 7.3 earthquake struck Nepal. The death toll topped out at more than 8000 souls, mostly from the collapse of unreinforced stone buildings, from avalanches, and from landslides (NB; Revision recommended to establish a connection between call-out and Figure 12.1 [which has nothing to do with the April 2015 earthquake in Nepal]). This quake is part of an active tectonic zone that extends from Yunnan, China, to Iran and is a result of the Indian tectonic plate crashing into the Asian crustal plate.

The 2004 Indian Ocean earthquake off the west coast of Sumatra, Indonesia, had a magnitude of 9.1–9.3. The undersea subduction-related earthquake triggered tsunamis along the coasts of the Indian Ocean, devastating coastal communities and killing 230,000 people with waves up to 30 m (100 ft) high (Figures 12.2 and 12.3).

FIGURE 12.1
The 2001 earthquake-triggered landslide near San Salvador, El Salvador. (Courtesy of US Geological Survey;
https://www.google.com/url?sa=i&rct=j&q=&esrc=s&source=images&cd=&ved=0ahUKEwiy-9Dqx8jRAh
Vo94MKHVMFD-gQjxwIAw&url=https%3A%2F%2Fwww2.usgs.gov%2Ffaq%2Fcategories%2F9752%2F2606
&psig=AFQjCNEdjeWt9HtNUDzdU5lEOfG4YBZbZw&ust=1484721226513862.)

A magnitude 7.0 quake leveled much of Port au Prince, Haiti, in 2010, killing between 100,000 and 250,000 people. This earthquake occurred along a left-lateral transform margin fault known as the Enriquillo–Plantain Garden fault zone.

A devastating magnitude 7.9 quake struck Tokyo and Yokahama in 1923, killing 105,000. Tens of thousands died, mostly in the firestorms that resulted as the city burned. The Kanto quake resulted from subduction of the Philippine Sea plate beneath the continental Amurian tectonic plate.

The 1906 San Francisco quake on the San Andreas Fault toppled buildings and broke water mains—allowing the fires it spawned to spread across the city unchecked. Modern estimates are that around 3000 lost their lives, mostly those trapped in collapsed buildings that then burned. As many as 225,000 were left homeless.

This is but a small sampling of the destructive power of earthquakes and the toll they take on civilization.

FIGURE 12.2
Banda Aceh, Indonesia, the most devastated region struck by the 2004 Indian Ocean tsunami. (Courtesy of United States Navy, image ID 050102-N-9593M-040; https://en.wikipedia.org/wiki/2004_Indian_Ocean_earth quake_and_tsunami.)

FIGURE 12.3
The December 26, 2004, tsunami striking Ao Nang, Krabi Province, Thailand. (Courtesy of David Rydevik, Stockholm, Sweden; https://commons.wikimedia.org/wiki/File:2004-tsunami.jpg.)

Where Do They Happen?

Active tectonic zones are the most likely places for earthquakes. But they are not the only places quakes can occur, as demonstrated by the 2011 quake that hit Washington, D.C. The magnitude 5.8 Virginia earthquake was felt in a dozen states and several Canadian provinces (Figure 12.4). The 7.8–8.2 magnitude Tangshan, China, earthquake in July 1976 killed close to 655,000 people, making it one of the most lethal earthquakes in history. Most people were killed by the collapse of unreinforced masonry homes. This quake occurred along a previously unknown strike-slip fault (one where slip is sideways rather than up/down) in an area where quakes of this magnitude were not expected and the buildings were not built to withstand large seismic events. Large quakes can happen in the tectonically quiet mid-continent.

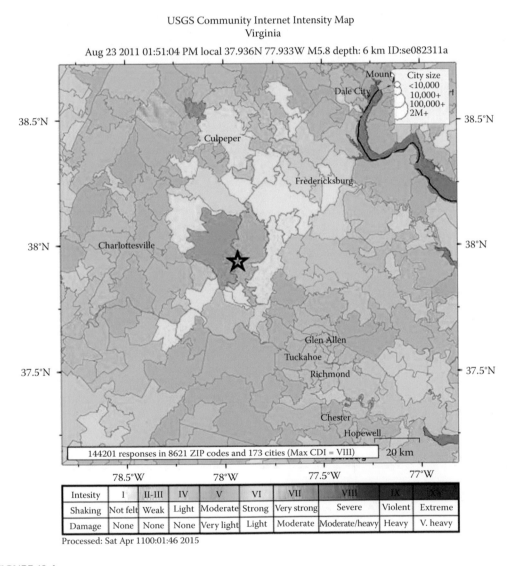

FIGURE 12.4
Community Internet Intensity map ("Did You Feel It?" map) for the magnitude 5.8 earthquake near Washington, D.C., August 23, 2011. (Courtesy of US Geological Survey; http://earthquake.usgs.gov/earthquakes/eventpage /se609212#dyfi.)

FIGURE 12.5
Engraving of the 1755 Lisbon earthquake. This copper engraving, made that year, shows the city in ruins and in flames. Tsunamis rush upon the shore, destroying the wharfs. (Original in Museu da Cidade, Lisbon; https://en.wikipedia.org/wiki/1755_Lisbon_earthquake.)

In 1811–1812, a group of magnitude 7.5 earthquakes centered near New Madrid, Missouri, affected an area between 200,000 and 334,000 km² (78,000 and 129,000 mi²) and was felt from Cairo, Illinois, to Memphis, Tennessee. The quakes were taken as a signal by Native Americans in the area to begin an uprising against the white American settlers urged by the Shawnee war chief Tecumseh. The resulting war led to the destruction of the Indian nations east of the Mississippi. An 8.5 to 9.0 quake hit normally quiet Lisbon in 1755, destroying the city and killing an estimated 100,000 inhabitants (Figure 12.5). A 6-m (20-ft) tsunami followed the quake by 40 minutes. Fires erupted and burned the city for the next 5 days.

The most destructive quake ever recorded, a magnitude around 8.0, occurred in Shaanxi, China, in the year 1556. About 830,000 people were killed and the effects were felt up to 800 km (500 mi) away. What most of these earthquakes have in common is being in active tectonic zones, mostly near tectonic plate margins. However, it is not the shuddering Earth that kills most people. The shaking destroys buildings, and it is the collapsing buildings and resulting fires that kill people.

Causes

Earthquakes occur suddenly and seemingly at random. They can be horribly destructive of lives and property, but what are they and what causes them? **Earthquakes** (also referred to as tremors, quakes, or temblors) are vibrations in the earth caused by tectonic forces that deform the earth. Earth's crust is under stresses of various magnitudes caused by plate tectonics, that is, by movement of large sections of the earth's brittle crust, and the release of stress causes sudden, sometimes violent shaking. **Tsunamis** (sometimes wrongly called "tidal waves") are sea waves usually caused by ocean-bottom earthquakes, but also by large landslides into water and by underwater volcanic eruptions. While they resemble a swell in the open ocean, as they approach shallow water, they slow down and their height increases. One of the most sinister aspects of tsunamis is that coastal water will run out to sea as the wave approaches, drawing the curious into its path.

The ultimate driver of earthquakes is convective heat transfer from the interior of the Earth toward the surface. The mantle, at depths between about 16 and 2900 km (10 and 1800 mi), is under tremendous pressure and is quite hot. The high temperature and high pressure allow the mantle to deform plastically. **Plastic deformation** means that there is a permanent change in shape without fracturing. As the plastic mantle rises under the influence of convection, it approaches the surface at mid-ocean ridges. At the ridges, the mantle stops rising and starts pushing the lithospheric plates away in opposite directions. Because this movement occurs at or near the surface, spreading centers tend to have shallow, near-surface earthquakes.

There are two kinds of tectonic plate boundary besides the spreading centers: convergent margins, where two plates collide, and transform margins, where two plates slide past each other. Examples of convergent boundaries include the India–Asia boundary along the Himalayas and the Pacific–South America boundary in the Andes. These boundaries are characterized by intermediate to deep earthquakes. Examples of transform plate boundaries include the San Andreas Fault that separates the Pacific and North American plates in California and the Falcon–El Pilar fault system that separates the Caribbean and South American plates in Colombia, Venezuela, and Trinidad. This type of boundary can have shallow, intermediate, or deep earthquakes.

Lithospheric plates undergo mostly brittle deformation. Brittle deformation means they break along faults. Faulting is what causes earthquakes.

Any time the crust experiences brittle deformation, the rocks break. The vibrations caused by that break spread outward as waves, much as ripples spread outward when a pebble is tossed into a pond. These waves are classified by geophysicists into one of three types: P, S, and L (Figure 12.6). The **P or Primary wave** is a **compressional body wave**; that is, it travels through a body in a push–pull (compression and dilation) movement of the rock particles. This is the fastest, therefore the first wave to hit a receiver (seismometer). The **S or Secondary wave** is a **shear wave** where the particles in the rock body move sideways at right angles to the direction of wave travel. A unique feature of the S wave is that it cannot move through liquids and thus cannot be transmitted through Earth's core. **Surface waves** only move on the surface of the earth. There are two types of surface waves: Love waves and Raleigh waves. **Love waves** (**L waves**) move the ground from side to side in a horizontal plane at right angles to the direction of propagation. This is particularly damaging to the foundations of buildings. **Rayleigh waves** move both vertically and horizontally in a vertical plane pointed in the direction in which the waves are traveling. P waves move faster than S waves, which move faster than L waves.

Seismometers are instruments that record these vibrations and their first arrival time. One can calculate the origin of the earthquake by using three or more seismometers, that is, you can triangulate its location at or below the earth's surface. The origin at depth is called the **focus** of the earthquake, and the location at the surface above the focus is called the **epicenter**.

The depth of the earthquake can be correlated to the destructiveness of the earthquake. Deep earthquakes are generally less destructive because the deep crust is slightly more plastic (less brittle) and vibrations are somewhat damped, dissipated, or minimized by the intervening crust. Shallow earthquakes (less than 5 to 10 mi, or 8 to 16 km) tend to be the most destructive because they occur in the most brittle parts of the crust and the vibrations originate near the surface and are quite intense.

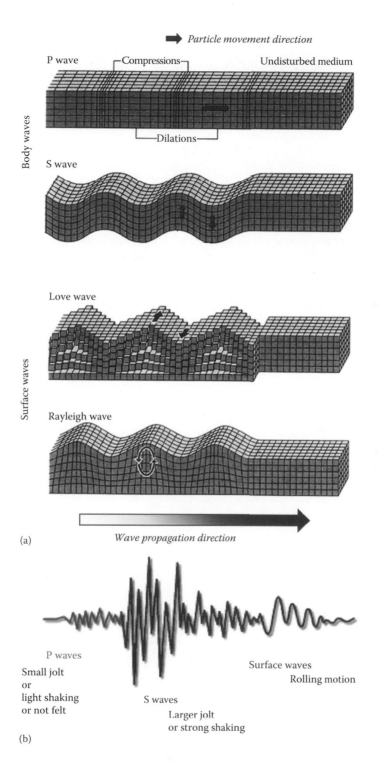

(a)

(b)

FIGURE 12.6
(a) Various seismic waves and corresponding particle motion. (b) Sequence of waves in time. P waves hit first, followed by S and surface waves. (Courtesy of US Geological Survey; http://earthquake.usgs.gov/learn/eqmonitoring/eq-mon-2.php.)

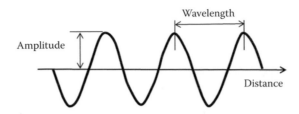

FIGURE 12.7
Relationship of amplitude and wavelength in a seismic wave.

Seismic waves are characterized by their wavelength and amplitude (Figure 12.7). Shorter-wavelength and higher-amplitude waves are the most destructive.

Earthquakes happen all the time everywhere on Earth. Most are so small that we don't feel them. The amount of energy released by an earthquake is measured using the **Moment Magnitude Scale** (MMS; formerly the Richter scale) or using the **Modified Mercalli Intensity scale** (Tables 12.1 and 12.2). **Intensity scales** measure the amount of shaking at a particular location. The Richter scale was superseded in the 1970s by the MMS. **Magnitude scales** measure the size or energy of the earthquake at its source. As with the Richter scale, an increase of one step on this scale corresponds to a 32-fold increase in the amount of energy released, and an increase of two steps corresponds to a 1000 times increase in energy. Tremors less than 3 on the MMS are rarely felt and are usually nondestructive, but they can be recorded. Values greater than 4 start damaging surface structures. The amount of damage is a function of the earthquake size, the distance from the focus, the surface materials, and the type of construction. Steel-reinforced concrete structures on a shock-absorbing foundation suffer less damage than unreinforced stone or adobe structures. Structures built on firm soil or bedrock suffer less damage than structures built on waterlogged soil or landfill. **Liquefaction** is when soil liquefies as a result of the vibrations and causes structures to tilt or sink (Figure 12.8). In addition to soil liquefaction, earthquakes cause surface displacements that uplift or downdrop large areas, and they can cause breaks (faults) that move the surface laterally, vertically, or open small local fissures. They also cause landslides, avalanches, and tsunamis.

Man-Made Earthquakes

Man-made earthquakes, also called **induced seismic events**, have been in the news lately. These tend to be smaller and less destructive than natural quakes, but have been felt and, in some cases, have caused damage. An early documented case involved injecting wastewater into a deep disposal well at the Rocky Mountain Arsenal near Denver, Colorado, between 1961 and 1966. Disposal was halted when the injection was tied to a resulting earthquake swarm.

Areas such as Kansas, Oklahoma, and Ohio, not generally known for quakes, have had a sudden increase in tremors over the past decade owing to wastewater disposal associated with hydraulic fracturing operations. The frac'ing process itself doesn't use enough fluid to cause obvious earthquakes, although any time rock is fractured, the resulting break causes a small shock. These induced seismic events are usually less than MMS magnitude 2.

However, produced oil and gas are almost always accompanied by groundwater. In the past, this **wastewater** was pumped into surface ponds or streams, but now regulations require this water be returned to the subsurface. The rather large volumes of reinjected water do sometimes cause earthquakes, either by breaking the rock during injection or by

TABLE 12.1

The Richter/Moment Magnitude Scale Used to Measure Earthquake Magnitude

Magnitude	Description	Mercalli Intensity	Average Earthquake Effects
Less than 2.0	Micro	I	Microearthquakes, not felt, or rarely felt by sensitive people. Recorded by seismographs.
2.0–2.9	Minor	I to II	Felt slightly by some people. No damage to buildings.
3.0–3.9		II to IV	Often felt by people, but very rarely causes damage. Shaking of indoor objects can be noticeable.
4.0–4.9	Light	IV to VI	Noticeable shaking of indoor objects and rattling noises. Felt by most people in the affected area. Slightly felt outside. Generally causes none to minimal damage. Moderate to significant damage very unlikely. Some objects may fall off shelves or be knocked over.
5.0–5.9	Moderate	VI to VIII	Can cause damage of varying severity to poorly constructed buildings. At most, none to slight damage to all other buildings. Felt by everyone.
6.0–6.9	Strong	VII to X	Damage to a moderate number of well-built structures in populated areas. Earthquake-resistant structures survive with slight to moderate damage. Poorly designed structures receive moderate to severe damage. Felt in wider areas; up to hundreds of miles/kilometers from the epicenter. Strong to violent shaking in epicenter area.
7.0–7.9	Major	VIII to X	Causes damage to most buildings, some to partially or completely collapse or receive severe damage. Well-designed structures are likely to receive damage. Felt across great distances with major damage mostly limited to 250 km from epicenter.
8.0–8.9	Great		Major damage to buildings, structures likely to be destroyed. Will cause moderate to heavy damage to sturdy or earthquake-resistant buildings. Damaging in large areas. Felt in extremely large regions.
9.0 and greater			Near or total destruction—severe damage to or collapse of all buildings. Heavy damage and shaking extends to distant locations. Permanent changes in topography.

Source: US Geological Survey; https://en.wikipedia.org/wiki/Richter_magnitude_scale.

lubricating nearby faults that are already under tectonic stress, causing them to slip. Factors like fault orientation relative to regional stress, pore pressure, and rock properties all play a part. Faults not well aligned with crustal stresses may not move even if large volumes are injected. Faults that are primed to slip can trigger earthquakes with small injected volumes. Recent studies suggest that reducing wastewater injection can reduce seismicity.

Earthquakes also occur as reservoirs are filled behind man-made dams. This **reservoir-induced seismicity** (RIS) is caused by the weight of the reservoir forcing water under pressure into cracks in the rocks beneath the lake. As with wastewater injection, this lubricates preexisting faults, causing them to slip.

The largest and most damaging of these was the magnitude 7.9 Sichuan quake that occurred in 2008. Some experts believe that this quake was triggered by filling the Zipingpu dam. Factors that influence seismicity include the depth and volume of the lake impounded by a dam and the presence of faults in the immediate vicinity of the reservoir. While no dams have failed (yet) because of RIS, several have come close.

TABLE 12.2

The Modified Mercalli Earthquake Intensity Scale

Intensity	Shaking	Description/Damage
I	Not felt	Not felt except by a very few under especially favorable conditions.
II	Weak	Felt only by a few persons at rest, especially on upper floors of buildings.
III	Weak	Felt quite noticeably by persons indoors, especially on upper floors of buildings. Many people do not recognize it as an earthquake. Standing motor cars may rock slightly. Vibrations similar to the passing of a truck. Duration estimated.
IV	Light	Felt indoors by many, outdoors by few during the day. At night, some awakened. Dishes, windows, doors disturbed; walls make cracking sound. Sensation like heavy truck striking building. Standing motor cars rocked noticeably.
V	Moderate	Felt by nearly everyone; many awakened. Some dishes, windows broken. Unstable objects overturned. Pendulum clocks may stop.
VI	Strong	Felt by all, many frightened. Some heavy furniture moved; a few instances of fallen plaster. Damage slight.
VII	Very strong	Damage negligible in buildings of good design and construction; slight to moderate in well-built ordinary structures; considerable damage in poorly built or badly designed structures; some chimneys broken.
VIII	Severe	Damage slight in specially designed structures; considerable damage in ordinary substantial buildings with partial collapse. Damage great in poorly built structures. Fall of chimneys, factory stacks, columns, monuments, walls. Heavy furniture overturned.
IX	Violent	Damage considerable in specially designed structures; well-designed frame structures thrown out of plumb. Damage great in substantial buildings, with partial collapse. Buildings shifted off foundations.
X	Extreme	Some well-built wooden structures destroyed; most masonry and frame structures destroyed with foundations. Rails bent.

Source: US Geological Survey; http://earthquake.usgs.gov/learn/topics/mercalli.php.

FIGURE 12.8
Liquefaction caused by the Niigata Earthquake, 1964. (From Japan National Committee on Earthquake Engineering, 1964; https://commons.wikimedia.org/wiki/File:Liquefaction_at_Niigata.JPG.)

Living in Earthquake Country

In summary, then, the earth will not open up and swallow villages; it is violent shaking that causes damage. So what can be done? First, avoid building on or near an active fault and on landfill or unstable (steep or waterlogged) soil. How do you recognize a fault? Many areas have fault maps that show where they are. See the "Sources of Earthquake and Tsunami Information" at the end of this section. If there are no maps, look for abrupt, right-angle bends and offsets of streams and valleys, alignments of springs, and natural escarpments (Figure 12.9). In developed areas, one can sometimes see offset fences and roads with dips that have been patched as the fault continues to creep (Figures 12.10 and 12.11). Landfill is shown on many hazard maps: avoid landfill because it can compact or liquefy. Check homes for cracked foundations.

FIGURE 12.9
Offset fence and escarpment along the San Andreas Fault as a result of the 1906 San Francisco earthquake south of Skinner Ranch barn. (Courtesy of J. C. Branner; http://earthquake.usgs.gov/earthquakes/states/events /1906_04_18_cow.php.)

FIGURE 12.10
Surface faulting in 1940 Imperial Valley earthquake offset (6 m) regular rows of orange trees. Fault displacement along this section of the Imperial fault was confined to a narrow zone. (Courtesy of William L. Ellsworth, US Geological Survey, 1991.)

FIGURE 12.11
View southwest at Long Point fault in Houston, Texas. The normal fault runs from the camera to a point to the left of the kiosk in the parking lot (dotted line). The upthrown side is to the right, and is marked by tire smudges on the road. This fault moves slowly but almost continuously. (Courtesy of Djmaschek; https://en.wikipedia.org /wiki/Long_Point%E2%80%93Eureka_Heights_fault_system#/media/File:Long_Point_Fault_SW_Houston.JPG.)

Buildings in earthquake-prone areas should use steel reinforcing to brace and strengthen all structures over two stories. Tall buildings should be built on pilings that go down to bedrock, or on shock-absorbing foundations that dampen vertical and lateral wave motion. Use steel rebar in masonry structures. Use shear panels in wood structures, and bolt structures to a foundation or bedrock. Secure bookcases and tall furniture to walls, brace chimneys, and tie down water heaters and gas appliances so they don't topple. Put latches on cabinets to prevent contents from spilling out.

Should you live in an active quake area, preparation is key. Once shaking starts, get clear of buildings or, if you cannot, take shelter under heavy furniture such as a table or desk and stay away from exterior walls, windows, and ceiling fixtures. Do not get into elevators. If you are outside, avoid falling debris and fallen power lines. If in a vehicle, pull off the road and stop, staying in the vehicle until shaking stops. Avoid bridges, overpasses, and fallen power lines. Anticipate that after a large event you will be cut off from outside help for several weeks, so have water and canned food in storage. Have warm clothing and vital medicine in a "go bag," sleep with flashlight and shoes by the bed in case of power outage and broken glass, and have a battery-operated radio and batteries for flashlights and lamps. Know where and how to turn off natural gas to your home. Put out any fires that may have started. Have a backup plan to communicate with and meet family members in case phone lines are tied up or not working. Be prepared for aftershocks.

Landslides can be triggered by earthquakes. If you live on a steep slope, be alert for falling debris and landslides and move to safer ground.

Tsunamis are triggered by earthquakes. If you are at elevations within 30 m (100 ft) of sea level and near the coast, move to higher ground and at least 3 km (2 mi) from the coast. Be aware of warning signs such as earthquakes lasting more than 20 s and seawater moving away from shore. Some coastal communities have tsunami warning sirens. You may

FIGURE 12.12

A view of tsunami deposits at the village of Lampuuk, Indonesia; 73 cm (30 in) of sand was deposited over soil (dark material in bottom center) in beds indicating multiple pulses of sand deposition. The lowermost 45 cm (19 in) of sand settled out of suspension in a single wave. (Courtesy of G. Gelfenbaum, US Geological Survey; http://soundwaves.usgs.gov/2005/03/.)

have less than 10–15 min warning before the wave hits the coast. Know evacuation routes from all low-lying areas.

Seawalls are the traditional defense against tsunamis. The wall has to be high enough to provide protection, and determining this height is critical. Most tsunamis are 3 m (10 ft) high or less, but at Fukushima, the 10-m (30-ft) seawall was just a speed bump for the 16-m (52-ft) wall of water that breached the reactors and reached as high as 39 m (128 ft) at Miyako city. The tsunami that washed over Banda Aceh in 2004 ranged from 15 to 30 m (50 to 100 ft) along a 100-km stretch of the Sumatra coast. The largest tsunami ever recorded was 524 m (1720 ft) high in Lituya Bay, Alaska. This enormous wave was generated by a large landslide at the end of a long, narrow bay.

We determine the height of expected tsunamis by mapping ancient tsunami deposits (Figure 12.12). If these deposits are found 50 m above sea level, this gives an indication of the height of tsunamis that have affected an area.

Sources of Earthquake and Tsunami Information

A National Seismic Hazard Map (https://earthquake.usgs.gov/hazards/hazmaps/) has been prepared by the US Geological Survey, and similar maps have been prepared by various state and national geological surveys to keep the public informed about areas with seismic risks.

The National Earthquake Information Center in Golden, Colorado (http://earthquake.usgs.gov/earthquakes/?source=sitenav), monitors and provides information to the general public regarding earthquakes.

The US Geological Survey is an important source of information on earthquake hazards:

http://earthquake.usgs.gov/learn/preparedness.php

The Department of Homeland Security has a website that is helpful:

https://www.ready.gov/earthquakes

Help preparing an earthquake survival kit can be found at the Red Cross and the Centers for Disease Control and Prevention (CDC):

http://www.redcross.org/get-help/prepare-for-emergencies/be-red-cross-ready/get-a-kit

https://www.cdc.gov/disasters/earthquakes/

The American Red Cross has a web page on earthquake preparedness:

http://www.redcross.org/prepare/disaster/earthquake

and for tsunami preparedness:

http://www.redcross.org/get-help/prepare-for-emergencies/types-of-emergencies/tsunami

The National Oceanic and Atmospheric Administration (NOAA) has a website dedicated to tsunami education:

http://www.tsunami.noaa.gov/

Landslides and Downslope Mass Movements

On March 22, 2014, a large landslide occurred just outside of Oso, Washington. An unstable hill collapsed, sending mud and debris across the North Fork of the Stillaguamish River, covering an area of approximately 2.6 km^2 (1 mi^2) and engulfing a rural neighborhood. Forty-three people were killed (Figure 12.13). This was an area of known previous slides (Figure 12.14).

The Vajont Dam was completed in 1959 in the valley of the Vajont River near Longarone, Italy, 100 km (60 mi) north of Venice. On October 9, 1963, while the reservoir was still being filled, a massive landslide caused a tsunami in the lake. Fifty million cubic meters of water overtopped the dam in a 250-m (820-ft) wave, leading to the complete destruction of several villages and towns and 1917 deaths in the Piave valley below the dam. This event occurred because the utility company and the Italian government dismissed evidence of previous and ongoing landslides on the south side of the reservoir. Numerous danger signs had been disregarded, and an attempt to safely control the landslide by lowering the lake level came too late to prevent the slide (Figure 12.15).

At 4:10 a.m. on April 29, 1903, a rockslide buried the east side of Frank, Alberta, under 90 million tons of rock and debris (Figure 12.16). Within 100 s, 90 residents of this coal mining town were buried alive. Mining operations may have weakened the mountainside, but the unstable slope was no doubt aggravated by a wet winter.

Landslides are sudden, rapid downslope movements of rock and soil (Figure 12.17). They include **rock falls** (usually along a cliff) and **debris flows** (slurries of saturated soil, mud, and rock). Landslides are responsible for great loss of life and destruction of property (Table 12.3). **Slumps** are slow-motion downslope movement, usually over periods of days or weeks. **Creep** is the gradual movement of soil downslope over long periods of

FIGURE 12.13
Aerial view of the Oso slide. (Courtesy of Washington Army National Guard, Spc. Matthew Sissel, Spc. Samantha Ciaramitaro; https://commons.wikimedia.org/wiki/File:Oso_Mudslide_29_March_2014_aerial_view_3.jpg.)

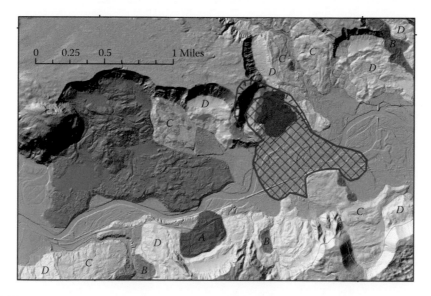

FIGURE 12.14
Oso landslide map by R. Haugerud, 2014. Red crosshatched area is the Oso slide; letters and colors denote previous slides (http://pubs.usgs.gov/of/2014/1065/pdf/ofr2014-1065.pdf).

time. Gravity is the ultimate driver, but the actual event may require a trigger such as an earthquake or a rainstorm.

Landslides are related to slope steepness, rock stability, and soil friction and cohesion. **Rock stability** is affected by moisture, the extent to which the rock is cemented or broken, and whether there are internal planes of weakness (such as bedding, cleavage, or faults) aligned with the slope. **Soil cohesion** is mainly a function of moisture

FIGURE 12.15
Aerial view of the Vajont dam and slide shortly after the disaster. The slide came down the hill on the right
(https://en.wikipedia.org/wiki/Vajont_Dam).

FIGURE 12.16
Aerial view of the Frank slide, Alberta. (Courtesy of Marek Ślusarczyk; https://en.wikipedia.org/wiki/List
_of_landslides.)

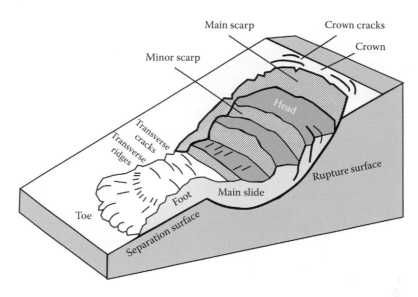

FIGURE 12.17
Parts of a landslide.

TABLE 12.3

Some of the Most Destructive Landslides of the Past 100 Years

Slide	Date	Number Killed	Triggering Event
Haiyuan, China	Dec. 16, 1920	>100,000	Haiyuan earthquake caused landslides
Vajont, Italy	Oct. 9, 1963	~2000	Unstable slope above a reservoir
Yungay, Peru	May 31, 1970	>22,000	Ancash earthquake. Slide buried Yungay
Armero, Colombia	Nov. 13, 1985	~23,000	Eruption of Nevado del Ruiz caused melting of its ice cap and mudflows
Vargas, Venezuela	Dec. 14–16, 1999	~30,000	Heavy rainstorm dropped 911 mm in few days causing deadly slides
Kedarnath, India	June 16, 2013	~5700	Flooding in northern India and resulting landslides

Source: Taken from https://en.wikipedia.org/wiki/List_of_landslides.

content, but can also be influenced by clay content, grain size, and whether the constituent grains are rounded or angular. Absence of vegetation means there are no roots to hold soil together. Erosion by wave action or rivers can cut into the toe of a slope and destabilize it. Man-made actions such as cutting trees and clearing brush, cutting into hillsides for roads and other developments, loading the top of a slope, and blasting can destabilize slopes.

Areas prone to landslides include areas with existing landslides, steep slopes, clay-rich rock or soil subject to a sudden increase in moisture, hillsides with internal planes of weakness parallel to the slope, and hillsides where development has added weight to the top of a slope or cut away material from the toe of a slope.

Living in Landslide Country

Warning signs of imminent slope failure include seeps or saturated ground in areas that have not been wet before. New cracks or elongated depressions appearing above the slope and movement of soil or fill material away from foundations are not good. Tilting of buildings, trees, powerlines, fence posts, or telephone poles in a downslope direction is a sign of downslope movement. Broken water lines or other underground utilities, cracked building foundations, offset fence lines, recent dips in roadbeds, and sticking doors and windows are all indications of slope movement (Figures 12.18 through 12.21). Low rumbling sounds, the sound of trees cracking, or boulders knocking together means it is time to leave. Quickly.

Avoid areas with previous slides. There are ways to increase the stability of slopes. Avoid cutting into the toe of slope. Decrease the load at the top of the slope, or decrease the slope angle. Install horizontal drain pipes or vertical wells to dewater the slope; plant trees, shrubs, and grasses to hold soil in place; divert moisture by using diversion ditches; cover the slope with water-repellant material such as plastic sheets; rock bolt broken rock slopes; and install retaining walls (Figures 12.22 and 12.23). High-pressure injection of grout, a cementing material, is sometimes used.

What can you do to avoid landslides? Check county hazard maps or get an engineering assessment before buying a property. Avoid areas with existing slides. Anywhere above or below a steep slope may be affected by landslides, particularly if there are signs that slips have occurred before. If your home is in a slide area, watch for signs of soil movement or

FIGURE 12.18
Gunn Lake incipient landslide, South Island, New Zealand. Arrows point to the main scarp. This slope may fail without further warning. (Courtesy of G. Prost.)

FIGURE 12.19
New dip in a road surface indicating downslope movement. (Courtesy of Burns, Hardin, and Andrew, 2008; http://www.oregongeology.org/sub/landslide/homeowners-landslide-guide.pdf.)

FIGURE 12.20
Cracked foundation resulting from downslope movement of soil. (Courtesy of Basement Questions; http://www.basementquestions.com/fdnmaintenance.php.)

FIGURE 12.21
These trees are bent downslope at the base and are vertical toward the top. The bending is evidence of downslope creep. (Courtesy of Western Washington University, Bellingham, WA; https://en.wikipedia.org/wiki /Downhill_creep.)

shifting and cracking of the foundation. Have flexible fittings on gas and water pipes. Plant ground cover to hold the soil in place, and install drainage pipes to keep soil moisture low. Consider building retaining walls and rock bolting or injecting cement into steep slopes.

Sources of Landslide Information

USGS Landslide Hazards Warning Program:

http://landslides.usgs.gov/learn/prepare.php

Red Cross Landslide Preparedness website:

http://www.redcross.org/prepare/disaster/landslide

The Federal Emergency Management Agency (FEMA) has a homeowner's guide to landslides:

http://www.oregongeology.org/sub/landslide/homeowners-landslide-guide.pdf

A good site for landslide education:

http://www.nature.com/scitable/topicpage/lesson-8-landslides-hazards-8704578

FIGURE 12.22
Retaining wall used to stabilize slopes. Note weep holes at base to allow drainage. (Courtesy of Eurico Zimbres; https://commons.wikimedia.org/wiki/File:Concrete_wall_drainage.jpg.)

Volcanoes

August 24, 79 CE, dawned normally in the Roman wine-growing region around Pompeii and Herculaneum. Then, around 1:00 p.m., the nearby mountain exploded. A column of ash rose into the sky above **Mt. Vesuvius** and then blanketed the surrounding countryside (Figure 12.24). People began to leave the area. Pliny the Younger described the eruption: *"Broad sheets of flame were lighting up many parts of Vesuvius; their light and brightness were the more vivid for the darkness of the night… it was daylight now elsewhere in the world, but there the darkness was darker and thicker than any night."*

Worse was to come. Early on the morning of August 25, pyroclastic flows raced down the slopes toward these towns, incinerating or suffocating all those who had not left (Figure 12.25). The remains of 1500 people have been found, but many more probably died.

Known as a **nuée ardente** (literally "burning cloud"), **pyroclastic flow**, or **ignimbrite**, these are fast-moving clouds of superheated gas, ash, and rock fragments ejected violently during an eruption (Figure 12.26). They are the most deadly type of eruption. In addition to the gas cloud, Vesuvius erupted lava flows and ejected tons of airborne pumice, ash, and lava bombs that buried the two cities.

(a)

(b)

FIGURE 12.23
(a) Fencing helps stabilize steep, broken rock of the Devonian Fredeburger Schiefer in a road cut in the Rothaar Mountains, Westphalia, Germany. (b) Rock bolt and fencing. (Courtesy of US Geological Survey; https:// en.wikipedia.org/wiki/Rock_bolt.) (Courtesy of Elop; https://commons.wikimedia.org/wiki/File:Fredeburger _Schiefer_bei_Silbach_(Elop).jpg.)

By the evening of the 25th, the mountain was quiet and the towns were gone, buried under several meters of ash and rock and erased from memory. They would not be uncovered for another 1700 years. Vesuvius, of course, remains active and overlooks the modern city of Naples (Figure 12.27).

A similar disaster befell the Caribbean island of Martinique in 1902. The town of St. Pierre is dominated by volcanic **Mt. Pelee**. Before the spring of 1902, this town was known as the "Paris of the West Indies." On April 23, there were some minor explosions at the summit of the volcano that were largely ignored in the town below. This was followed over the next few days by small earthquakes and ash eruptions that drove hordes of ants and poisonous snakes into town. As many as 50 people died of snake bites. By May 5, the summit lake was boiling when the crater rim collapsed, sending torrents of scalding water down the mountainside. The water mixed with ash to form a **lahar**, a muddy debris flow that buried 23 people and caused a 3-m tsunami. The wave flooded low-lying areas along the St. Pierre waterfront. Still, people remained in town.

On the morning of May 8, a nuée ardente rushed down the mountain and enveloped the town (Figure 12.28). In less than a minute, the town was gone, destroyed by searing heat and hurricane force winds (Figure 12.29). A 3-ton statue was moved 16 m, and ships in the harbor were set ablaze. There were 2 survivors out of a population of 28,000.

A volcanic fissure in southern Iceland called **Lakagigar**, or **Laki**, erupted for 8 months between June 1783 and February 1784. The fissure, along with the adjacent Grímsvötn volcano, emitted an estimated 14 km^3 (3.4 mi^3) of basalt, 8 million tons of hydrofluoric acid vapor, and 120 million tons of sulfur dioxide gas, creating what became known as the "**Laki haze**." These gases and aerosols rose 15 km (10 mi) into the atmosphere where they spread over Europe and the Middle East. On Iceland, the "móðuharðindin" (mist hardships) killed 80% of the sheep, 50% of the cattle, and 50% of the horses on the island. Between 20% and 25% of the population died in the resulting famine.

The deadly Laki fogs extended from Ireland to Prague, causing crop failures and famine. The sun was blood colored; the "dry fog" was so thick that boats were unable to navigate (Figure 12.30). Benjamin Franklin wrote of "a constant fog over all Europe, and a great part of North America." While the summer of 1783 was the hottest on record, the erupted aerosols blocked incoming solar radiation, dropping global temperatures and causing the severe winter of 1783–1784. Sulfur dioxide reacted in people's lungs to create sulfurous acid, causing them to slowly choke to death. An estimated 23,000 people were poisoned in Britain alone, while the extremely cold winter killed another 8000. The naturalist Gilbert White reported 28 days of continuous frost in Hampshire in southern England, and the Rev John Cullum wrote to the Royal Society that barley "became brown and withered … as did the leaves of the oats; the rye had the appearance of being mildewed." The extreme weather in Europe lasted several years, causing crop failures and famine that increased poverty and, some speculate, leading to the French revolution in 1789. Weak monsoons caused droughts and famine in Egypt, Arabia, and India. In North America, the winter of 1784 was one of the coldest on record, with the longest freezing of Chesapeake Bay, the Mississippi River freezing at New Orleans, and ice on the Gulf of Mexico. Globally, the eruption and its effects are estimated to have killed upward of 6 million people, making it one of the deadliest in history.

Three centuries later, another Icelandic volcano, Eyjafjallajökull, put fine ash into the atmosphere that led to the shutdown of air traffic over Europe and the North Atlantic for 8 days in 2010. More than 107,000 flights were canceled, costing the airline industry over $1.7 billion and stranding millions of passengers.

FIGURE 12.24
The last major eruption of Mt. Vesuvius in 1944. (Courtesy of John Reinhardt, B24 tail gunner, US Army Air Force; https://en.wikipedia.org/wiki/Mount_Vesuvius.)

The eruption of **Mount Tambora** on the Indonesian island of Sumbawa in 1815 was one of the largest eruptions in recorded history. The Volcanic Explosivity Index (VEI) describes the amount of erupted material, with each level 10 times the volume of the previous level. Tambora is classified a VEI-7, with roughly 100 km³ ejected. A caldera 7 km (4.3 mi) across and 700 m (2300 ft) deep was formed.

Tambora became active in 1812, generating tremors and ash clouds. On April 5, 1815, the mountain exploded with thunderous booms heard as far away as Batavia (now Jakarta) 1260 km (780 mi) away, and Sumatra more than 2600 km (1600 mi) away (Figure 12.31). Pyroclastic flows wiped out the village of Tambora at the base of the mountain and extended at least 20 km (12 mi) from the summit. All vegetation on the island was destroyed. Trees and pumice washed into the sea and formed rafts up to 5 km (3.1 mi) across. A tsunami up to 4 m (13 ft) hit the Indonesian islands. An ash column reached the stratosphere, extending over 43 km (27 mi) in elevation. High-level winds spread aerosols around the globe, creating brilliantly colored sunsets in 1815 and 1816. Between 10 and

FIGURE 12.25
Plaster casts of victims of the eruption of Vesuvius in 79 CE. (Courtesy of Lancevortex; https://en.wikipedia
.org/wiki/Pompeii.)

120 million tons of sulfur dioxide were put into the atmosphere. The spring and summer of 1815 saw a persistent "dry fog" in the northeastern United States that reddened and dimmed sunlight. Neither wind nor rainfall dispersed the fog. It has since been identified as a sulfate aerosol.

Atmospheric particulates from Tambora lowered global temperatures, leading to global cooling and worldwide harvest failures in 1816, the "Year without a Summer." Average global temperature dropped 0.5°C. On June 4, 1816, frosts were reported in Connecticut, and on the following day, most of New England was freezing. On June 6, 1816, snow fell in Albany, New York. The cold lasted at least 3 months and ruined most crops in North America. Thirty centimeters (12 in) of snow fell near Quebec City on June 6 to 10, 1816. Europe suffered cooler summers and stormier winters. The cooler weather weakened the Indian monsoons, leading to an outbreak of cholera in the Bay of Bengal in 1816. Cholera spread over much of Asia and Europe in the coming years. Lt. Philips, an eyewitness reporting to the Lieutenant Governor of Java, stated "Since the eruption, a violent diarrhoea has prevailed in Bima, Dompo, and Sang'ir [Java],

FIGURE 12.26
1984 nuée ardente at Mayon Volcano, Philippines. (Courtesy of C. G. Newhall, US Geological Survey; https://en.wikipedia.org/wiki/Pyroclastic_flow.)

FIGURE 12.27
Vesuvius forms the backdrop for the modern city of Naples. (Courtesy of Adelle and Rick Palmer.)

which has carried off a great number of people. It is supposed by the natives to have been caused by drinking water which has been impregnated with ashes; and horses have also died, in great numbers, from a similar complaint." Failure of wheat, oat, and potato harvests led to famine and riots in Europe and south Asia. Swiss glaciers actually advanced during 1816 and 1817, giving rise to the theory of ice ages championed by local Swiss engineer Ignace Venetz and naturalist Louis Agassiz. Multiyear failures

FIGURE 12.28
Nuée ardente moving down the flank of Mt. Pelee. (From Heilprin, 1908. The Eruption of Mt. Pelee: Philadelphia Geographic Society; http://www.geology.sdsu.edu/how_volcanoes_work/Pelee.html.)

of the rice crop in Yunnan led to a switch to opium poppies in what is now called the "golden triangle," a key source of world opium.

The estimated number of deaths associated with Tambora varies depending on what is included. The number of direct deaths caused by pyroclastic flows on Sumbawa Island was around 10,000; about 38,000 deaths were attributed to starvation, and 10,000 deaths were a result of disease and starvation on Lombok Island. The total number of direct and indirect deaths in Indonesia may have approached 100,000.

Another deadly effect of volcanoes is silent, invisible, and odorless. On August 21, 1986, 1700 people and 3500 livestock were discovered dead in villages on the flank of an inactive volcano northwest of Yaondé, Cameroon. It took some time to figure out what happened, as there were no signs of violence. The villages lay in topographic depressions near **Lake Nyos**, a small volcanic crater lake. After several years of study, it was determined that the volcano is leaking carbon dioxide into the lake waters. Turnover of the stratified waters in this lake does not occur periodically and carbon dioxide levels can get quite high. For some reason, perhaps an underwater landslide or gradual heating from magma below the lake, the lake burped and released a large amount of gas. Carbon dioxide, a colorless and odorless gas that is heavier than air, flowed out of the lake and ponded in depressions where the villages lay. All living things within 25 km (15 mi) of

FIGURE 12.29
The remains of St. Pierre after the eruption of Mt. Pelee on May 8, 1902. (From Heilprin, 1908. The Eruption of Mt. Pelee: Philadelphia Geographic Society; http://www.geology.sdsu.edu/how_volcanoes_work/Pelee.html.)

FIGURE 12.30
Volcanic haze seen from atop the Eiffel Tower is related to the eruption of Eyjafjallajökull in Iceland on April 17, 2010. (Courtesy of US Geological Survey; https://volcanoes.usgs.gov/publications/2010/20100417eiffel.jpg.)

FIGURE 12.31
Painting of the eruption of Tambora, Indonesia, in 1815. (Artist unknown; https://learnodo-newtonic.com/mount-tambora-facts.)

the lake were killed (Figure 12.32). This is the first documented large-scale asphyxiation caused by a volcanic event. Since that time, tubes have been installed to siphon water from the bottom layers of the lake to allow the carbon dioxide to leak off slowly and safely.

What do all these disasters have in common? First, people settled and built farms on the slopes of volcanoes or in the path of volcanic flows. Second, because they were invested in their

FIGURE 12.32
Dead livestock near Lake Nyos, Cameroon, August 1986. (Courtesy of US Geological Survey; http://www.atlas obscura.com/places/lake-nyos-the-deadliest-lake-in-the-world.)

properties, the inhabitants chose to ignore warning signs of imminent eruptions. This was at least partly because of many false alarms, but ignoring the signs led to wholesale destruction of life and property. Third, the larger eruptions had global consequences: they affected weather patterns, crop yields, and the spread of disease and famine; they interrupted transportation; promoted revolutions; and gave birth to new scientific theories including that of the ice ages.

Living in Volcano Country

Volcanoes affect people when they explode, when they erupt lava, and when they spew ash and gas. Ash is erosive to machinery, notably jet engines, is a breathing hazard, and, in extreme cases, can collapse buildings. Lava flows threaten structures, but people usually have time to escape. Pyroclastic flows and lahars rush down valleys, destroying everything in their path. Floods caused by eruptions under glaciers have washed out roads and bridges in Iceland. Eruptions can trigger landslides and tsunamis, as when Unzen erupted in Japan in 1792. Killer gases such as carbon dioxide and hydrogen sulfide have asphyxiated livestock and people and cause lung damage. Volcanic explosions and the ignimbrites and lahars they generate are the deadliest. Even volcanic bombs can kill. The eruption of Mount St. Helens lobbed a large rock onto a vehicle 16 km (10 mi) from the vent, killing the driver. Many of the world's largest cities sit at the base of explosive volcanoes. Others lie in the path of lava and debris flows.

The Pacific **"ring of fire"** describes the nearly continuous chain of volcanoes that surround the Pacific Ocean from the tip of South America to Alaska, through Kamchatka to Japan and Indonesia. These volcanoes are all a result of subduction and melting of oceanic tectonic plates around the margins of the Pacific. **Mount Fuji** and its predecessors in Japan have been active for the past 700,000 years. The most recent series of eruptions from Mount Fuji ended in 1707. The last eruption deposited ash a far as the town of Edo (now known as Tokyo) 100 km (60 mi) away.

Geologists determine the activity of volcanoes by measuring gas emissions (rate and composition), monitoring seismic activity at and near the volcano, and measuring deformation at the surface (swelling and tilting, indicating near-surface magma movement). Seismic tomography indicates that magma under Mount Fuji today is at depths around 10 km (6 mi). The internal pressure of the magma chamber suggests that an eruption may be imminent. Steam and gas are pouring from the crater, and hot gas and water are discharging at **fumaroles** (vents). In the past few years, many earthquakes occurred beneath the mountain. Earthquakes, mudslides, lahars, and ash falls from Mount Fuji threaten roughly 8 million people living in the Tokyo area (Figure 12.33).

On the other side of the Pacific is **Mount Rainier**, the tallest volcano in the Cascade Range. This volcano looms over the 3.6 million inhabitants of the Seattle–Tacoma metropolitan area 100 km (60 mi) away. It last erupted in 1894–1895. Volcanic hazards in this area include earthquakes, pyroclastic flows, and ash falls, but the event feared most by experts is eruption-triggered lahars. These are debris flows caused by instantaneous melting of the glaciers capping the mountain.

Lahars can travel tens of kilometers at up to 80 km/h (50 mph) and have the consistency of wet concrete. In the past, they left deposits 30 m (100 ft) thick in narrow valleys on the slopes of Mount Rainier. They slow down, thin, and spread out in the wide lowland valleys. Over the past several thousand years, large lahars have reached the Puget Sound lowlands an average of once every 500 to 1000 years. The west flank of Mount Rainier has the highest risk of lahars: they threaten the Puyallup and Nisqually River valleys and as many as 80,000 people living downstream.

FIGURE 12.33
Mt. Fuji viewed from Tokyo. (Courtesy of Morio; https://commons.wikimedia.org/wiki/File:Skyscrapers_of_Shinjuku_2009_January.jpg.)

The USGS and Pacific Northwest Seismic Network at the University of Washington constantly monitor seismic activity near Mount Rainier. The USGS and Washington State Emergency Management Division have lahar warning systems comprising arrays of vibration monitors. These early warning systems issue automatic alerts to agencies that can then respond appropriately. In a worst-case scenario, the residents of these valleys would have only 40 min warning (transmitted by siren, Internet, radio, and television) to get to higher ground. Moving to high ground is the only way to escape these flows. Individuals should be alert to the warning signs of an approaching lahar: shaking ground and loud rumbling or roaring sounds. Success would depend on effective notification, prompt response, and prearranged evacuation routes.

Lava flows regularly threaten homes and villages in Sicily, Hawaii, and Iceland. These rivers of 1000°C (1832°F) molten rock move slowly but inexorably down valleys. If you live in the path of a lava flow, can you stop it?

In 1669, the residents of Catania armed themselves with shovels and picks and built a barrier to redirect a lava flow from **Mount Etna**. Mt. Etna, on the island of Sicily, has erupted throughout recorded history. This time, they would do something about it. But their neighbors in Paterno didn't like the fact that the lava would be directed toward them. So they breached the barrier and a large part of Catania was destroyed by the flow. Fast-forward to 1993: Etna is erupting again and the lava is flowing toward the town of Zafferana. Workers built barriers and trenches to stop the lava and direct it away from the town. When that proved only partly successful, they dropped concrete blocks into the flow, fully diverting it. Etna is still active today (Figure 12.34).

A successful attempt to stop a lava flow occurred on the island of Heimey off the coast of Iceland in 1973–1974. Lava from the **Eldfell** volcano threatened the town of Vestmannaeyiar. For almost 5 months, seawater was pumped through fire hoses onto the advancing lava to cool it (Figures 12.35 and 12.36). When lava cools, it becomes more viscous, slowing the flow. About 80% of the town was saved along with the harbor. The combination of slow-moving lava and an inexhaustible supply of water probably saved the town.

FIGURE 12.34
Mt. Etna on the island of Sicily continues to spew steam and lava. (Courtesy of Adelle and Rick Palmer.)

FIGURE 12.35
View to the south from Vestmannaeyjar's outer harbor on May 4, 1973. Seawater is being sprayed directly onto the lava flow front to arrest infilling of the harbor entrance. (From Williams and Moore, 1983. *Man against Volcano: The Eruption on Heimaey, Vestmannaeyjar, Iceland*. US Geological Survey General Interest Publication, 33 pp.)

FIGURE 12.36

The ships *Sandey* and *Lóðsinn* pumping seawater onto the forward margin of the lava at the breakwater in Vestmannaeyjar, Iceland, during May of 1973. (Courtesy of Sigurgeir Jónasson and US Geological Survey; http://pubs.usgs.gov/of/1997/of97-724/lavaoperations.html.)

Some "Dormant" Volcanoes Aren't

Human understanding is limited by our experience. If we don't see a volcano erupting, we assume it is long dead. This is not necessarily the case. Looking at a volcano or lava flow can tell you if it was recently active. If trees are growing on it, it has been several hundreds to thousands of years since it last erupted. If, on the other hand, you see fresh rock with no trees, shrubs, or grass, it was probably active in the last couple of hundred years.

California is not usually thought of as volcanically active. The last eruption was at **Mount Lassen**, the southernmost volcano of the Cascade Range, during 1914–1917. Before that, the last eruption was 1100 years ago. The area around Lassen Peak still has fumaroles and hot springs. But what about the rest of California? What about North America?

Mono Lake in the Owens Valley of eastern California sits in a volcanic landscape. The last eruption in the area, at Putnam Crater, was only 700 years ago. The volcanoes in the area date back 40,000 years, suggesting that activity in this area is not over yet.

On the east side of the Sierra Nevada, near the resort town of Mammoth, is one of the largest calderas on Earth. The **Long Valley Caldera** measures 32 km (20 mi) long by 18 km (11 mi) wide. This eruptive center was formed 760,000 years ago by a massive explosion that covered 3900 km^2 (1500 mi^2) under lava and much of the western United States under ash. It is estimated that a total of 600 km^3 (140 mi^3) of material was ejected.

In 1980, an earthquake swarm hit the Mammoth area. The Long Valley Caldera was uplifted as much as 25 cm (10 in), and uplift continues. There are hot springs in the area, and new gas vents appeared on the surface (Figure 12.37). The USGS Volcano Hazards Program is monitoring the area, and an "escape road" has been added to the local road network.

FIGURE 12.37
Example of a fumarole near the top of Mount Baker, Washington. Geologists are sampling gases. (Courtesy of W. Chadwick, US Geological Survey; https://commons.wikimedia.org/wiki/File:Baker_Fumarole.jpg.)

Most of us don't think of Arizona as having recent volcanoes. Yet 4700 km² (1800 mi²) of northern Arizona are covered by lava flows and as many as 600 volcanoes of the **San Francisco volcanic field**. Humphreys Peak, a large stratovolcano above Flagstaff, is 3851 m (12,633 ft) high. Eruptions began about 6 million years ago. Given that the youngest, Sunset Crater, is less than 1000 years old, it would be fair to say that volcanic activity is not over in this area (Figure 12.38). Some geologists suspect there is a hot spot, or area of localized melting of the mantle, below northern Arizona. The USGS is monitoring the area and says that the small eruptions that are likely to occur in the future pose little threat because of the remoteness of the area.

Idaho is not usually considered volcanically active. Yet, southern Idaho along the Snake River contains lava flows that range in age from 16 million years to about 2000 years old (Figure 12.39). Craters of the Moon National Monument preserves about 1600 km² (618 mi²) of spectacular volcanic features including flows and cinder cones (Figure 12.40). Tectonic extension across the region releases pressure at depth, allowing magma to move to the surface. Volcanic events have averaged 2000 years between eruptions, and the last event was about 2000 years ago. Eruptions in this area are expected to be small flows or cinder and ash discharges.

Although the last full-scale eruption was 630,000 years ago, volcanic activity in Yellowstone National Park, Wyoming, is ongoing and recent. The ground literally vibrates with volcanic energy: geysers, fumaroles, hot springs, bulging of the ground surface, and earthquake swarms. The only question is whether another eruption is imminent. Between 2004 and 2008, the ground around the **Yellowstone Caldera** moved upward about 20 cm (8 in), suggesting magma was approaching the surface. By the end of 2009, however, uplift had slowed or stopped. Scientists from the USGS, the Park Service, and the University of Utah are monitoring the situation. The Yellowstone Volcano Observatory in December 2013 issued the following statement regarding danger in the area: "Although fascinating, the new findings do not imply increased geologic hazards at Yellowstone, and certainly do

FIGURE 12.38
SP Crater and associated lava flow (black), near Sunset Crater in the San Francisco volcanic field, Arizona.
(Courtesy of US Geological Survey; https://pubs.usgs.gov/fs/2001/fs017-01/.)

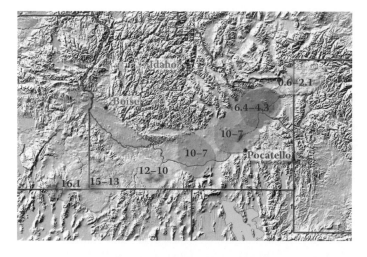

FIGURE 12.39
Age of volcanism in millions of years, Snake River plain to Yellowstone National Park, Idaho and Wyoming.
(Courtesy of Kelvin Case, National Park Service web posting; https://en.wikipedia.org/wiki/Yellowstone_Caldera.)

FIGURE 12.40
Recent volcanism, Craters of the Moon National Monument, Idaho. (Courtesy of G. Prost.)

not increase the chances of a 'supereruption' in the near future. Contrary to some media reports, Yellowstone is not 'overdue' for a supereruption."

Sources of Volcano Hazard and Safety Information

The Red Cross has a website that explains the risks and how to be prepared when living in volcano country:

http://www.redcross.org/prepare/disaster/volcano

The USGS Volcano Hazards Program monitors volcanic activity in the United States:

http://volcanoes.usgs.gov/index.html

The CDC website has information on emergency preparedness at:

http://emergency.cdc.gov/disasters/volcanoes/before.asp

The Department of Homeland Security also has information for the public:

https://www.ready.gov/volcanoes

Flooding

Mark Twain quipped about flooding of the Mississippi, *"The Mississippi River will always have its own way; no engineering skill can persuade it to do otherwise…"* Natural flooding falls into six categories: (1) annual runoff owing to snowmelt or rainy season; (2) flash flooding attributed to intense rainfall in a small area; (3) flooding caused by ice dams; (4) coastal flooding attributed to high tides, onshore winds, or a combination of the above along with rainfall; (5) storm surges; and (6) land subsidence as a result of dewatering and compaction along the coast. Flooding can also be caused by the collapse of dams because of earthquakes, landslides, or overtopping attributed to heavy rainfall and snowmelt.

River flooding of agricultural lowlands are among the worst natural disasters in history. The Central China Floods on the Yangtze River during July to August 1931 killed an estimated 3.7 million people by drowning, disease, and starvation and left 80 million more people homeless. In September 1887, the Yellow River overran its levees and dikes in China's Henan Province. The flood covered 130,000 km^2 (50,000 mi^2), leaving millions homeless and killing an estimated 900,000 to 2,000,000 people.

The worst dam collapse in history resulted from 122 cm (4 ft) of rain in 1 day along the Ru River watershed, China. A small dam burst, sending a wall of water downstream, destroying one dam after another. A total of 62 dams failed, and the resulting floods killed an estimated 230,000.

Along the coast, a combination of runoff, heavy rain, and storm surges can do a lot of damage. In the year 1099, high tides and storms were responsible for up to 100,000 deaths in coastal lowlands in Holland and England. A storm surge killed as many as 80,000 in Holland in 1287 after a dike collapsed. The city of Galveston, Texas, is built on a barrier island. The average elevation is 2 m (6.5 ft). The Galveston hurricane of 1900 brought a storm surge that moved over the barrier island. The water moving inland did damage, but the water returning to the sea was devastating, obliterating the city and sweeping between 6000 and 8000 people out to sea. More recently, Hurricane Sandy, which hit the northeast coast of the United States in 2012, brought with it a storm surge that flooded coastal areas of New York and New Jersey. Homes were flooded, tunnels and subways were under water, and power was cut throughout the area. As many as 233 people were killed in the eight countries along its path, and damage in the United States alone amounted to $71.4 billion.

Flooding of coastal lowlands is usually a slow process that results from subsidence owing to natural causes such as compaction of river delta sediments. Every so often, the flooding is rapid, as during Hurricane Katrina, which hit the area around New Orleans on the US Gulf Coast in 2005. Most of New Orleans is below sea level: it is protected from flooding of the Mississippi River by natural levees and man-made dikes built up over the centuries. It was especially prone to flooding once the levees failed because of the storm surge. At least 1400 people died as a direct cause of the flooding, as many as 2000 altogether, millions were left homeless, and damage has been estimated at $108 billion. However, New Orleans was just one area hit by the storm: the storm surge reached between 10 and 19 km (6 and 12 mi) inland, flooding 90% of coastal communities in Louisiana and Mississippi. The surge was up to 9 m (30 ft) high.

"There are two easy ways to die in the desert: thirst and drowning." So said Craig Childs in his book *Secret Knowledge of Water.* Flash floods happen rarely, but are impressive when they do. In 1976, a storm system stalled over the Big Thompson watershed in Colorado and dumped between 30 and 36 cm (12 and 14 in) of rain in a 4-h period. The water all converged on Big Thompson Canyon. The river, which usually runs 46 cm (18 in) deep, reached depths of 6.1 m (20 ft) and, moving at 9 m/s, displaced boulders weighing up to 250 tons. Homes were ripped away, camp sites were scoured, and 143 people lost their lives. In areas like the mountains or desert, where rain can fall quickly and tends to run off rather than soak into the ground, a storm in one area can cause flooding many kilometers downstream, especially in narrow canyons.

Living in Flood Country

The US government National Flood Insurance Program was designed to reduce the effects of flooding on public and private structures. It does this by providing affordable flood insurance and by encouraging responsible development in flood-prone areas. As part of this program, the FEMA had to define what areas are prone to flooding in any given period of time.

Thus, the concept of the **one-hundred-year flood**: a flood event that has a 1% probability of occurring in any given year. This kind of flood would theoretically occur on average once every 100 years. However, people buying homes in the 100-year-flood zone shortly after such a flood should not be fooled into thinking it won't happen again for a hundred years. It could, for a number of reasons. First, this is just a statistical average. Statistically, you could have two of these floods in 10 years and then none for the next 190 years. Another way to have more frequent large floods is to change the context. Development in a floodplain (roads, homes, and industry) changes runoff patterns such that what had been a hundred-year flood in farm country now occurs on average every 20 years in a suburban setting. The only way to prevent flood damage, as opposed to flooding, is to keep development out of the floodplain. Since that is not always practical, development should be controlled so that floodplains contain parks or structures that can withstand periodic flooding.

Communities try to prevent flooding by building dikes, by increasing the height of natural levees, and by building dams and holding basins for runoff. Clearly, these work in normal years. In years of unexpected weather, they do not. During the spring snow run-off of June 2013, a high-pressure weather system stalled over the city of Calgary, Alberta, Canada. The city has two flood control dams upstream on the Bow River and another dam on the Elbow River, and berms along the banks of the Bow River through town. The rivers normally are high with snowmelt at this time of year. After a week of rain, an additional 200 mm (7.9 in) fell in Calgary in less than 2 days. Upstream in Canmore, over 220 mm (8.7 in) fell in 36 h. In High River, 325 mm (12.8 in) was recorded in less than 2 days. Dams upstream were full and water had to be released to keep them from overtopping and failing. The Bow River flowed through Calgary at five times its normal level for this time of year (Figure 12.41). The Elbow and Highwood Rivers were flowing at 10 times their averages for this time of year. A total of 100,000 people were displaced from their homes in southern Alberta, losses exceeded $1.7 billion, and five deaths resulted from the flooding.

FIGURE 12.41
Sunnyside neighborhood under water. Flooding in Calgary, Alberta, in the spring of 2013. (Courtesy of G. Prost.)

Flash floods are caused by heavy rainfall, failure of a dam or levee, or by breaking of an ice or debris jam. Usually, there is little warning, so if you live in areas prone to flash floods, you should be prepared. Streets can become rivers, underpasses will flood, and basements and first floors will be inundated. A trickle of water in a dry riverbed can become a muddy torrent in minutes. Water levels can rise many meters, especially when in confined areas such as canyons. Walking, swimming, or driving through a flood can be fatal. Two-thirds meter (2 ft) of water can carry away a normal car. Fourteen centimeters (6 in) of fast-moving water can sweep you off your feet.

Know how high your property is and whether you are in a flood-prone area. Check local disaster maps for flood risk before you buy a home. If you are in a floodplain, elevate and reinforce your home. Consider building a berm around your home. Move electrical panels, furnace, and water heater to a second floor. A water heater may be your only source of freshwater after a flood. Install check valves or backflow preventers on sewer lines to prevent sewage from backing up into your home.

Have an evacuation plan and discuss it with your family. Listen to weather radio when rainfall is intense or long-lasting. If you see water start to rise, stop what you are doing and get to high ground.

Sources of Flood Information

Find out about flood preparedness and flood safety at:

http://www.redcross.org/get-help/prepare-for-emergencies/types-of-emergencies /flood

Learn about floods at the National Severe Storm Lab:

http://www.nssl.noaa.gov/education/svrwx101/floods/types/

Get weather information from the National Weather Service at:

http://www.weather.gov/

Learn what causes flooding at:

http://www.disastercenter.com/guide/flood.html

Subsidence, Sinkholes, and Expanding Soils

Natural subsidence is a slow process and a nuisance to property but rarely causes death. Sudden subsidence caused by collapsing sinkholes can be lethal.

Subsidence is caused either by compaction of near-surface soil and rock, or by dissolving away rock near the surface. In the first case, subsidence is likely to be slow and gradual and may only be noticed because of cracked foundations or ponding of surface water. This occurs in areas where soil and rock is slowly compacting and dewatering, usually on the shore of lakes or near wetlands along the coast. Developers may put in a layer of fill to raise the area before building, or may build on existing landfill as in the Millenium Tower example. However, the slow process of subsidence will continue. The coastal wetlands of Louisiana have been subsiding owing to natural compaction and dewatering of delta sediments since they were deposited by the Mississippi River. Mexico City was built on soft sediment of a lake bed and has been slowly subsiding for the past 500 years. The most famous example of a building affected by subsidence is the leaning tower of Pisa

(Figure 12.42). Pisa got its name from the Greek word for "marsh." The tower at the church of St. Michele, the bell tower of St. Nicola church, and the cathedral and baptistery in Pisa are also sinking.

Then there is the city of Venice. Venice, built on a series of islands within a coastal lagoon, has been sinking for centuries. The wet sediments have been compacted by the weight of building stone. More recently, pumping of groundwater has increased the rate of subsidence. Once this was realized, groundwater removal was stopped. It was thought that the subsidence had also stopped, but recent studies indicate that the city is again subsiding at a rate of 1 to 2 mm/year (0.04–0.08 in/year). In fact, the entire lagoon and all 117 islands are still sinking (Figure 12.43). This is attributed to a combination of continued compaction of sediments and subduction of the Adriatic tectonic plate (on which Venice sits) beneath the Apennines Mountains of the Eurasian plate.

Southeast Louisiana consists of a number of overlapping ancient deltas of the Mississippi River. As one delta is built out, the river shifts to an easier, shorter, or lower path, abandoning the old delta. The old delta sediments then begin to compact under their own weight, driving water out and sinking. As the land floods, vegetation changes from trees to grass and then dies off as the marsh is replaced by open water (Figure 12.44). Channeling of the

FIGURE 12.42
The leaning tower of Pisa. (Courtesy of Alkarex Malin äger; https://commons.wikimedia.org/wiki/File:Leaning_tower_of_pisa_2.jpg.)

FIGURE 12.43
Usually dry, at high water the Piazza San Marco, Venice, is submerged. (Courtesy of Wolfgang Moroder; https://commons.wikimedia.org/wiki/File:Acqua_alta_in_Piazza_San_Marco-original.jpg.)

FIGURE 12.44
An aerial view of subsiding marshes in the Mississippi River Delta. (Courtesy of US Geological Survey; http://www.usgs.gov/blogs/features/2011/03/28/celebrate-american-wetlands-month-and-wade-into-usgs-wetlands-research/.)

Mississippi by building levees to prevent flooding and maintain navigation within the present river channel has only enhanced the process of subsidence in the surrounding marshes. Man-made levees prevent the coastal wetlands from receiving seasonal loads of sediment that would slow the subsidence process. The city of New Orleans was originally built on a natural river levee. That levee has been built up by the Army Corps of Engineers to keep the Mississippi within its banks while the entire area has been subsiding as a result of compaction and dewatering. At this time, the city is completely surrounded by levees and the average elevation is half a meter (between 1 and 2 ft) below sea level, with some parts as low as 2 m (7 ft) below sea level.

When subsidence is caused by dissolving of near-surface rock the results can be unexpected and fatal. One night in March 2013, Jeffrey Bush had gone to bed in his suburban Tampa home. Suddenly and without warning, the bedroom collapsed into a 6 by 9 m (20 by 30 ft) sinkhole. His brother Jeremy rushed over to help but all he could see was the top of the bed and no sign of Jeffrey. He tried digging his brother out but had to be pulled out of the hole himself. Rescue teams found no signs of life. The hole was filled in and the body was never recovered.

Sinkholes, also called **cenotes** or **karst collapse** features, are caused by slightly acid rainwater slowly dissolving away underground limestone. This is how most underground caverns are formed. The sinkhole opens up when the roof of the underground cavern collapses under its own weight (Figure 12.45). In areas with thick underground salt or gypsum, two minerals that dissolve in freshwater, the movement of groundwater can dissolve these rocks and cause caverns and sinkholes as well. The trigger can be drought followed by rain. Drought lowers the water table such that caverns previously filled with water, which helped support the cave roof, no longer have that support. Rain then adds weight to the rocks and soil above the cavern and the roof collapses.

Caverns and sinkholes are a fact of life any place in the world where limestone is close to the surface. In the United States, this would include parts of every state (Figure 12.46), but

FIGURE 12.45
A sinkhole opened suddenly just meters away from a home in Florida in 2010. (Courtesy of US Geological Survey; https://www.usgs.gov/media/images/sinkholes-west-central-florida-freeze-event-2010-0.)

FIGURE 12.46
USGS map of areas prone to karst and sinkholes (green) in the United States (http://water.usgs.gov/ogw/karst /kig2002/jbe_map.html).

sinkholes are particularly common in Florida, Georgia, west Texas, Alabama, Missouri, Iowa, Kentucky, Tennessee, and Pennsylvania.

It would not do to leave this topic without mentioning human-induced subsidence. This subsidence occurs when groundwater is pumped faster than it can recharge, as in the Central Valley of California (Figure 12.47), or when oil and gas are withdrawn from underground reservoirs, such as at San Pedro, California. Whereas subsidence can be measured by a careful survey, the signs homeowners may notice include soil pulling away from foundations, cracks appearing in foundations and walls, doors and windows that don't close properly, breaks in underground pipes and utilities because of shifting soil, the appearance of local depressions, or trees and utility poles tilting at odd angles. Abandoned mines, such as coal mines along the foothills south and west of Denver, Colorado, will collapse if not backfilled. If they are close to the surface, a depression will form. Sometimes, these will fill with water and become ponds that roughly conform to the location and shape of the underground mine workings. These ponds and depressions are distinct from natural stream drainage patterns.

Expanding or **swelling clay** is a nuisance for those constructing roads, buildings, and other infrastructure (Figure 12.48). Swelling clays include **smectite, bentonite, montmo- rillonite, illite, chlorite**, and a few others. These clays absorb water and swell, or dry out and shrink. Soils containing these clays can expand 10% or more. Both swelling and shrinking cause damage to foundations, walls, and buried pipes. It is not the presence or lack of moisture that causes problems, but rather the change from wet to dry and dry to wet. Perhaps as many as a quarter of all homes in the United States have suffered some damage from swelling clay soil. Although damage can run into billions of dollars in a typical year, most people are not aware of it because the damage occurs slowly and is often blamed on shoddy construction. All construction projects should include a soil survey to identify expanding clays so that measures can be taken to eliminate the effects of swelling on structures.

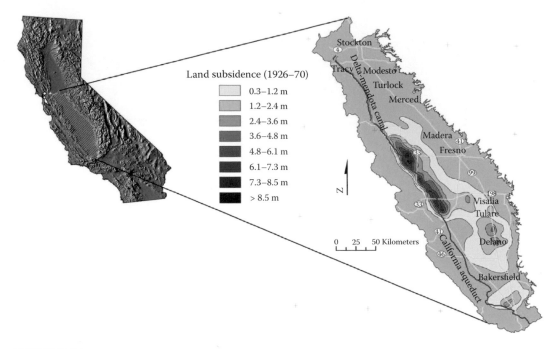

FIGURE 12.47
Land subsidence in the San Joaquin Valley, California, due to groundwater withdrawal, 1926–1970. (Modified from Ireland et al., 1984. Land Subsidence in the San Joaquin Valley, California, as of 1980. US Geological Survey Professional Paper 437-I, 103 pp.; http://ca.water.usgs.gov/projects/central-valley/delta-mendota-canal.html.)

FIGURE 12.48
This "roller-coaster road" on the eastern slope of the Rockies in Colorado results from uneven swelling of alternating clay-rich and clay-poor bedrock. (Courtesy of Dave Noe, Colorado Geological Survey; http://colorado geologicalsurvey.org/geologic-hazards/heaving-bedrock/.)

Living in Subsidence Country

Home buyers should check with local government for maps showing where land fill has been used to build up wetlands that may continue to subside because of dewatering and compaction. Local governments usually have maps showing areas prone to subsidence and locations of old subsurface mines. Find out what has been done to alleviate subsidence in an area, or just don't buy a home there.

Stay away from areas prone to sinkholes. If you live in such an area, prevent sinkholes by ensuring that your property has good drainage. Fill any depressions and direct runoff away from the property. Sometimes, filling a sinkhole can stop its growth or recurrence.

It is possible to build in areas with swelling soil. Sound construction in these areas calls for constant, stable moisture conditions and may require that a foundation be floated on the soil or be tied to deep pilings unaffected by volume changes in near-surface soil. Another method is to inject grout or other material that cements the soil beneath a structure.

Sources of Subsidence, Sinkhole, and Swelling Soil Information

The USGS has a web page explaining subsidence:

http://water.usgs.gov/ogw/subsidence.html

Information about subsidence risk management can be found at:

http://www.subsidencesupport.co.uk/downloads/Property%20Assure%20 Guide%20to%20Subsidence.pdf

A USGS website explains land subsidence due to pumping groundwater at:

http://water.usgs.gov/ogw/pubs/fs00165/

Arizona has information on subsidence due to groundwater withdrawal:

http://www.azwater.gov/AzDWR/Hydrology/Geophysics/LandSubsidenceIn Arizona.htm

Other states have similar websites.

Colorado School of Mines has a bulletin explaining what homeowners should know and what they can do about subsidence over abandoned coal mines:

http://inside.mines.edu/fs_home/tboyd/Coal/homeowner/

The USGS has an information page on sinkholes:

http://www.usgs.gov/blogs/features/usgs_top_story/the-science-of-sinkholes/

Maps of sinkholes in the United States provided by the USGS:

http://water.usgs.gov/ogw/karst/kig2002/jbe_map.html

Sinkhole maps of Florida can be obtained at:

http://fcit.usf.edu/florida/maps/galleries/sinkholes/index.php

A USGS swelling clays map of the United States can be found at:

http://ngmdb.usgs.gov/Prodesc/proddesc_10014.htm

Natural Pollution

Pollution is the presence of substances that are unsafe or harmful or poisonous to the environment. Natural pollution exists in many forms. Volcanic gases contain sulfur dioxide and hydrogen sulfide. Radon gas, which occurs naturally in many regions, seeps into basements and becomes concentrated if there is no ventilation. Arsenic occurs naturally in some bedrock and soil and leaches into groundwater or surface streams. Natural oil seeps occur on the seafloor or along rivers (http://walrus.wr.usgs.gov/seeps/; http://www.earth.columbia.edu/articles/view/3272; http://www.whoi.edu/oilinocean/page.do?pid=52296&tid=201&cid=54634&ct=362). Methane is produced by decomposing organic matter and occurs in large concentrations in coal seams, swamps, bogs, and as hydrates on the seafloor and in permafrost.

Volcanoes are a source of deadly **hydrogen sulfide** and smelly sulfur dioxide. Hydrogen sulfide causes eye irritation at levels above 5 parts per million (ppm); it is toxic at levels greater than 250 ppm. Luckily, most people do not live near volcanic vents. Some, however, attempt to mine the sulfur that is deposited around the vents (Figure 12.49).

Radon is an invisible, odorless radioactive gas derived from the decay of uranium in the earth. It is usually found in igneous rocks and soil, but occasionally can be found in groundwater. It escapes easily from rocks, soil, and water and collects in enclosed spaces such as caves, mines, tunnels, and basements. Underground uranium miners have the highest levels of radon exposure. Studies have shown a clear link between breathing radon and lung cancer. The US Environmental Protection Agency (EPA) ranks radon as the second leading cause of lung cancer, contributing to about 21,000 deaths per year in the United States. It is the leading cause of lung cancer deaths among nonsmokers.

The risk posed by radon is preventable. It can be detected using a simple low-cost test and can be eliminated using barriers and ventilation. The EPA estimates that 1 out of every 15 homes in the United States has radon levels at or above the recommended maximum of 4 picoCuries per liter (pCi/L) of air.

FIGURE 12.49
Sulfur vapor rising from Kawah Ijen, Java, Indonesia. Sulfur miners can be seen in the foreground for scale. (Copyright Sémhur/Wikimedia Commons/CC-BY-SA-3.0; https://commons.wikimedia.org/wiki/File:Sulfur _mining_in_Kawah_Ijen_-_Indonesia_-_20110608.jpg.)

Before buying a home, check the area's radon potential. The EPA has maps showing areas with the greatest risk (Figure 12.50). Test your home with kits available at most hardware stores, or order a kit from the National Radon Services Program at Kansas State University (http://sosradon.org/test-kits). If you find high levels of radon gas, you can open your windows to ventilate, or install a venting system. If you are building a home in a high radon-risk area, the EPA has a number of recommendations:

1. Place a gas-permeable layer beneath the floor to allow gas to move from under the house to the sides.
2. Place impermeable plastic sheets between the permeable layer and the floor.
3. Seal and caulk all cracks in the foundation.
4. Install venting pipe in the permeable layer to assist movement of the gas away from the home.
5. Install venting fans in homes with crawlspace foundations.

Methane is natural gas. It is colorless, odorless, and combustible. It forms an explosive mixture in air at levels greater than 5% to 15% by volume. In addition to being found in oil and gas fields, it is the main component of swamp gas, the product of bacterial decomposition of organic matter. It is almost always associated with coal seams, one of the factors that make coal mining so hazardous. It is also bound up with water molecules in **methane hydrates**, the natural ice deposits that occur under special conditions on the seafloor and in permafrost. Natural methane seeps have been known for centuries (Figures 12.51 and 12.52).

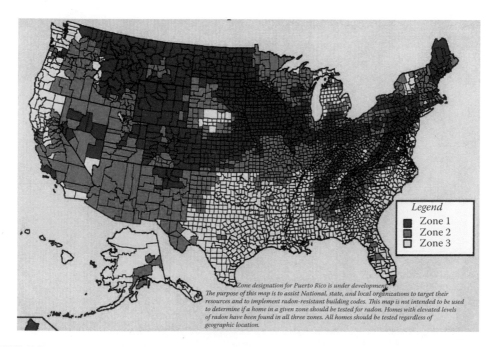

FIGURE 12.50
Radon map of the United States shows varying risk levels. (Courtesy of Environmental Protection Agency; https://www.epa.gov/sites/production/files/2015-07/documents/zonemapcolor.pdf.)

FIGURE 12.51
The eternal fire of Baba Gurgur, Iraq. Visitors to Kirkuk can see the flames from the city (12 km, or 7 mi). (Courtesy of Chad.r.hill; https://commons.wikimedia.org/wiki/File:P3110004.jpg.)

FIGURE 12.52
Eternal Flame Falls is a waterfall in the Shale Creek Preserve of Chestnut Ridge Park, western New York. (Courtesy of Mpmajewski; https://en.wikipedia.org/wiki/Eternal_Flame_Falls.)

Of particular interest is **methane-contaminated groundwater**, as featured in the 2010 documentary film *Gasland*. Yet, landowners with water wells encounter natural gas in their well water any time that water is in contact with coal beds or if the water-bearing layer is connected to a gas field by faulting. The gas slowly seeps out of the coal or along the fault and accumulates dissolved in the groundwater much as natural carbonation (carbon dioxide) does. When the temperature is increased and pressure is lowered by bringing the water to the surface, the gas comes out of solution. Dissolved methane escapes easily from water. Have an accredited laboratory test the water to see if it contains methane. Gas detectors can also identify methane and other gases such as carbon monoxide in your home. Some of these detectors can be found at hardware stores. Since methane is lighter than air, the detectors should be installed near the ceiling.

Water wells should have a sealed well cap and vent that allows accumulated gas to escape harmlessly. This is especially important if a wellhead is located in a basement, since otherwise the gas would escape into the house. Methane can also be stripped from water by aeration, either in an enclosed tank or in tower aerators. Contact your state or local Department of Health for help.

Naturally occurring **arsenic** is primarily derived from arsenopyrite and iron oxides, but is a component of more than 200 minerals. It is found in volcanic rocks and geothermal systems. Areas with high natural arsenic have elevated levels of soluble arsenic in groundwater. Inorganic arsenic can be highly toxic. Arsenic-contaminated water poses a health risk to drinking water, food crops, and food preparation. Long-term exposure causes skin lesions and cancer. It has also been associated with developmental effects, neurotoxicity, cardiovascular disease, and diabetes.

Countries with naturally occurring arsenic include Argentina, Bangladesh, Chile, China, Ghana, India, Mexico, Thailand, Taiwan, the United Kingdom, and the United States (Figure 12.53). *USAToday* (August 30, 2007) reported that over 137 million people in more than 70 countries are probably affected by arsenic poisoning of drinking water. The maximum safe limit as defined by the World Health Organization (WHO) and the US Environmental Protection Agency (EPA) is 0.01 mg/L. For some reason the city of Fallon, Nevada, in the United States has natural levels exceeding 0.08 mg/L.

There are several methods for removing arsenic from groundwater. Small-scale techniques include "under-the-sink" filters that use adsorption (Bayoxide E33, GFH, or titanium dioxide) or the reverse osmosis process. Large-scale water treatment is accomplished through coagulation/filtration, which uses iron or alum to precipitate the arsenic, iron oxide adsorption filters, activated alumina, ion exchange, reverse osmosis, and electrodialysis. Arsenic may be removed from groundwater without a water treatment plant by injecting aerated water to create a zone of oxygen-rich groundwater that allows arsenic-oxidizing microorganisms to trap iron and arsenic within the soil.

Natural acid drainage containing iron oxides and iron sulfides occurs in areas where geothermal systems or mineral deposits are exposed to the atmosphere. Streams originating in these areas have rocks coated with orange-colored mineral deposits (Figure 12.54). The Rio Tinto in Spain, Red River in Cornwall, and Iron Creek in Colorado all have natural acid drainage. Natural acid drainage decreases surface water quality by lowering the pH (making it more acidic) and increasing the amount of dissolved metals in the water. The impact is a function of the amount of neutralization and dilution that occurs: it has a greater impact when concentrated in a small stream than when mixed into a large river. The combination of dissolved metals and acidity can be toxic to fish, amphibians, and aquatic insects and plants.

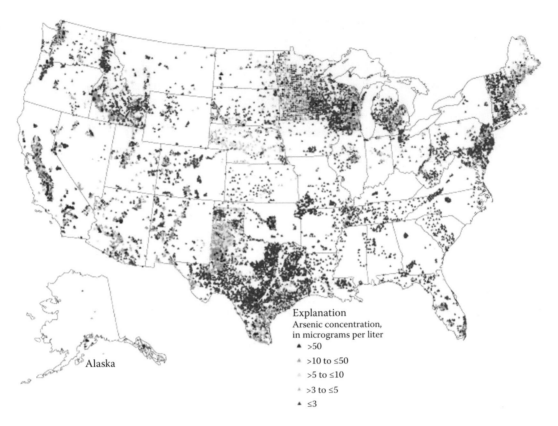

FIGURE 12.53
Map showing arsenic concentrations in the United States. (Courtesy of US Geological Survey; http://water.usgs .gov/nawqa/trace/arsenic/.)

Remediation methods for natural acid drainage include dilution, inhibiting oxidation at the source by using wet or dry covers, and chemically treating water to neutralize acids by adding calcium oxide or calcium carbonate to water. Biological methods include creating aerobic and anaerobic wetlands, employing iron-oxidizing bacteria, and biogenic production of hydrogen sulfide to generate alkaline conditions that make the dissolved metals insoluble and thus remove them from water.

Fortunately, natural acid drainage is not a major problem in most populated areas.

Natural **oil seeps** occur in oil basins such as the Los Angeles basin (La Brea tar pits), basins offshore California, the Gulf of Mexico, the Athabasca region of the Western Canada Sedimentary Basin, and the San Joaquin Valley of central California (Figure 12.55). These seeps leak directly out of oil-bearing formations, or leak along faults that connect the oil formations to the surface. Along the Athabasca and Christina rivers in Alberta, tar-saturated McMurray Formation sandstone seeps tar down riverbanks during warm weather, forming sheens on the water surface (Figure 12.56). In the case of California, fault-related oil seeps on the seafloor are responsible for tarring the feet of beachgoers when the resulting tar balls float ashore (Figures 12.57 and 12.58).

Tube worms, mussels, and clams appear to do well around hydrocarbon seeps on the ocean floor. **Chemosynthetic communities** of marine organisms grow around

FIGURE 12.54
Iron and aluminum minerals precipitating in the South Fork of Lake Creek at the confluence of a natural acidic drainage. The source is the mineral-rich Oligocene Grizzly Peak Caldera between Leadville and Aspen, Colorado. (Courtesy of D. Coulter, 2006, Colorado School of Mines PhD., with permission.)

hydrocarbon seeps and derive much of their nutrients directly from the seep (Figure 12.59). Over 600 natural seeps have been identified in the Gulf of Mexico that leak between 2700 and 13,700 barrels of oil per day. A recent study estimates that natural seeps in the Santa Barbara Channel offshore California introduce on the order of 100 barrels per day of oil into the marine environment. These seeps are a potential source of chronic, low-level stress to invertebrates, fish, marine mammals, and birds. Yet, the environment is able to handle these natural seeps, and they do not result in major mortality of marine animals.

On the other hand, controversy surrounds damage caused to people living downstream from hydrocarbon seeps along rivers. Studies indicate that higher than normal levels of carcinogenic polycyclic aromatic hydrocarbons (PAHs) as well as high levels of heavy metals such as arsenic, mercury, selenium, and cadmium are found in kidney and liver samples from moose, ducks, beavers, and muskrats harvested downstream from the Canadian oil sands. The Canadian government has issued consumption advisories regarding fish caught downstream from seeps and oil sands mining operations, especially around the First Nations community of Fort Chipewyan. It has been difficult to determine whether health issues are a result of high natural levels of these compounds, a result of mining activities, or both. Various adverse effects of PAHs, including sleeping disorders, migraines, increased stress, allergies, asthma, and hypertension have been reported. The greatest concern, however, is with increasing cancer rates. Studies linking these effects to

FIGURE 12.55
Natural tar seep at the McKittrick Oil Field, California. (Courtesy of Lldenke; https://en.wikipedia.org/wiki
/McKittrick_Tar_Pits.)

FIGURE 12.56
Oil seeping out of McMurray Formation sandstone, Christina River, Alberta. (Courtesy of G. Prost.)

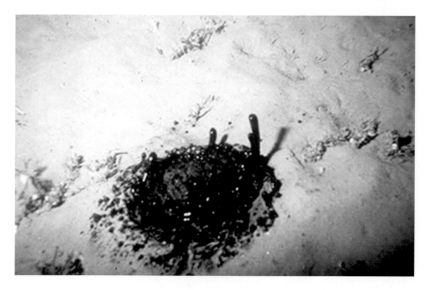

FIGURE 12.57
Oil seep rising from the seafloor near Coal Oil Point, southern California. (Courtesy of A. Allen/NOAA; http://response.restoration.noaa.gov/oil-and-chemical-spills/oil-spills/resources/what-are-natural-oil-seeps.html.)

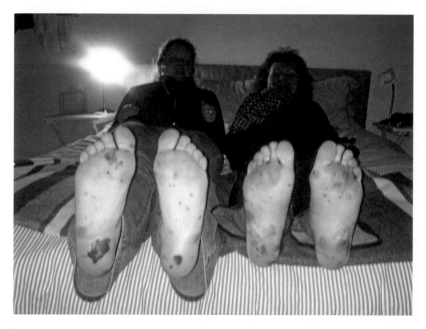

FIGURE 12.58
Tarred feet after a day on Venice Beach, southern California. (Courtesy of Monika Falkenberg.)

FIGURE 12.59
Chemosynthetic community on Viosca Knoll, Gulf of Mexico. Dense clusters of tubeworms grow where hydrocarbons are actively seeping. (Courtesy of Ken Sulak, USGS; http://fl.biology.usgs.gov/images/pictures /CHEMO_TUBEWORM_BUSH.jpg.)

hydrocarbon contamination have so far been inconclusive. People most at risk of adverse health effects are those with a traditional (not store-bought) diet and who drink untreated surface water. Water treatment and a less traditional diet might be a short-term solution to risk concerns.

Sources of Information on Natural Pollution

For more information about radon, go to EPA's website:

www.epa.gov/radon

or call a national toll-free hotline at 1-800-SOS-RADON (1-800-767-7236).

Information on Radon can be found at the EPA site:

https://www.epa.gov/radon

and at the WHO:

http://www.who.int/ionizing_radiation/env/radon/en/

The Water Research Center has information on removing methane from well water:

http://www.water-research.net/index.php/methane

The Alberta Department of Agriculture and Forestry has information on detecting and removing methane from groundwater at:

http://www1.agric.gov.ab.ca/$department/deptdocs.nsf/all/agdex10840

The WHO has a health sheet on arsenic at:

http://www.who.int/mediacentre/factsheets/fs372/en/

The USGS has a web page dedicated to arsenic in groundwater at:

http://water.usgs.gov/nawqa/trace/pubs/gw_v38n4/

It Came from Outer Space!

At 9:20 p.m. local time on February 15, 2013, a 20-m (60-ft) meteor flashed over the town of Chelyabinsk, Russia (Figure 12.60). Traveling at 19 km/s (40,000 mph), the meteor exploded at a height of 30 km (18 mi) with the force of 20 to 30 Hiroshima atom bombs. The fireball was brighter than the sun and could be seen up to 100 km (60 mi) away. About 1500 people were injured, mainly from windows that were shattered by the shock wave. Some 7200 buildings in six cities in the region were also damaged by the shock wave.

A little over 100 years earlier, a meteor or comet exploded over the Stoney Tunguska River in Krasnoyarsk Krai, Russia. On the morning of June 30, 1908, the 100-m-wide (300-ft-wide) object burst at an altitude between 5 and 10 km (3 to 6 mi) with the power of roughly 1000 Hiroshima atom bombs. Twenty years after the Tunguska event, a Russian expedition documented the destruction, noting that downed trees were oriented outward from the blast center (Figure 12.61). Estimates are that 80 million trees were knocked down in an area 2150 km^2 (830 mi^2) and that the shock wave would have registered a 5.0 on the Richter scale. While an explosion of this magnitude could level a city, it is more likely to hit in the ocean and generate a tsunami.

There are only a handful of cases where people are known to have been hurt by meteors. An Italian manuscript published in 1677 tells of a Milanese friar who was killed by a meteorite. On November 30, 1954, a meteor fell in Oak Grove, Alabama. After crashing through the roof of a house and bouncing off a large radio, a grapefruit-sized fragment of the Sylacauga meteorite struck Ann Fowler Hodges. She was badly bruised but alive. In 1992, a meteorite fragment hit a Ugandan boy in Mbale. He was uninjured because the 3-g (0.11-oz) object had been slowed down by a tree.

FIGURE 12.60
The 2013 Chelyabinsk meteor trail and explosion. (Courtesy of Aleksandr Ivanov, Ogg Theora video; https://en.wikipedia.org/wiki/File:%D0%92%D0%B7%D1%80%D1%8B%D0%B2_%D0%BC%D0%B5%D1%82%D0%B5%D0%BE%D1%80%D0%B8%D1%82%D0%B0_%D0%BD%D0%B0%D0%B4_%D0%A7%D0%B5%D0%BB%D1%8F%D0%B1%D0%B8%D0%BD%D1%81%D0%BA%D0%BE%D0%BC_15_02_2013_avi-iCawTYPtehk.ogv.)

FIGURE 12.61
Trees felled by the 1908 Tunguska event. (Courtesy of Leonid Kulik Expedition; http://apod.nasa.gov/apod
/ap071114.html.)

FIGURE 12.62
City killer: Barringer Crater (Meteor Crater), Arizona, is 1.6 km wide and 200 m deep. The impact energy is estimated at 150 times the atomic bomb that leveled Hiroshima. (Courtesy of Shane.torgerson; http://en.wikipedia
.org/wiki/File:Meteorcrater.jpg.)

Meteor Crater, near Winslow, Arizona, is one of the most recent and best preserved impact craters on Earth with an age of around 50,000 years (Figure 12.62). The meteorite that made this crater weighed 272 million kg and was composed of nickel and iron. It is estimated to have been about 40 m across and traveling over 42,000 km/h on impact. For comparison, the asteroid or comet that created the Chicxulub crater in Mexico 65 million years ago, and is thought to have caused the extinction of the dinosaurs, was between 9 and 19 km in diameter. The effects of a large impact include vaporization of a large area of the earth's surface, tsunamis, firestorms, impact winter, and extinction.

Scientists are investigating how to anticipate where and when a meteor will impact Earth and how to avoid such impacts. These efforts fall into three categories: identification, destruction, and deflection. The first requirement is to identify which objects are likely to hit Earth. It would take a year to decades of warning to prepare an effective response, so early identification is essential. If the response is to destroy the target, then a nuclear or kinetic impact (dense) device is likely to be launched to try to break up the object and disperse the fragments. If deflection is the objective, then all that is needed is to change the velocity of the body slightly so that it crosses Earth's orbit before or after Earth gets there.

Sources of Information on Impacts

NASA has a website describing their near-Earth object program at:

http://neo.jpl.nasa.gov/neo/report2007.html and http://neo.jpl.nasa.gov/

The Armagh Observatory in Armagh, Ireland, has a site describing near Earth object impact hazards, risks, and deflection technologies:

http://star.arm.ac.uk/impact-hazard/

The European Space Agency has a site dedicated to a near-Earth object directory at:

http://newton.dm.unipi.it/neodys2/

References and General Reading

Burns, S. F., T. M. Harden, and C. J. Andrew. 2008. *Homeowners Guide to Landslides-Recognition, Prevention, Control, and Mitigation.* Portland State University and FEMA, 12 pp. http://www.oregongeology.org/sub/landslide/homeowners-landslide-guide.pdf

Childs, C. 2000. *The Secret Knowledge of Water.* Back Bay Books, New York. 288 pp.

D'Arcy Wood, G. 2014. *Tambora—The Eruption that Changed the World.* Princeton University Press, Princeton. 293 pp.

Ellsworth, W. L. 1990. Earthquake history, 1769–1989, in R. E. Wallace (ed.), USGS Professional Paper 1515.

Hamblyn, R. 2009. *Terra—Tales of the Earth—Four Events that Changed the World.* Picador, New York. 268 pp.

Haugerud, R. 2014. Preliminary Interpretation of Pre-2014 Landslide Deposits in the Vicinity of Oso, Washington. US Geological Survey Open-File Report 2014–1065, 6 pp.

Heilprin, 1908. *The Eruption of Mt. Pelee*: Philadelphia Geographic Society.

Ireland, R. L., J. F. Poland, and F. S. Riley. 1984. Land Subsidence in the San Joaquin Valley, California, as of 1980. US Geological Survey Professional Paper 437-I, 103 pp.

Japan National Committee on Earthquake Engineering. 1964. *Proceedings of the 3rd World Conference in Earthquake Engineering*, Volume III, pp. 78–105.

Lynch, D. K., K. W. Hudnut, and D. S. P. Dearborn. 2009. *Seismological Research Letters* 81 (3) May/June 2010, doi: 10.1785/gssrl.81.3.453.

McPhee, J. 1989. *The Control of Nature*. Noonday Press, New York. 272 pp.

Twain, M. 1940. *Mark Twain in Eruption—Hitherto Unpublished Pages about Men and Events*. Harper Brothers Publishing, New York. 402 pp.

USAToday, 2007. Arsenic in Drinking Water Seen as Threat. August 30. http://usatoday30.usatoday.com/news/world/2007-08-30-553404631_x.htm

Williams, R. S. Jr., and J. G. Moore. 1983. *Man against Volcano: The Eruption on Heimaey, Vestmannaeyjar, Iceland*. US Geological Survey General Interest Publication, 33 pp.

Winchester, S. 2003. *Krakatoa—The Day the World Exploded, August 27, 1883*. Harper Collins, New York. 416 pp.

Winchester, S. 2005. *A Crack in the Edge of the World—America and the Great California Earthquake of 1906*. Harper Collins Publishers, New York. 463 pp.

Witze, A., and J. Kanipe. 2014. *Island on Fire—The Extraordinary Story of a Forgotten Volcano that Changed the World*. Pegasus Books, New York. 224 pp.

13

Groundwater

Whiskey is for drinking and water is for fighting.

Attributed to Mark Twain
Humorist

Much groundwater is parceled out under the Rule of Capture, otherwise known as "the biggest pump wins."

Anonymous

Much of our drinking water and the water that irrigates our farms comes from wells. Most surface water issues from springs. How does this water occur in the ground? Where did it come from? How do you find it? Who owns it? These are essential questions for those who depend on groundwater. And most of us depend on groundwater.

The Hydrologic Cycle

Water moves in a continuous flow from the atmosphere to the surface and subsurface of the earth to lakes and oceans and back to the atmosphere. This is known as the **hydrologic cycle** or **water cycle** (Figures 13.1 and 13.2). The physical processes include evaporation, condensation, precipitation, infiltration, and runoff. The process where surface water percolates into the ground is called **infiltration**. Infiltration can be rapid in sandy soil, and it can be very slow in clayey or rocky areas. Infiltration **recharges** the water table. Infiltration can also occur from melting snowpack. It generally takes 10 vertical centimeters of snow to make 1 cm of water at temperatures close to freezing (the ratio is greater than 10:1 when it is colder).

Recall that rocks are usually not entirely solid: they have fractures and spaces between grains called **pores**. Saturated ground is much like a water-filled sponge. The **water table** is the boundary between water-filled, **saturated zone** and the air-filled, unsaturated zone. It moves up and down slightly depending on the rainy season and amount of recharge. **Springs** occur where the water table comes to the surface.

Where did the water ultimately come from? Water is thought to have entered the atmosphere from the degassing of Earth's interior by way of volcanic eruptions. Once in the atmosphere, the water vapor cools and precipitates as rain. The rain infiltrates into the ground or runs off the surface and collects in low spots. Groundwater is either picked up by plant roots and transpired back to the atmosphere, or seeps into rivers, lakes, and the ocean. From there, evaporation puts moisture back into the atmosphere and the cycle continues.

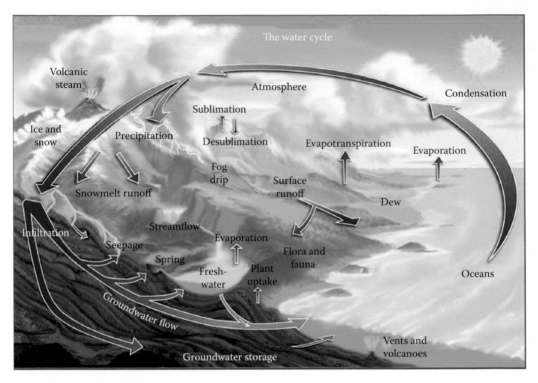

FIGURE 13.1
The hydrologic cycle. (Courtesy of John Evans, Howard Perlman, U.S. Department of the Interior, USGS; http://water.usgs.gov/edu/watercycle.html.)

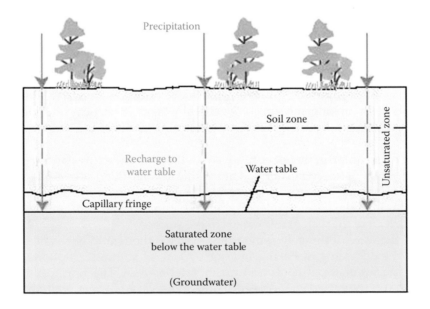

FIGURE 13.2
The water table separates the saturated zone from the unsaturated zone. (Courtesy of USGS; http://water.usgs.gov/edu/watercyclegwstorage.html.)

Underground rock layers with lots of water-filled pores are called **aquifers**, from the Latin "water-bearing." This saturated rock contains freshwater or brine that is thousands to millions of years old. Underground water is also on the move, although usually very slowly (Figure 13.3). A velocity of 0.3 m/day (1 ft/day) is fast for groundwater. More common is flow around 0.3 m/year or even 0.3 m per decade. Groundwater moves downslope under the influence of gravity from areas of recharge to areas of **discharge**, where it comes to the surface as springs or seeps along streams, lakes, and wetlands. Groundwater also discharges directly into bays and the ocean.

Rocks with connected pores are **permeable**: water is stored in and moves through the pores. A well drilled into an aquifer will encounter water at the water table and a pump may be required to lift the water to the surface. There are also layers of rock that do not allow the movement of groundwater. These **aquitards** either do not have pores, or the pores are not connected: they are **impermeable** or **impervious**. When an aquifer is confined between aquitards, the groundwater can be under considerable pressure from the overlying rock. A well drilled into a confined aquifer will release the pressure and the water may rise up in the well or even flow at the surface. These are **artesian** wells (Figure 13.4). Artesian wells have hydraulic head above the local water table. **Hydraulic head** is the level to which water rises in a well. It is measured in distance and is a proxy for the pressure exerted on the column of water. Groundwater always moves from areas with high hydraulic head to areas with low hydraulic head.

Not all groundwater is fresh, or **potable** (drinkable). About 54% of it is brackish or salty, containing minerals dissolved from the surrounding rocks. Unless it is purified, this water is not good to drink, and may not be good for agriculture.

The water table can move up and down naturally over time as a result of changing precipitation and surface water recharge. It can also be lowered by human pumping (Figure 13.5). Sometimes, the drawdown is local; sometimes, if there are lots of wells over a large area, the drawdown is regional and substantial, as in the San Joaquin Valley of central California.

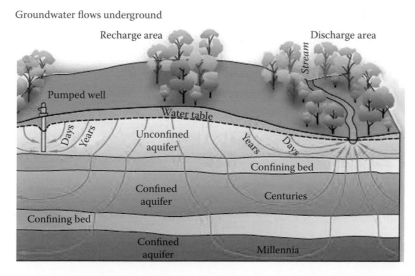

FIGURE 13.3
How groundwater moves. (Courtesy of USGS; http://water.usgs.gov/edu/watercyclegwdischarge.html.)

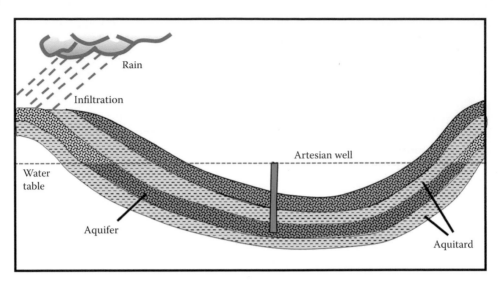

FIGURE 13.4
Conditions needed for an artesian well.

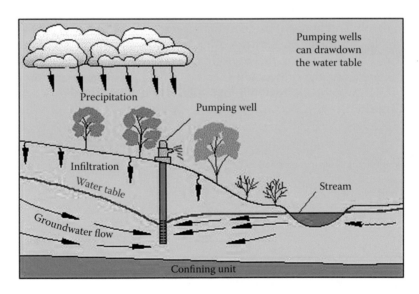

FIGURE 13.5
Pumping draws down, or lowers the local water table. (Courtesy of USGS.)

Information on groundwater resources is available from the government. In the United States, contact the US Geological Survey (USGS) or state water agencies. The Groundwater atlas of the United States can be found at: http://pubs.usgs.gov/ha/ha730/. Other groundwater data for the United States can be found at: http://waterdata.usgs.gov/nwis/gw.

Finding Groundwater

Groundwater may be just below the surface, or it can be thousands of meters deep. The depth to the water table is of interest to you if your home depends on well water or your farm needs wells to irrigate. Where do you find groundwater?

Just about everywhere, really. The only questions are (1) how deep is it? (2) at what rate will it flow into the well? and (3) is it fresh?

Groundwater is closer to the surface in valleys than on hills. If you are on a flat plain, groundwater will be closer to the surface near springs, lakes, or swamps. If there is no surface water, it can still be near the surface if you see **phreatophytes**, water-loving plants like cottonwood trees, willows, or tamarisk.

Sandstones are the best aquifers because they often have good interconnected pores (both porosity and permeability), and because the layers can extend for many kilometers. A good example is the **Ogallala aquifer** that extends under a large part of the Great Plains in the United States (Figure 13.6).

Sandstone aquifers provide the best water flows over time. Igneous and metamorphic rocks can yield groundwater if they have interconnected fractures, and flow can be good at first, but flow rates decline rapidly as fractures are drained (Figure 13.7).

The best information about water-bearing layers comes from existing wells. **Hydrologists**, people who specialize in groundwater, take information from wells in the region, especially depth to the water table. The resulting **potentiometric surface** maps show groundwater flow and how deep wells have to go to encounter water (Figure 13.8). By looking at the amount of drawdown in a well, they can calculate the volume of water in the system. **Chemical tracers** can be added at a well and tracked from well to well to determine how fast and in what direction water is moving underground.

A **perched water table** is a local, shallow, saturated layer, usually of limited extent, that exists in the unsaturated zone above a regional water table. River channel sandstone surrounded by shale is a common perched aquifer (Figure 13.9).

This section would not be complete without mentioning **water dowsing** (Figure 13.10). This is the practice of using a forked stick or other device to locate groundwater (and minerals). The dowser, or **water witch**, holds one fork in each hand and the other end of the "Y" is pointed upward or outward. As they walk back and forth over an area, the presence of water or minerals near the surface is supposed to exert a pull on the end of the stick and it points downward.

Is there any scientific basis to water witching? No. In some places, they are just plain lucky, as water is just about everywhere in the subsurface. It is just a matter of how deep you drill. In other cases, the water witch knows from experience what clues to look for. They know that groundwater is closer to the surface in valleys than on hills. They look for areas with sandstone near the surface. They look for water-loving plants. They know where plants suffer least from water stress, such as areas that turn green earliest in the spring and where trees lose their leaves last in the fall. As far back as 1556, this practice

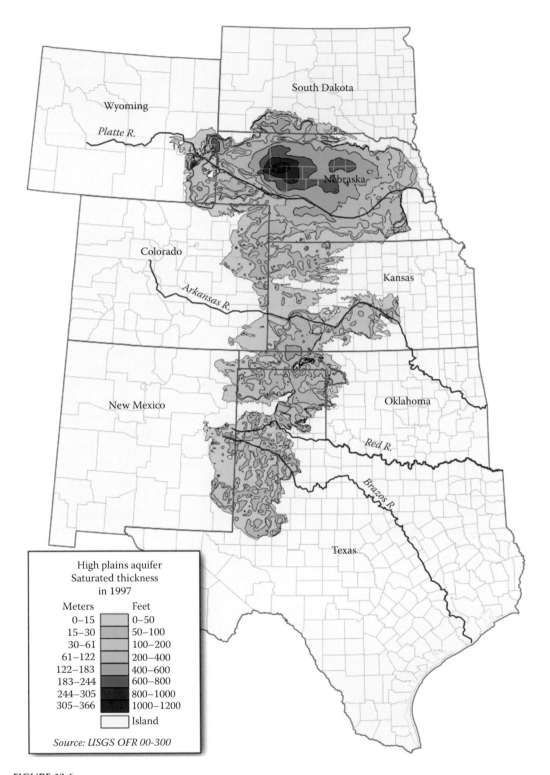

FIGURE 13.6

Distribution and saturated thickness of the Ogallala Aquifer, central United States. (Courtesy of USGS; https://commons.wikimedia.org/wiki/File:Ogallala_saturated_thickness_1997-sattk97-v2.svg.)

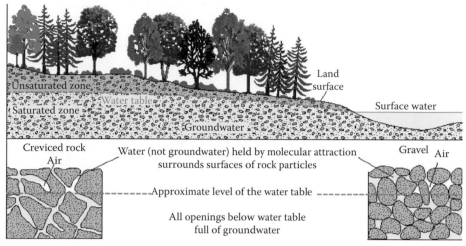

FIGURE 13.7
Groundwater is stored in fractures and pores. (Courtesy of USGS; http://pubs.usgs.gov/gip/gw/how_a.html.)

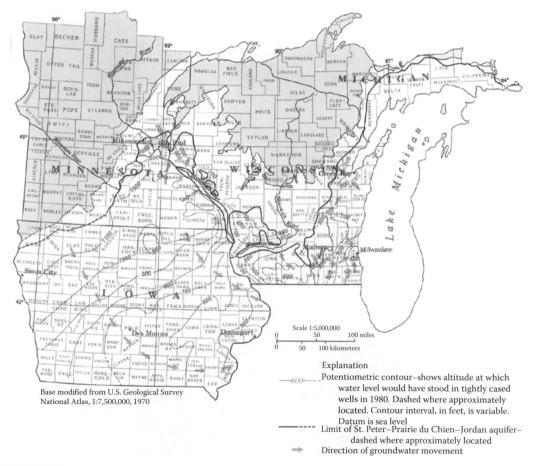

FIGURE 13.8
Potentiometric surface map of Minnesota, Wisconsin, and Michigan. Map shows depth to the water table and flow direction of subsurface water. (Courtesy of USGS; http://pubs.usgs.gov/ha/ha730/ch_j/jpeg/J116.jpeg.)

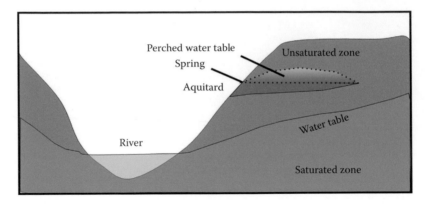

FIGURE 13.9
Perched water tables are above and separate from most groundwater. They tend to be limited in size and distribution.

was recognized as fraudulent: the primary mining text of the day, *De Re Metallica* (On the Nature of Metals) states "the manipulation is the cause of the twig's twisting motion…. Therefore a miner, since we think he ought to be a good and serious man, should not make use of an enchanted twig, because if he is prudent and skilled in the natural signs, he understands that a forked stick is of no use to him, for as I have said before, there are the natural indications of the veins [or water] which he can see for himself without the help of twigs."

FIGURE 13.10
Dowser. (From Agricola, G. 1556. *De Re Metallica*.)

Groundwater Ownership

Groundwater can belong to the public or be privately owned. Publicly owned groundwater is usually managed on the principle that it is "common to all" and that the government has a responsibility to regulate the quantity and quality of groundwater in the public's best interest. Groundwater is allocated based on an aquifer's **sustainable yield**, that is, an amount that can be maintained year over year. Rights are allotted through a system of permits. A regulator ensures that withdrawals stay within limits set by the state. This means the state can set the number of wells, the spacing of wells, and the amount of water produced from a well.

On reservations or treaty lands, aboriginal customs govern the use of surface water and groundwater. In most cases, these rights cannot be infringed upon by state governments.

Private groundwater may or may not be regulated, depending on local laws and traditions. There are several common systems of management.

The **Rule of Capture** allows a landowner to capture as much groundwater as can be used, but only to the extent that it is beneficial and not malicious. Landowners are not guaranteed a set amount of water. A property owner is not liable to pay a neighbor for drawing water from under the neighbor's land. This system of water allocation is prone to overproduction as each landowner drills bigger or deeper wells so as not to have their water produced by a neighbor.

The **Reasonable Use Rule** (also known as the American Rule) allows unlimited water extraction as long as it does not damage the aquifer or other wells. This rule gives precedent to historical uses and limits new uses for the water if they interfere with prior use. State agencies or the courts determine who gets a well and how much can be pumped.

Prior Allocation (also called "first-in-time, first-in-right") gives exclusive rights to groundwater based on a seniority system whereby older permits have first claim over any newer license holder. During times of shortage, the older permits can take their full share before newer licensees. Water can only be used for socially accepted purposes as defined by legislation.

Correlative groundwater rights allocate groundwater on the basis of the size of the surface area owned. Each landowner gets a proportional share of available groundwater. The amounts are set by the state or by courts. The amounts may be decreased during droughts. Landowners can sue neighbors for encroaching on their rights. This system favors ranchers and others with large tracts of land but low water usage and hurts those with smaller lands but high demand, such as farmers or cities.

Groundwater Mining and Sustainability

Recharge constantly replenishes groundwater, but groundwater can be removed faster than it is replenished. Both usage and recharge vary with the seasons. The **water budget** is the total of all water entering, leaving, and stored in the groundwater system (Figure 13.11). Before development, this system is in long-term equilibrium. The amount of water in storage is relatively constant, changing only as a result of annual rainfall or multiyear climate variations. Human pumping draws down the water budget by increasing the outflow. If

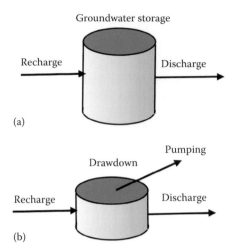

FIGURE 13.11
Simple water budgets for a groundwater system (a) predevelopment and (b) during development. (Courtesy of USGS.)

the maximum pumped is less than the natural inflow, this is called a **safe yield**. If the inflow is less than the outflow, the amount in storage decreases and the water table drops. In the long run, that is not sustainable.

Humans also change the inflow by irrigating (adding to inflow), by paving and other urban developments (decreasing inflow), and by changing vegetation cover (can increase or decrease inflow).

Pumping also has an effect on outflow. Changing the groundwater flow patterns and lowering the water table can cause springs and wetlands to dry up.

Droughts should be considered in any long-term water budget planning. Droughts reduce recharge, resulting in less groundwater in storage. Human water needs generally increase during a drought. Managing groundwater withdrawal through a drought requires tough decisions regarding who gets water, how much water they get, how much it costs, and how to maintain the resource for the long haul.

Groundwater mining refers to a progressive decrease in the amount of groundwater in storage. The amount removed exceeds the recharge rate of the aquifer. **Overdraft** refers to excessive withdrawal of groundwater, that is, groundwater mining that leads to irreparable drawdown of the aquifer. Whereas pumping a single well always has a local effect on groundwater movement and the water table, pumping many wells over a large area has regional effects on the groundwater system.

Groundwater Contamination

Groundwater **contamination** is the term used for the natural presence of minor, unwanted constituents, contaminants, or impurities. Groundwater **pollution** usually refers to man-made pollutants that are released on or just under the ground and make their way into groundwater. The movement of groundwater and dispersion of the pollutants lead to a chemical plume in the aquifer. The **pollution plume** can be tapped by wells or can come to the surface at springs and seeps.

Pollution is the result of effluent from landfills, wastewater treatment plants, leaking sewer pipes and septic tanks, leaking gas stations and pipelines, burial of industrial waste, leakage from underground storage tanks, overapplication of fertilizers and pesticides, leakage from stock yards, mine waste, road salt, and other sources (Table 13.1). Groundwater can appear clean because of the natural filtering of rocks and soils, but chemicals and bacteria can still be present. Polluted groundwater is a public (and wildlife) health hazard through poisoning or spread of disease. Anyone whose water supply comes from a well should have it tested for contaminants.

Frac'ing and Groundwater

Gasland, a 2010 documentary by Josh Fox, shocked the public by showing a homeowner turning on a water tap, then lighting the water on fire. This was attributed to frac'ing. What is going on?

Frac'ing, fracking, or **hydraulic fracture stimulation**, is a process that has been used since the 1940s to enhance production where oil or gas in rock pores is difficult to recover (the permeability is low, so fluid cannot flow from pore to pore). A combination of water and chemicals with sand is pumped into the reservoir layer under pressures high enough to fracture the rock and connect the pores. The sand props open the fractures, allowing the hydrocarbons to flow to the well. Generally less than half the water used to fracture a formation is recovered during production, meaning that some frac'ing water and chemicals are left in the subsurface.

People's concerns about health risks associated with hydraulic fracturing and groundwater fall into three categories: (1) natural gas in drinking water, (2) water produced along with oil has to be disposed of, and injection of this wastewater may contaminate groundwater, and (3) the chemicals used in the fracturing process may pose health risks.

Groundwater aquifers are usually within a few hundred meters of the surface. The layers being fractured are almost always thousands of meters below the aquifer: it is almost impossible for the relatively small volume of water injected during fracturing to break through to near-surface layers. Thus, it is unlikely that this is an avenue for gas to move from the deep subsurface to near-surface aquifers. It is possible, if the well is not drilled properly, that oil or gas could migrate along the wellbore to shallower layers. In a poorly completed well (not properly cemented around the wellbore), this could happen regardless of whether frac'ing was used or not.

Wastewater is disposed of by subsurface injection in much greater volumes than are used for frac'ing. If there are old, played out wells in the area that have not been properly abandoned (cemented from base to the surface), these can act as a conduit for fluids under pressure to move from the low permeability horizon to the near-surface aquifer. If there is a fault near the injection well, the fault can act as a conduit for injected fluids to move up to the aquifer.

There are other ways to get gas in groundwater, some man-made and some natural. When Ohio and Texas investigated gas in groundwater over the 16-year period from 1993 to 2008 (Kell 2011) they found that, for more than 16,000 horizontal shale gas wells, there was not a single incident where groundwater contamination resulted from site preparation, drilling, well construction, well completion, hydraulic fracture stimulation, or production operations at any of these wells. Many of the contamination incidents were

TABLE 13.1

Common Groundwater Contaminants

Contaminant	Sources to Groundwater	Potential Health and Other Effects
Inorganic Contaminants Found in Groundwater		
Aluminum	Occurs naturally in some rocks and drainage from mines.	Can precipitate out of water after treatment, causing increased turbidity or discolored water.
Antimony	Natural weathering, industrial production, municipal waste disposal, and manufacturing of flame retardants, ceramics, glass, batteries, fireworks, and explosives.	Decreases longevity, alters blood levels of glucose and cholesterol in laboratory animals exposed at high levels over their lifetime.
Arsenic	Natural processes, industrial activities, pesticides, and industrial waste, smelting of copper, lead, and zinc ore.	Causes acute and chronic toxicity, liver and kidney damage; decreases blood hemoglobin. A carcinogen.
Barium	Occurs naturally in some limestones, sandstones, and soils in the eastern United States.	Can cause a variety of cardiac, gastrointestinal, and neuromuscular effects. Associated with hypertension and cardiotoxicity in animals.
Beryllium	Occurs naturally in soils, groundwater, and surface water. Often used in electrical equipment and components, nuclear power and space industry. Mining operations, processing plants, and improper waste disposal. Found in low concentrations in rocks, coal, and petroleum.	Causes acute and chronic toxicity; can cause damage to lungs and bones. Possible carcinogen.
Cadmium	Found in low concentrations in rocks, coal, and petroleum. From industrial discharge, mining waste, metal plating, water pipes, batteries, paints and pigments, plastic stabilizers, and landfill leachate.	Replaces zinc biochemically in the body and causes high blood pressure, liver and kidney damage, and anemia. Destroys testicular tissue and red blood cells. Toxic to aquatic biota.
Chloride	Often from saltwater intrusion, mineral dissolution, industrial and domestic waste.	Deteriorates plumbing, water heaters, and municipal waterworks equipment at high levels.
Chromium	From old mining operations runoff and leaching into groundwater, fossil-fuel combustion, cement-plant emissions, mineral leaching, and waste incineration. Used in metal plating and as a cooling-tower water additive.	Chromium III is essential for nutrition. Chromium VI is more toxic than Chromium III and causes liver and kidney damage, internal hemorrhaging, respiratory damage, dermatitis, and ulcers on the skin at high concentrations.
Copper	From metal plating, industrial and domestic waste, mining, and mineral leaching.	Stomach and intestinal distress, liver and kidney damage, anemia in high doses. Adverse taste and staining to clothes and fixtures. Toxic to plants and algae at moderate levels.
Cyanide	Often used in electroplating, steel processing, plastics, synthetic fabrics, and fertilizer production; from improper waste disposal.	Poisoning is the result of damage to spleen, brain, and liver.

(Continued)

TABLE 13.1 (CONTINUED)

Common Groundwater Contaminants

Contaminant	Sources to Groundwater	Potential Health and Other Effects
Fluoride	Occurs naturally or as an additive to municipal water supplies; widely used in industry.	Decreases incidence of tooth decay but high levels can stain or mottle teeth. Causes crippling bone disorder (calcification of the bones and joints) at very high levels.
Iron	Occurs naturally as a mineral from sediment and rocks or from mining, industrial waste, and corroding metal.	Imparts a bitter astringent taste to water and a brownish color to laundered clothing and plumbing fixtures.
Lead	From industry, mining, plumbing, gasoline, coal, and as a water additive.	Affects red blood cell chemistry; delays normal physical and mental development in young children. Causes slight deficits in attention span, hearing, and learning in children. Can cause slight increase in blood pressure in some adults.
Manganese	Occurs naturally as a mineral from sediment and rocks or from mining and industrial waste.	Imparts brownish stains to laundry. Affects taste of water; causes dark brown or black stains on plumbing fixtures. Toxic to plants at high levels.
Mercury	Occurs as an inorganic salt and as organic mercury compounds. From industrial waste, mining, pesticides, coal, electrical equipment (batteries, lamps, switches), smelting, and fossil-fuel combustion.	Causes acute and chronic toxicity. Targets the kidneys and can cause nervous system disorders.
Nickel	Occurs naturally in soils, groundwater, and surface water. Often used in electroplating, stainless steel and alloy products, mining, and refining.	Damages the heart and liver of laboratory animals exposed to large amounts over their lifetime.
Nitrate (as nitrogen)	Occurs naturally in mineral deposits, soils, seawater, freshwater systems, and biota. More stable form of combined nitrogen in oxygenated water. From fertilizer, feedlots, and sewage.	Toxicity results from the body's natural breakdown of nitrate to nitrite. Causes "bluebaby disease," or methemoglobinemia, which threatens oxygen-carrying capacity of the blood.
Nitrite (combined nitrate/nitrite)	From fertilizer, sewage, and human or farm-animal waste.	Toxicity results from the body's natural breakdown of nitrate to nitrite. Causes "bluebaby disease," or methemoglobinemia, which threatens oxygen-carrying capacity of the blood.
Selenium	Enters environment from naturally occurring geologic sources, sulfur, and coal.	Causes acute and chronic toxic effects in animals—"blind staggers" in cattle. Nutritionally essential at low doses, but toxic at high doses.
Silver	From ore mining and processing, product fabrication, and disposal. Often used in photography, electric and electronic equipment, sterling and electroplating, alloy, and solder. Silver recovery practices are typically used to minimize loss.	Can cause argyria, a blue-gray coloration of the skin, mucous membranes, eyes, and organs in humans and animals with chronic exposure.
Sodium	Derived geologically from leaching of surface and underground salt deposits and decomposition of various minerals. From road deicing and washing products.	Can be a health risk factor for those individuals on a low-sodium diet.

(Continued)

TABLE 13.1 (CONTINUED)

Common Groundwater Contaminants

Contaminant	Sources to Groundwater	Potential Health and Other Effects
Sulfate	Elevated concentrations may result from saltwater intrusion, mineral dissolution, and domestic or industrial waste.	Forms hard scales on boilers and heat exchangers; can change the taste of water, and has a laxative effect in high doses.
Thallium	Naturally in soils; used in electronics, pharmaceuticals manufacturing, glass, and alloys.	Damages kidneys, liver, brain, and intestines in laboratory animals when given in high doses over their lifetime.
Zinc	Found naturally in water, most frequently in areas where it is mined. Enters environment from industrial waste, metal plating, and plumbing.	Aids in the healing of wounds. Causes no ill health effects except in very high doses. Imparts an undesirable taste to water. Toxic to plants at high levels.
Organic Contaminants Found in Groundwater		
Volatile organic compounds	Enter environment when used to make plastics, dyes, rubbers, polishes, solvents, insecticides, inks, varnishes, paints, disinfectants, gasoline products, pharmaceuticals, preservatives, spot removers, paint removers, degreasers.	Can cause cancer and liver damage, anemia, gastrointestinal disorder, skin irritation, blurred vision, exhaustion, weight loss, damage to the nervous system, and respiratory tract irritation.
Pesticides	Enter environment as herbicides, insecticides, fungicides, rodenticides, and algicides.	Cause poisoning, headaches, dizziness, gastrointestinal disturbance, numbness, weakness, and cancer. Destroys nervous system, thyroid, reproductive system, liver, and kidneys.
Plasticizers, chlorinated solvents, benzo[a] pyrene, and dioxin	Used as sealants, linings, solvents, pesticides, plasticizers, components of gasoline, disinfectant, and wood preservative. From improper waste disposal, leaching, leaking storage tanks, and industrial runoff.	Causes cancer. Damages nervous and reproductive systems, kidney, stomach, and liver.
Microbiological Contaminants Found in Groundwater		
Coliform bacteria	Occurs naturally from soils and plants and in the intestines of humans and other warm-blooded animals. Is an indicator for the presence of pathogenic bacteria, viruses, and parasites from domestic sewage, animal waste, or plants or soil.	Bacteria, viruses, and parasites can cause polio, cholera, typhoid fever, dysentery, and infectious hepatitis.

Source: Some information from Waller, R. M. 1982. *Ground Water and the Rural Homeowner.* US Geological Survey; https://water.usgs.gov/edu/groundwater-contaminants.html.

related to "legacy" issues, where wells had been abandoned before current standards were in place. The companies that operated the wells had walked away without **plugging** them (filling them with cement), and many of the operators were now bankrupt or out of business (Kell 2011).

Gas accumulations are categorized as **thermogenic** when they are the result of deep burial and heating of organic matter. They are considered **biogenic**, **microbial**, or **swamp gas**, when they are the result of bacterial decomposition of organic material at shallow depths. The USGS (Breen et al. 2005) published the result of an investigation into gas in groundwater in Pennsylvania and determined that it was mainly derived from a nearby gas storage facility:

> Proximity to the axis of the Sabinsville Anticline and the eastern margin of the gas-storage field correspond to the presence of thermogenic gas in water wells…. The weight of the evidence … points to storage-field gas as the likely origin of the natural gases found in water wells near Tioga Junction.

Further, when they examined gas in groundwater in New York, they found that it was all from naturally occurring thermogenic or microbial sources (USGS 2013):

> Results of sampling indicate that occurrence of methane in groundwater of the region is common…. Methane in valley wells was predominantly thermogenic in origin…. Water samples from wells in a valley setting indicate a mix of thermogenic and microbial methane….

The point is that hydraulic fracture stimulation of wells almost never leads to gas in groundwater. Gas in groundwater has been found to be (1) a result of natural seepage of gas, (2) a result of, in rare cases, abandonment of old wells before current safeguards were in effect, (3) a result of poor well completion (cementing of the wellbore), and (4) a result of leakage from gas storage facilities.

The developers of frac'ing chemicals have been reluctant to disclose the substances used in the process, thinking of them as proprietary and as providing a competitive advantage. The public, however, is concerned that these chemicals might be carcinogenic or pose other health risks and has been clamoring to make this information public. A 2014 review by Public Health England of potential health risks owing to exposure to chemical pollutants used in shale gas production stated that

> An assessment of the currently available evidence indicates that the potential risks to public health from exposure to the emissions associated with shale gas extraction will be low if the operations are properly run and regulated. Most evidence suggests that contamination of groundwater, if it occurs, is most likely to be caused by leakage through the vertical borehole. Contamination of groundwater from the underground hydraulic fracturing process itself (i.e. the fracturing of the shale) is unlikely. However, surface spills of hydraulic fracturing fluids or wastewater may affect groundwater.

Countries such as Scotland and states such as New York have placed a moratorium on frac'ing because of public health concerns. England and South Africa have lifted their bans and rely on regulation to handle health issues. Countries like Germany allow frac'ing except in wetland areas. The European Union has issued minimum standards for hydraulic fracturing that includes full disclosure of all additives. In the United States, hydraulic fracturing is excluded from the provisions of the Safe Drinking Water Act's underground injection regulations except when diesel fuel is used as an additive. An online voluntary disclosure database for frac'ing fluids, funded by the oil and gas industry and the US Department of Energy, can be found at FracFocus.org.

Hot Springs and Geothermal Energy

Hot springs, or **geothermal springs**, are where groundwater has been heated by a near-surface source of heat, usually magma. However, if the source is deep enough, it can be heated by the earth's normal geothermal gradient. If the water is heated to boiling or becomes superheated, the steam pressure can cause it to erupt as a **geyser**. There is no accepted definition of how hot the water has to be in order to be called a hot spring. If only steam comes to the surface, it is a **fumarole**.

Hot springs are distributed around the world in both volcanic and nonvolcanic areas. Historically, the water has been thought to have therapeutic value, either because of its warmth or because of the minerals dissolved in it. Famous baths and spas have been built around these springs, as at Baden Baden, Germany, Warm Springs, Georgia in the United States, and Bath, England (Figure 13.12). The healing value is probably more in the mind of the bather than in the water, but entire tourist industries have been built around these baths.

Geothermal energy taps into thermal springs and uses the heated water to turn turbines that generate electricity. This requires water heated to greater than 57°C (135°F), which usually requires volcanic areas. Geothermal energy is produced in 24 countries around the world, notably in Iceland, The Geysers (California), and the Taupo Volcanic Zone of New Zealand.

FIGURE 13.12
Restored Roman baths in Bath, England. (Courtesy of G. Prost.)

References and General Reading

Agricola, G. 1556. *De Re Metallica*.

Breen, K. J., K. Révész, F. J. Baldassare, and S. D. McAuley. 2005. Natural Gases in Ground Water near Tioga Junction, Tioga County, North-Central Pennsylvania—Occurrence and Use of Isotopes to Determine Origins, 2005. U.S. Geological Survey Scientific Investigations Report 2007-5085. 75 pp. https://pubs.usgs.gov/sir/2007/5085/pdf/sir2007-5085.pdf

Johnson Division, Universal Oil Products. 1972. *Ground Water and Wells*. St. Paul, Minnesota. 440 pp.

Kell, S. 2011. State Oil and Gas Agency Groundwater Investigations and their Role in Advancing Regulatory Reforms A Two-State Review: Ohio and Texas. Prepared for The Ground Water Protection Council, 165 pp. http://www.gwpc.org/sites/default/files/State%20Oil%20%26%20Gas%20Agency%20Groundwater%20Investigations.pdf

Public Health England. 2014. Shale gas extraction: Review of potential public health impacts of exposures to chemical and radioactive pollutants. Report PHE-CRCE-009, ISBN978-0-85951-752-2, 60 pp. www.gov.uk.

U.S. Geological Survey. 2013. Occurrence of Methane in Groundwater of South-Central New York State, 2012—Systematic Evaluation of a Glaciated Region by Hydrogeologic Setting. Scientific Investigations Report 2013–5190, 42 pp.

Waller, R. M. 1982. *Ground Water and the Rural Homeowner*. US Geological Survey.

14

Oceans

We are tied to the ocean. And when we go back to the sea, whether it is to sail or to watch—
we are going back from whence we came.

John F. Kennedy
Politician and U.S. President

For though we are strangers in your silent world, to live on the land we must learn from
the sea.

John Denver
Singer and songwriter, in *'Calypso'*

We don't know much about the sea, and most of what we do know we have learned in
the past 100 years. It took the advent of sonar and of gravity satellites to learn the shape
of the ocean floor. Deep sea sampling told us its age. Magnetometers told us that the seafloor is spreading away from the mid-ocean ridges. Remotely operated vehicles revealed
black smokers, those deep sea volcanic vents that have their own exotic life communities.
Much of the technology used to explore the sea comes from military applications, weather
forecasting, and the energy and mining industries, each of which has an interest in better
understanding the oceans.

In the Beginning...

Water bound in minerals was released by volcanism on the early Earth and created an atmosphere rich in water vapor. As we have seen, the oceans formed by condensation of this
atmospheric vapor starting in the Hadean, when Earth was only around 150 million years
old, and by the Archean there is good evidence, mainly from oxygen isotopes of zircon, that
the surface was cool enough and wet enough for worldwide oceans. There were no continents at first, as the crust was entirely dense basalt, which does not float as high on the mantle as lighter granite. The early ocean was more acidic than now: the early atmosphere had
thousands of times more carbon dioxide than exists in the air today. The carbon dioxide–rich
atmosphere mixed with rain to form carbonic acid. The oceans were probably warm or hot
because of the near-surface magma, and the combination of hot water (a good solvent), no
continents (continents contain large deposits of ocean-derived salt), and acidic seawater (dissolves minerals better) meant that the oceans may have been twice as salty as they are now.

The composition of the oceans changed gradually as granitic continents began to form
and as plants changed the proportion of carbon dioxide and oxygen in the atmosphere.

How Deep Are They?

On average, the ocean is about 3600 m (12,000 ft) deep. But the bottom isn't flat: there are volcanic seamounts, mid-ocean ridges that extend like the Rocky Mountains for thousands of kilometers (Figure 14.1), there are deep sea trenches, and there is the **continental shelf**, the submerged part of the continents. The shelf extends down to depths around 200 m (650 ft), whereas trenches get as deep as 11,000 m (36,200 ft) in the Challenger Deep of the western Pacific.

What's Down There? The Ocean Floor

Oceanic crust is thinner and denser than continental crust, which is why it floats lower on the mantle. It averages about 10 km (6 mi) thick and has a density of about 2.9 g/cm³ as opposed to 35 to 40 km thick and a density of 2.7 g/cm³ for continents. Where continents are largely granitic, the oceans are mainly basaltic. There are four main layers to oceanic crust: from top to base these are (1) unconsolidated and semiconsolidated sediments, (2) basaltic pillow lavas and sheet flows, (3) basaltic sheeted dikes, and (4) gabbro, rock with the same composition as basalt but that cooled slowly at depth.

Sedimentary cover is thin to absent over the mid-ocean ridges and increases in thickness away from the ridges. **Deep sea sediments** consist of the small shells of plankton that slowly rain down on the seafloor, volcanic ash, and fine-grained continental sediments

FIGURE 14.1
Topography of the Atlantic seafloor. The continental shelf is shown in light blue; the deep sea (abyssal) plains are dark blue; the Mid-Atlantic Ridge snakes down the center. (Courtesy of NOAA; http://www.virginiaplaces .org/boundaries/ocs.html.)

that extend far out into the sea as turbidity currents. **Turbidity currents** are the result of sediment accumulating along the continental margin until it becomes unstable and slumps into the deep basins like a submarine landslide.

Below the sediment cover is **pillow basalt**, so called because it forms pillow-shaped blobs as a result of being erupted under water (Figure 14.2).

The **sheeted dikes zone** contains the feeder dikes that brought lava from a magma chamber in the mantle or lower crust to the surface along mid-ocean ridges (Figure 14.3).

(a)

(b)

FIGURE 14.2

(a) Recently erupted pillow basalts off Hawaii. (Courtesy of NOAA and Colin Vosper; https://en.wikipedia.org/wiki/Pillow_lava.) (b) Outcrop of pillow basalt at Chipley Quarry, UK (https://commons.wikimedia.org/wiki/File:Pillow_Lava_SSSI_Chipley_Quarry_closeup_-_geograph.org.uk_-_1187277.jpg).

FIGURE 14.3
Quaternary sheeted dikes, Kupreanof and Zarembo Islands, southeast Alaska. (From Karl et al. 1999. Reconnaissance Geologic Map of the Duncan Canal-Zarembo Island Area, Southeastern Alaska. U.S. Geological Survey Open-File Report 99-168, Version 1.0; http://pubs.usgs.gov/of/1999/of99-168/of99-168-pamphlet.html.)

The deepest layer, comprising the lower 5 km or so of crust, is gabbro. This was the magma chamber that erupted to form the overlying lavas and dikes. The upper part of the gabbro is more or less uniform. The lower portion, called a **cumulate gabbro**, is layered as a result of separation of minerals by density before crystallization. Heavier minerals settled to the bottom of the magma and lighter minerals rose to the top of the chamber.

Sections of ocean floor that have been pushed and hoisted onto the margins of continents by plate tectonic forces are called **ophiolites**.

The Continental Shelf

According to the United Nations Convention on the Law of the Sea, "The continental shelf of a coastal State comprises the seabed and subsoil of the submarine areas that extend beyond its territorial sea … to the outer edge of the continental margin, or to a distance of 200 nautical miles … where the outer edge of the continental margin does not extend up to that distance."

The geological definition of the continental shelf is simpler: it is "that part of the continent that extends under the sea." The continental shelf extends from the coast to the shelf break at the continental slope (Figure 14.4). The average water depth on the shelf is about 60 m (200 ft). The width of the shelf varies from 1 km (0.6 mi) off the California coast to more than 1290 km (800 mi) off Siberia. On average, it is about 65 km (40 mi) wide. These broad, gently sloping plains are interrupted by **submarine canyons** that carry sediment from continental rivers and shorelines to the deep ocean (Figure 14.5). Much of the continental shelf was emergent and exposed to erosion during the last ice age when sea level was up to 120 m (390 ft) lower than at present.

The relatively shallow water of the continental shelves is home to abundant sea life and holds important fisheries as well as some of the world's largest remaining oil and gas accumulations.

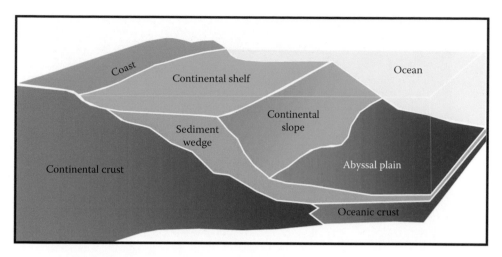

FIGURE 14.4
Relationship of continental shelf, slope, and abyssal plain.

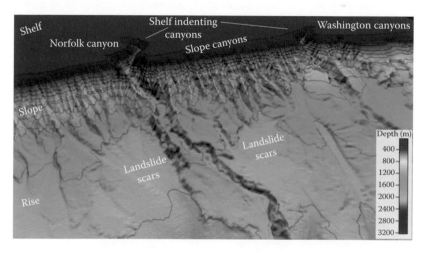

FIGURE 14.5
Sonar image of the continental shelf and slope, offshore Virginia, USA. (Courtesy of NOAA; http://oceanexplorer
.noaa.gov/explorations/11midatlantic/hires/seafloormappingfig3_hires.jpg.)

The Deep Seafloor

The deep seafloor lies between the mid-ocean rise and the continental slope and is known
as the **abyssal plain** (Figure 14.6). The average depth is 3000 to 6000 m (10,000 to 20,000 ft).
These plains are relatively flat as a result of clay and silt deposited by turbidity currents,
and the slow but constant deposition of **pelagic sediments** (mid-ocean sediments consisting mainly of microscopic shells of plankton). Traces of meteorite dust and occasional volcanic ash also occur. These sediments average about 1 km (0.6 mi) in thickness.

An interesting feature of the deep ocean plains are metallic nodules that precipitate
on the ocean floor (Figure 14.7). These nodules contain manganese, iron, nickel, cobalt,

FIGURE 14.6
The abyssal plain is relatively flat and featureless and is in total darkness. Grenadier fish are the most common fish in the deep sea. This *Coryphaenoides leptolepis* is at 3158 m (10,360 ft) water depth at the Davidson Seamount, California. (Courtesy of NOAA; https://en.wikipedia.org/wiki/Grenadiers_(fish).)

FIGURE 14.7
Manganese nodules forming on the seafloor are rich in metals like manganese, iron, nickel, copper, and cobalt. (Courtesy of USGS; https://commons.wikimedia.org/wiki/File:Manganese_nodules.gif.)

and copper. The process that forms them is uncertain: they may be a result of precipitation from seawater, precipitation from submarine hot springs, or precipitation through the activity of microorganisms. These potato-sized **manganese nodules** and are of interest for future mining because they can contain up to 30% manganese, 6% iron, 3% aluminum, 1.5% nickel, 1.4% copper, and 0.25% cobalt.

Trenches

Deep sea trenches are a result of oceanic crust moving down into the mantle at subduction zones (Figures 9.7 and 9.13). They are oceanward of volcanic island arcs (like the Aleutians) or continental volcanic arcs (like the Andes). Trenches form long, narrow depressions in the seafloor that range in depth from 7300 to 11,000 m (24,000 to 36,000 ft). The longest trench is the Peru–Chile trench at 5900 km (3700 mi). They are usually less than 100 km (60 mi) wide.

Most trenches have little or no sediment fill since they are far from sources of sediment and since most sediment in the trench is either subducted or scraped off the downgoing plate and **accreted** (pasted onto) to the overriding plate. Accreted sediment is called a **mélange**, the French word for mixture. The Franciscan Mélange along the central coast of California is a good example (Figures 14.8 and 14.9).

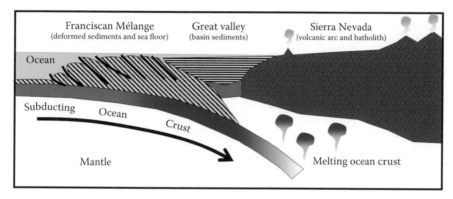

FIGURE 14.8
The Franciscan Mélange was accreted to coastal California during the Jurassic period.

FIGURE 14.9
Franciscan Mélange, Glen Canyon Park, San Francisco, California. The intense folding is a result of the tectonic forces applied to these layers. (Courtesy of Eric Schiff; https://en.wikipedia.org/wiki/Franciscan_Assemblage.)

Trenches are the site of deep earthquakes that occur as the two lithospheric plates grind past each other. At the trench, they occur at depths to 55 km (34 mi); as one goes landward, they occur at ever greater depths, up to 500 km (300 mi) or more. These earthquakes are thought to reveal the location of the brittle downgoing tectonic plate.

The pull of gravity is lower over trenches. This suggests that less dense lithosphere is being forced down into the denser mantle. Heat flow is also lower at trenches than in the abyssal plains, suggesting that a cool lithospheric slab is descending into the mantle.

Mid-Ocean Ridges

Mid-ocean ridges are formed by upwelling magma at oceanic spreading centers. They consist of a central rift and flanking mountains. This is where oceanic crust is made, and these spreading centers are characterized by sheet dikes and pillow lavas in the near surface. These are the youngest rocks in the ocean floor and have the highest heat flow of any of the regions on the ocean bottom. The ridge is the source of many small to moderate earthquakes caused by volcanic eruptions and splitting of the ocean floor. Where spreading is fast, such as in the East Pacific Rise (6 to 16 cm/year), the topographic relief on the mid-ocean ridge is relatively gentle. Where the spreading is slow, as in the Mid-Atlantic Ridge (2 to 5 cm/year), the mountains are fairly rugged. The average depth to the top of the mid-ocean ridge is around 2500 m (8200 ft), but they can be as deep as 4000 m (13,000 ft). The Mid-Atlantic Ridge is exposed at the surface and you can actually walk across it at Thingvellir, Iceland (Figure 14.10).

FIGURE 14.10
You can walk into the central rift of the Mid-Atlantic Ridge at Thingvellir National Park in Iceland. On the west side you are standing on the North American tectonic plate; on the east side you are on the Eurasian plate. (Courtesy of Pmarshal; https://en.wikipedia.org/wiki/Mid-Atlantic_Ridge.)

Mid-ocean ridges are offset by fracture zones. **Triple junctions** are where three spreading centers come together at angles roughly 120° apart. An example is the Rodriguez triple junction in the Indian Ocean (Figure 14.11). Eventually, one of the spreading centers goes dormant. This then becomes a **failed rift**.

An intriguing feature of mid-ocean ridges is the existence of **hydrothermal vents**. Ocean water that has seeped into deep fissures gets heated to 400°C (750°F) or more and gushes from the vents. **White smokers** are mineral chimneys where carbon dioxide–rich water issues from vents carrying dissolved barium, calcium, and silica (Figure 14.12). **Black smokers** contain iron sulfide–rich waters that issue from mineral mounds that have built up around the vents (Figure 14.13). In addition to iron sulfide (pyrite), these waters contain zinc, copper, and trace amounts of gold. **Chemosynthetic communities** of bacteria and archaea thrive in the zone between the superhot fluids and ice-cold seawater. They live totally independent of sunlight and get their energy by oxidizing hydrogen sulfide and converting carbon dioxide into organic material in a process called **chemosynthesis**. These microbes in turn support a community of tube worms, clams, mussels, crabs, and shrimp that feed on them. Some biologists believe life may have begun around these vents.

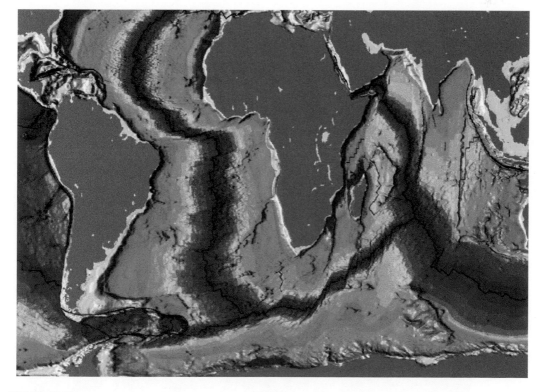

FIGURE 14.11
A map showing the age of the oceanic crust reveals the Rodriguez triple junction in the Indian Ocean. Wide red (young) areas indicate fast spreading rates. (Courtesy of NOAA; https://commons.wikimedia.org/wiki/File: Earth_seafloor_crust_age_1996_-_2.png.)

FIGURE 14.12
White smokers emit high-temperature liquid carbon dioxide at the Champagne vent, Northwest Eifuku vol-
cano, Marianas Trench Marine National Monument. (Courtesy of NOAA; https://en.wikipedia.org/wiki
/Hydrothermal_vent.)

FIGURE 14.13
Black smoker at Sully Vent in the Main Endeavour Vent Field, NE Pacific. Tube worms cover the base of the
vent. An acoustic hydrophone and resistivity-temperature-hydrogen probe surround the vent. (Courtesy of US
Department of Commerce and NOAA; http://www.pmel.noaa.gov/eoi/gallery/smoker-images.html.)

Reference

Karl, S. M., P. J. Haeussler, and A. McCafferty. 1999. Reconnaissance Geologic Map of the Duncan Canal-Zarembo Island Area, Southeastern Alaska. U.S. Geological Survey Open-File Report 99-168, Version 1.0.

15

Mineral Resources

There are strange things done in the midnight sun
By the men who moil for gold;
The Arctic trails have their secret tales
That would make your blood run cold.

Robert Service
Author, from *The Cremation of Sam McGee*

Mining is the art of exploiting mineral deposits at a profit. An unprofitable mine is fit only for the sepulcher of a dead mule.

T.A. Rickard
Mining engineer and editor

The ages of mankind are named after minerals and metals: the Stone Age, the Bronze Age, and the Iron Age. It is fair to say that the development of human culture is intimately tied to mineral extraction.

The world's earliest mines were sites for obtaining flint for tools. Flint tools included spear points, arrow points, scrapers, and flints for striking fire. Egyptian flint mines at Nazlet Khater and Nazlet Sabaha on the Nile were being worked between 33,000 and 100,000 years ago. Salt was mined in Austria's Hallstatt underground mine around 4500 BCE.

Copper is found as a native metal in certain locations and is **malleable**: it can be beaten into useful shapes without breaking. Copper artifacts around 7000 years old have been found in Turkey as well as in Wisconsin in the United States. Someone may have dropped a copper nugget or implement into a campfire and noticed that it melted and formed a new shape when it cooled. It didn't take long to figure out that copper could be made into weapons, pots, and other useful items. Soon, it was noticed that the green and blue rocks found near native copper, when heated, flowed molten copper. Mining these rocks soon followed: 3800-year-old objects smelted from copper ore have been found in Iran. **Gold**, collected as nuggets from streams, was soft enough to work into useful shapes, and the fact that it does not corrode or oxidize made it special: it was used for jewelry, burial relics, cups, figurines, and crowns.

Tin and copper are sometimes found together. It was found that melting copper with tin created an alloy, **bronze**, that was actually harder than either of the two original metals. Bronze weapons took a sharper edge and lasted longer than copper. The earliest bronze tools were found at Ur, in present-day Iraq, dating to around 2800 years ago.

Iron was mined as long as 43,000 years ago at Lion Cavern (Bomvu Ridge) in Swaziland for the red pigment (ochre) derived from hematite. Ochre was mined 12,000 years ago at the San Ramon 15 mine in Chile. Iron as a metal appears around 1500 BCE in Turkey. The Hittites were the first to work with iron. Its higher melting point required the invention of the bellows to create a hotter oven for smelting the metal. Around 1100 BCE, someone in the Middle East discovered that if you add charcoal to molten iron, it becomes harder, and if you quench it in water, it becomes harder still. Thus, steel was discovered, and it soon replaced other metals for tools and weapons.

Metals

Ultimately, all metals come from magma deep in the earth. As the magma cools, some minerals crystallize early and others late. Many metals remain with the late-crystallizing magma and, along with quartz and water vapor, undergo late-phase high-pressure injection into fractures in the surrounding country rock. These cool to form **lode** deposits such as gold–silver–quartz veins (Figure 15.1). Massive, randomly oriented veins are known as **stockworks**.

Other metals crystallize throughout a magma, such as copper and molybdenum disseminated in granite or granodiorite **porphyry** ore deposits (Figure 15.2). Porphyries are named for the **porphyritic texture** of the host rock, a fine-grained igneous rock with large feldspar or quartz crystals (Figure 15.3). Porphyry deposits contain economically important amounts of gold and silver, in addition to copper and molybdenum.

Classifications of metallic ore deposits usually describes the genesis, environment, or geologic setting in which the ore was deposited. **Hydrothermal deposits** are concentrations of metals carried by hot magmatic fluids or groundwater. Volcanic surface vents and fumaroles, both on land and on the sea bottom, are the site of massive deposits of sulfide minerals including iron (**pyrite**), lead (**galena**), zinc (**sphalerite**), and copper (**chalcopyrite** and **bornite**), among others. These are known as **volcanogenic massive sulfide** deposits (Figure 15.4). **Hot spring deposits** form where minerals, chiefly mercury and gold, precipitate as the thermal spring water cools.

Slightly acidic metal-rich hydrothermal fluids frequently come into contact with reactive rocks such as limestone and precipitate metals such as iron, lead, and zinc in **skarn**

FIGURE 15.1
Gold-quartz hydrothermal vein, Red Mountain District, Ouray County, Colorado. (Courtesy of James St. John; https://commons.wikimedia.org/wiki/File:Gold-quartz_hydrothermal_vein_(Red_Mountain_Mining _District,_Ouray_County,_Colorado,_USA)_(16862850998).jpg.)

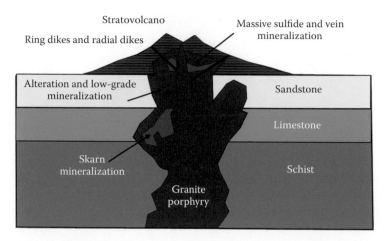

FIGURE 15.2
Deposits associated with porphyries. Hydrothermal fluids from the magma carry metals in solution to wall rocks where they react chemically with limestone and precipitate metals as skarn deposits.

FIGURE 15.3
Porphyritic texture here is a result of large plagioclase crystals (white) in granodiorite. Porphyry deposits have disseminated metals in an intrusive igneous host. Width approximately 10.7 cm. (Courtesy of William Bullock Clark, Maryland Geological Survey; https://commons.wikimedia.org/wiki/File:Granite_Porphyry_PlateXI _MD_Geological_Survey_Volume_2.jpg.)

deposits (Figure 15.5). **Carlin-type black shale disseminated gold** deposits form when hot water circulates through silty carbonate rocks (limestone and dolomite). The original rock is altered to **jasperoid** (a silica–pyrite–hematite rock), and gold, antimony, and mercury are precipitated (Figure 15.6).

Once metals have crystallized from magma or precipitated from hydrothermal fluids, they are available to be reworked and concentrated by other processes. Erosion of a gold–quartz vein moves gold into streams and to beaches where nuggets of the dense, heavy

FIGURE 15.4
Massive sulfide from a seafloor hydrothermal vent, northern East Pacific Rise. The golden-brown material in the upper left is chalcopyrite (a copper–iron sulfide). The remaining dark material is chalcopyrite and sphalerite (zinc sulfide). The white mineral is anhydrite (calcium sulfate). (Courtesy of James St. John; https://commons .wikimedia.org/wiki/File:Sample_of_massive_sulfide_deposit_from_EPR.jpg.)

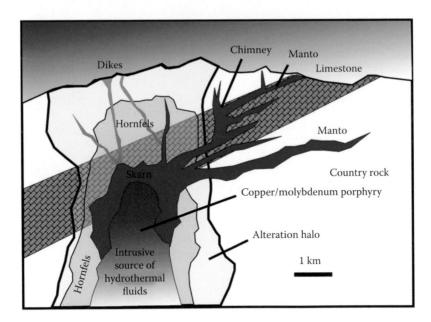

FIGURE 15.5
Schematic diagram of a carbonate replacement deposit with base and precious metal mineralization.

gold are concentrated in **placer deposits** by the winnowing action of currents or waves (Figure 15.7). Other placer deposits include the heavy minerals **magnetite** (iron), **ilmenite** (titanium), **chromite** (iron–chromium), **wolframite** (iron–manganese–tungsten), **cassiterite** (tin), diamond, and garnet.

Groundwater can leach metals from their igneous host rock and concentrate them in blanket-shaped deposits called **mantos** at the water table. Mantos usually follow

FIGURE 15.6
Carlin-type ore. The silicified Lower Ordovician Comus Formation siltstone contains invisible gold. Twin Creeks mine, Nevada (https://en.wikipedia.org/wiki/Carlin%E2%80%93type_gold_deposit).

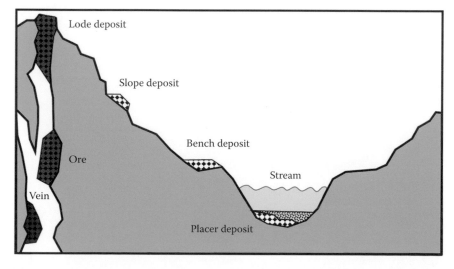

FIGURE 15.7
A deposit found in place is a lode deposit. When it has been eroded, moved, and concentrated it is a placer deposit.

sedimentary bedding (Figure 15.8). In tropical climates, ultramafic rocks rich in nickel and aluminum weather deeply. The dispersed metals leach out of the rock and become concentrated in a residual soil known as aluminum or nickel **laterites** (Figure 15.9).

Slightly acidic groundwater is capable of leaching metals and carrying them until they come into contact with slightly basic carbonate rocks like limestones, where they precipitate

FIGURE 15.8
Copper manto at Mantos Grandes project, Limari province, Chile. Mineralization consists of primary bornite and chalcopyrite with secondary copper oxides (blue and green rocks). (Courtesy of Cobre Montana; http://www.resourcesroadhouse.com.au/_blog/Resources_Roadhouse/post/cobre-montana-moves-chilean -copper-mountain/.)

FIGURE 15.9
Bauxite, an aluminum-laterite. (Courtesy of NASA; http://mars.nasa.gov/mer/classroom/schoolhouse/rock library/source/bauxite.html.)

as **carbonate-hosted** or **Mississippi Valley–type** copper–lead–zinc deposits (Figure 15.10). **Stratiform sandstone copper** deposits precipitate in high permeability sediments where metal-rich oxygenated groundwater comes into contact with an oxygen-poor environment around coal or organic-rich shales (Figure 15.11). These deposits contain copper and frequently silver, lead, and zinc as well.

One classification scheme for ore deposits is provided in Table 15.1.

FIGURE 15.10
Mississippi Valley–type zinc specimen from the Joplin field, Tri-state district, Missouri, USA. Sphalerite, dolomite, and chalcopyrite sample from the Charles Hansen Collection (https://en.wikipedia.org/wiki/Sphalerite).

FIGURE 15.11
Stratabound copper ore from the Mansfeld (Germany) copper deposit. Upper Permian Kupferschiefer ore consists of bornite. Size of the polished section is about 25 × 80 mm. (Courtesy of Ion Tichy; https://commons.wiki media.org/wiki/File:Mansfelder_Kupferlineal.JPG.)

TABLE 15.1

Classification of Mineral Deposits

Hydrothermal epigenetic deposits	• Mesothermal lode gold deposits—Golden Mile, Kalgoorlie • Archaean conglomerate-hosted gold-uranium deposits—Elliot Lake, Canada, and Witwatersrand, South Africa • Carlin-type gold deposits, including epithermal stockwork vein deposits
Granite-related hydrothermal	• Iron oxide copper gold (IOCG) deposits—supergiant Olympic Dam Cu–Au–U deposit • Porphyry copper ± gold ± molybdenum ± silver deposits • Intrusive-related copper-gold ± (tin–tungsten)—Tombstone, Arizona • Hydromagmatic magnetite iron ore deposits and skarns • Skarn ore deposits of copper, lead, zinc, tungsten, and others

Magmatic deposits	• Magmatic nickel–copper–iron–PGE deposits including the following:
	   • Cumulate vanadiferous or platinum-bearing magnetite or chromite concentrated by settling within magma
	   • Cumulate hard-rock titanium (ilmenite) deposits
	   • Komatiite (ultramafic, mantle-derived lava)-hosted Ni–Cu–Platinum Group Elements (PGE)
	   • Subvolcanic (medium to shallow depth intrusive) feeder subtype—Noril'sk-Talnakh and the Thompson Belt, Canada
	   • Intrusive-related Ni–Cu–PGE—Voisey's Bay, Canada and Jinchuan, China
	   • Lateritic nickel ore deposits—Goro and Acoje (Philippines) and Ravensthorpe, Western Australia

Volcanic-related deposits	• Volcanic-hosted massive sulfide (VHMS) Cu–Pb–Zn including Teutonic Bore and Golden Grove, Western Australia; Besshi and Kuroko, Japan
Metamorphically reworked deposits	• Podiform serpentinite-hosted paramagmatic iron oxide-chromite deposits—Savage River, Tasmania iron ore, Coobina chromite deposit • Broken Hill Type Pb–Zn–Ag, a class of reworked SEDEX deposits
Carbonatite-alkaline igneous related	• Phosphorus–tantalite–vermiculite—Phalaborwa, South Africa • Rare earth elements—Mount Weld, Australia and Bayan Obo, Mongolia • Diatreme-hosted diamond in kimberlite, lamproite, or lamprophyre
Sedimentary deposits	• Banded iron formation iron ore deposits, including the following: • Channel-iron fluvial channel sedimentary deposits or pisolite-type iron ore • Heavy mineral sands ore deposits and other sand dune–hosted deposits • Alluvial gold, diamond, tin, platinum, or black sand deposits • Alluvial oxide zinc deposit type: sole example Skorpion Zinc
Sedimentary hydrothermal deposits	• Sedimentary Exhalative (SEDEX)—from release of ore-bearing hydrothermal fluids into the ocean resulting in the precipitation of stratiform ore • Lead–zinc–silver—Red Dog, McArthur River, Mount Isa • Stratiform arkose-hosted and shale-hosted copper—Zambian copperbelt • Stratiform tungsten—Erzgebirge deposits, Czechoslovakia • Exhalative spilite-chert–hosted gold deposits • Mississippi valley type (MVT) zinc–lead deposits • Hematite iron ore deposits of altered banded iron formation
Astrobleme (large impact)-related ores	• Sudbury Basin nickel and copper, Ontario, Canada

Source: Modified after Wikipedia, "Ore."

Exploring for Mineral Deposits

The obvious mineral deposits have all been found. These are the ones where the metal-bearing rocks are exposed by erosion and are right at the surface. Modern prospectors use geophysical techniques, geochemical surveys, and remote sensing to locate indicators of mineralization. Indicators of mineralization include bleaching caused by mineralizing fluids that alter feldspars to clay, and the weathering of pyrite that turns the landscape red (Figure 15.12). **Remote sensing** examines the surface using various sensors, including airborne and satellite cameras, charge-coupled devices, and radar to locate surface alteration zones and host rocks (Figures 15.13 and 15.14). **Geochemical surveys** collect samples at or near the surface, send them to a laboratory to be assayed, and then make maps that show where concentrations of metals are high or low. Geophysical techniques investigate the subsurface using a number of methods.

Electromagnetic (EM) surveys look for metallic conductors. The best EM response comes from massive sulfides. Graphite, pyrite, and pyrrhotite are responsible for most observed responses. EM works by creating an alternating magnetic field by passing a current through a coil. This generates strong currents in conductive deposits and in turn creates a measurable secondary magnetic field.

Magnetic surveys use magnetometers to measure variations in the strength and direction of Earth's magnetic field. They can be used to locate ferrous (iron-bearing) mineral deposits, identify faults, and map igneous intrusions (Figure 15.15).

Resistivity surveys are used to explore for sand and gravel deposits, oil sands, and groundwater, among others. The instrument uses an antenna to produce an electric current in the ground, which then induces a magnetic field. The magnetic field is measured to determine the resistivity of subsurface materials. Metallic deposits are more conductive and less resistive; sand and gravel have higher resistivity.

Induced polarization surveys measure the electrochemical response of subsurface sulfides and clays to an imposed electrical current. A current is introduced using electrodes

FIGURE 15.12
Red Mountain, Colorado. The rusty red color is a result of weathering of iron minerals, such as pyrite, to the iron oxides limonite, goethite, and hematite. (Courtesy of Andreas Borchert; https://en.wikipedia.org/wiki/Red_Mountain_(Ouray_County,_Colorado).)

FIGURE 15.13
True color air photo of phyllic alteration (quartz, sericite, and limonite) around the Grizzly Peak Caldera, Colorado. Mineralization consists of sulfides of molybdenum, copper, and iron in quartz stockworks and veins. (From Dave Coulter, Colorado School of Mines PhD, 2006. Reproduced with permission.)

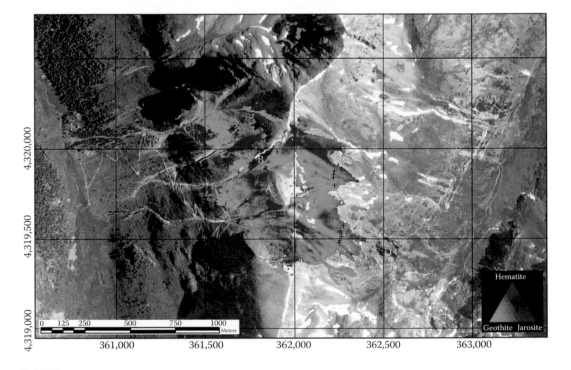

FIGURE 15.14
HyperSpecTIR (HST) image shows distribution of iron minerals at Grizzly Peak, Colorado. Natural acid seeps are located at A, B, and C. Mineral key is in lower right. (From Dave Coulter, Colorado School of Mines PhD, 2006. Reproduced with permission.)

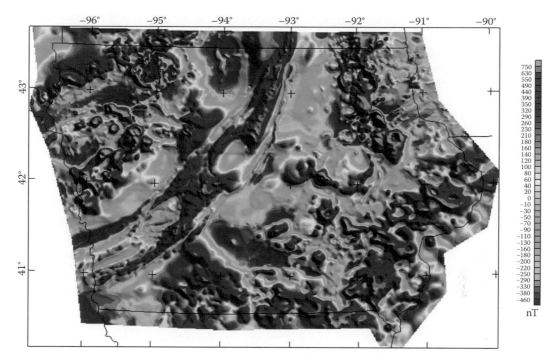

FIGURE 15.15
Magnetic anomaly map of Iowa. The red arc in the center represents deeply buried magnetic basalts in the mid-continent rift, unknown here until discovered by magnetic and gravity surveys. (Courtesy of Robert P. Kucks and Patricia L. Hill, US Geological Survey; https://en.wikipedia.org/wiki/File:Iowa_magnetic_map.jpg.)

and the voltage is monitored through receiver electrodes. Either the voltage decay rate is measured, or the resistivity is measured at different current frequencies.

Radiometric surveys use **gamma-ray spectrometers** or **scintillometers** to detect gamma rays emitted from Earth's surface. Gamma radiation at Earth's surface comes from the decay of uranium, thorium, and potassium in rocks and soil. Radiometric surveys have been used to explore for uranium and to map the extent of granitic intrusions, which are rich in potassium feldspar (Figure 15.16).

When explorers have located a prospective area, they bring in the drills. Carefully placing the drills in an attempt to intersect the subsurface deposits, they use diamond-studded drill bits to cut and recover rock **core** (Figure 15.17). The core is described and sampled for assaying. If metal values are high enough, and the deposit is extensive enough to be economic, a mine may be developed.

Energy Minerals

"Energy minerals" include uranium and coal. While coal is technically *not* a mineral (because it is organic), it is an important energy resource and is commonly called an energy mineral.

FIGURE 15.16
Radiometric image map of part of New South Wales, Australia. This map shows proportions of the isotopes of potassium (red), thorium (green), and uranium (blue) in rocks or soil at or near the surface. (From Geological Survey of New South Wales. Copyright State of New South Wales through Department of Trade and Investment, Regional Infrastructure and Services, Australia; http://www.resourcesandenergy.nsw.gov.au/miners-and -explorers/geoscience-information/products-and-data/maps/statewide-maps/radiometric-map-of-nsw.)

Uranium

Uranium is used primarily to generate heat to run turbines in electrical power plants. The most important uranium minerals are the oxides **uraninite** and **pitchblende**. Less common minerals like **carnotite** (uranium potassium vanadium oxide) and **coffinite** (uranium silicate) are sometimes mined as well.

Igneous uranium deposits occur in granites and **pegmatites** (granitic dikes with very large crystals), disseminated, or as veins. Major deposits include Rössing and Husab in Namibia, Palabora in South Africa, Bancroft in Canada, Radium Hill in South Australia, and Jachymov in the Czech Republic. Granite breccia uranium deposits, such as Olympic Dam, Australia, occur in a hematite-and-copper–rich granite **breccia** (angular broken rock) with minor amounts of gold and silver.

Shear zone uranium deposits are associated with volcanic calderas, such as at Xiangshan, China; Dornod and Gurvanbulag, Mongolia; and Streltovska, Russia.

Sandstone uranium and uranium–vanadium deposits occur in porous and permeable sandstones at **roll fronts**, places where oxygenated groundwater encounters oxygen-poor (reducing) conditions in the presence of organic matter, hydrocarbons, or sulfides (pyrite, hydrogen sulfide; Figure 15.18). Uranium minerals, soluble in oxygen-rich groundwater, precipitate at the contact with a reducing environment. Major deposits include the Mount Taylor

FIGURE 15.17
Core from a mining operation. (Courtesy of Blastcube; https://commons.wikimedia.org/wiki/File:Diamond
_Core.jpeg.)

FIGURE 15.18
This uranium roll front is hosted in Cretaceous Dakota Sandstone near Morrison, Colorado, USA. The dark
rim contains uraninite; the yellow material is the uranium ore tyuyamunite. (Courtesy of James St. John; www
.jsjgeology.net/Home-page.htm.)

and Uravan districts in New Mexico and Colorado; Matoush, Canada; the Lodève District, France; Dalur, Budenovskoye, Tortkuduk, and Khiagda in Russia; Hamr-Stráž pod Ralskem in the Czech Republic; Akouta, Arlit, and Imouraren in Niger; the Franceville basin, Gabon; and deposits in the Karoo Supergroup in South Africa, Zambia, Malawi, and Tanzania.

Metamorphic uranium deposits occur in metasediments and metavolcanics. Examples include Forstau, Austria; Shinkolobwe in the Democratic Republic of Congo; Rozna in the Czech Republic; and Port Radium, Canada. **Metasomatic uranium** is deposited by hydrothermal fluids that alter a host rock by adding sodium and potassium. Examples of this type of deposit occur at Elkon, Russia; Lagoa Real-Caetite, Brazil; Valhala, Australia; and Michelin, Canada.

Proterozoic unconformity uranium deposits, such as those found in Saskatchewan, Canada, and the Alligator Rivers region of the Northern Territory, Australia, exist under, at, and just above unconformities. Uraninite and pitchblende are the primary minerals.

Minor accumulations of uranium occur in phosphate beds, in lignite (low-grade coal), and in karsted carbonate rocks, among others. Uranium can also occur in soils and unconsolidated sediments, usually where uranium-rich granites are deeply weathered and the weathered zone is cemented with calcite. The main mineral is **carnotite**.

Nuclear Waste Repositories

We would be remiss if we didn't address the geologic aspects of **nuclear waste disposal**. There are two categories of radioactive waste: **low-level waste** (LLW), which includes contaminated tools and clothing and some medical and industrial products, and **high-level waste** (HLW), which includes spent fuel rods and nuclear weapons waste. In the United States, nuclear weapons waste is disposed of at the Waste Isolation Pilot Plant in New Mexico. In 2017, there are 100 nuclear power plants operating and 11 new reactors under construction. There is currently 70,000 metric tons of nuclear waste in temporary storage, and the operating plants generate another 2200 metric tons of spent fuel annually.

At the end of their useful life, nuclear fuel rods are stored in pools of water for at least 5 years while they cool and lose most of their radioactivity. Then, they are transported to a storage facility. The ideal facility is isolated from human activity (remote and deep), is in impermeable rock (so it cannot leak), should not be in contact with groundwater, and is tectonically stable (no earthquakes). These facilities are designed to limit public exposure to radiation for 1 million years. Security is also an issue. In the United States, deep salt deposits in New Mexico have been considered because they are easy to mine and are dry, and salt flows rather than fractures. Yucca Mountain, at the Nevada Test Site near Las Vegas, has also been considered. It is in a secure area, in volcanic tuff, and waste can be buried 300 m (1000 ft) below the surface and still be 300 m above the water table. Similar facilities exist in several countries in rocks ranging from mudstone to granite (Table 15.2).

At this time (2017), there is no operating nuclear waste repository for LLW and HLW in the United States. It is not a geological problem. The issue seems to be political: no one wants nuclear waste stored in their backyard, and no one wants trains full of waste moving through their neighborhood. As a mining engineering professor at Colorado School of Mines used to say, burying nuclear waste almost anywhere would be safer than storing it at Hanford in rusting steel drums that are leaking into the Columbia River. This was the mid-1980s, and the site is still a matter of concern (Columbia Riverkeeper 2009; Durden 2017).

TABLE 15.2

Nuclear Waste Facilities

Country	Facility Name	Location	Geology	Depth	Status
Belgium	HADES Underground Research Facility	Mol	Clay	223 m	In operation 1982
Canada	AECL Underground Research Lab	Pinawa	Granite	420 m	1990–2006
Canada	OPG DGR	Ontario	Argillaceous limestone	680 m	License application 2011
Finland	Onkalo	Olkiluoto	Granite	400 m	Under construction
Finland	VLJ	Olkiluoto	Tonalite	60–100 m	In operation 1992
Finland		Loviisa	Granite	120 m	In operation 1998
France	Meuse/Haute Marne Underground Research Lab	Bure	Mudstone	500 m	In operation 1999
Germany	Schacht Asse II	Lower Saxony	Salt dome	750 m	Closed 1995
Germany	Morsleben	Saxony-Anhalt	Salt dome	630 m	Closed 1998
Germany	Schacht Konrad	Lower Saxony	Sedimentary rock	800 m	Under construction
Japan	Horonobe Underground Research Lab	Horonobe	Sedimentary rock	500 m	Under construction
Japan	Mizunami Underground Research Lab	Mizunami	Granite	1000 m	Under construction
Korea	Korea Underground Research Tunnel		Granite	80 m	In operation 2006
Korea	Gyeongju			80 m	Under construction
Sweden	Äspö Hard Rock Lab	Oskarshamn	Granite	450 m	In operation 1995
Sweden	SFR	Forsmark	Granite	50 m	In operation 1988
Sweden		Forsmark	Granite	450 m	License application 2011
Switzerland	Grimsel Test Site	Grimsel Pass	Granite	450 m	In operation 1984
Switzerland	Mont Terri Rock Laboratory	Mont Terri	Claystone	300 m	In operation 1996
USA	Waste Isolation Pilot Plant	New Mexico	Salt bed	655 m	In operation 1999
USA	Yucca Mountain Project	Nevada	Volcanic tuff	200–300 m	Proposed, canceled 2010

Source: https://en.wikipedia.org/wiki/Deep_geological_repository.

Coal

Coal is used to heat homes and cook food in much of the undeveloped world. It is burned to generate steam that turns turbines and generates electricity in much of the developed world. In fact, coal is the biggest source of energy to generate electricity worldwide. **Coking coal**, or **coke**, is derived from coal by heating it in an oven without oxygen to drive off the volatile gases. It is then used as a reducing agent in a blast furnace for the smelting of iron ore. When mixed with iron, coal is a key ingredient in the making of steel.

Coal can be converted to combustible hydrogen gas by combining it with water vapor and oxygen under heat and pressure. Coal can be converted to liquid fuels like gasoline or diesel using the **Fischer–Tropsch process**. This process converts solid coal to a mixture of carbon monoxide and hydrogen gas and ultimately into liquid fuel.

Coal is the organic remains of plants that grew in forests, swamps, and marshes. The coal was preserved by being buried before it could be oxidized. During burial, the plant material goes through a slow metamorphosis by dewatering, compaction, and heating that converts it from plants to **peat**. With continued exposure to Earth's heat and pressure, it becomes **lignite**, **bituminous coal**, and, ultimately, **anthracite**. Geologists estimate that it takes around 10 vertical meters of peat to make a 1-m-thick bituminous coal.

The first important coal beds appear in the Carboniferous period (literally "coal-bearing") between 359 and 299 Ma. Coal occurs as sedimentary layers, usually next to sandstone or shale. The depositional environments include swamps on alluvial plains and river deltas or in coastal lagoons and tidal marshes. Coal beds can be hundreds of meters thick and extend for tens of kilometers, or they can be thin and discontinuous.

Coal was originally collected at the surface. When surface deposits were depleted, it was mined underground. More recently, it has been **strip mined**: if the deposit is thick enough, it becomes economic to strip off the overlying rock and soil up to around 70 m in order to get at the coal (Figure 15.19). Mining regulations in most areas require that the original topsoil be stockpiled and, when the mine is abandoned, the mine be smoothed out and the topsoil replaced (Figure 15.20).

FIGURE 15.19
Eocene Fort Union Formation coal, Wyoming, USA. (Courtesy of US Bureau of Land Management; https://en .wikipedia.org/wiki/Surface_mining.)

FIGURE 15.20
Reclaimed coal strip mine in Ohio, USA. (Courtesy of Ohio Department of Natural Resources. Copyright 2016 Ohio DNR; http://minerals.ohiodnr.gov/.)

Nonmetallic Economic Minerals

Gemstones

Gems are the celebrities of the mineral world, glamorous, flashy, and expensive. Some are large crystals of pure minerals. Others are just brightly colored stone. Their properties such as hardness, color, clarity, size, and rarity make them special. Some well-known precious gems are diamonds, corundum (sapphire and ruby), beryl (emerald and aquamarine), topaz, opal, and jade. Semiprecious stones include tourmaline, turquoise, lapis lazuli, garnet, peridot, amethyst, and citrine, among others. Diamond, corundum, and garnet have industrial uses as well: their hardness makes them well suited for drilling, grinding, sanding, and polishing. Rubies are used in lasers to generate red light.

Diamonds are a form of pure carbon formed in Earth's mantle. They get to the surface during an explosive eruption of **kimberlite**, a mica peridotite or **lamproite**, a silica-deficient magma with rare mantle-derived minerals. The gas-charged magma is ejected at supersonic velocity. The cone-shaped near-vertical pipe and vent is known as a **diatreme** (Figure 15.21). Diatremes weather to topographic depressions that are frequently filled by a pond or lake (Figure 15.22). Explosion breccia forms the crater rim. The surface around the crater is usually littered with broken bits of the crust and mantle.

After mining, the rock is crushed or milled to release the diamonds. Diamonds are separated from the remaining material using grease (since diamonds stick to grease), or by their luminescence (a machine picks them out if they light up under an x-ray). Diamonds are also found as placer deposits in streams and on beaches because of their hardness (10 on the Mohs scale) and density (3.51 g/cm^3). In one case, at Crater of Diamonds State Park near Murfreesboro, Arkansas, diamonds have weathered out of a 95 Ma lamproite volcano and are found by tourists in a plowed field (Figure 15.23). The main diamond-producing countries are South Africa, Russia, and Botswana. Australia produces the most industrial-grade diamonds. Significant amounts also come from India, Brazil, Canada, the United States, and China.

FIGURE 15.21
Oblique aerial view of Moses Rock diatreme (dark rock, center), Cane Valley, Utah, USA. Because the bedding here has been tilted, this is effectively a cross-sectional view through the Earth. (Courtesy of Doc Searles; https://en.wikipedia.org/wiki/Diatreme#/media/File:Moses_Rock_Dike.jpg.)

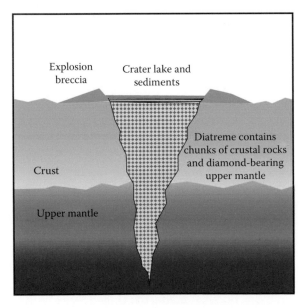

FIGURE 15.22
Cross section of a typical diamond pipe.

FIGURE 15.23
A gem-quality 0.74-carat diamond found at Crater of Diamonds State Park, Pike County, Arkansas, USA. Size: 0.5 × 0.5 × 0.4 cm. (Courtesy of Rob Lavinsky/iRocks.com; https://commons.wikimedia.org/wiki/File :Diamond-260144.jpg.)

"**Blood diamonds**" are diamonds mined in countries with civil wars between rebels and an internationally recognized government. They are exactly the same mineral as other diamonds: it is just that they are produced by rebels through slave labor and are used to buy arms. Their name derives from the fact that they destroy lives.

Corundum, an aluminum oxide, is found in metamorphic gneiss and schist, and in low-silica igneous rock and pegmatites (Figure 15.24). Because of its hardness (9 on the Mohs scale) and density (4.02 g/cm^3), it frequently occurs as a heavy mineral placer deposit in streams and on beaches. The gem forms include **ruby** (red color is attributed to chromium) and **sapphire** (blue and other colors are attributed to titanium and iron). Some of the best deposits of rubies are in Thailand, Cambodia, Burma, India, Afghanistan, Australia, Namibia, and Colombia. Sapphires are mined in Australia, Thailand, Sri Lanka, China, Madagascar, and Montana in the United States.

Beryl is a beryllium aluminum silicate usually found in granite pegmatites, but also in mica schists and carbon-rich shales or limestones that have been intruded by granitic magma or subjected to regional metamorphism. **Emerald** (green variety) contains trace amounts of chromium and vanadium; **aquamarine** (blue and blue-green beryl) is a result of iron impurities (Figure 15.25); **morganite** (pink beryl) is attributed to minor amounts of manganese in the crystal. Emeralds are found in Colombia, Brazil, Zambia, Zimbabwe, Madagascar, Pakistan, India, Afghanistan, Russia, the United States (North Carolina), and Canada (Yukon). Most aquamarines are from Brazil, Sri Lanka, Madagascar, and the United States (Colorado). Morganite comes from Sweden, Brazil, Madagascar, Afghanistan, Mozambique, Namibia, Russia, and the United States (California and Maine).

FIGURE 15.24
Red corundum (ruby) crystal from Luc Yen, Yenbai Province, Vietnam. Size: 3.4 × 3.3 × 2.3 cm. (Courtesy of Rob Lavinsky, irocks.com; https://commons.wikimedia.org/wiki/File:Corundum-180102.jpg.)

FIGURE 15.25
Aquamarine crystal from Nagar, Hunza Valley, Gilgit District, Northern Areas, Pakistan. Size: 12.9 × 10.5 × 10.5 cm. (Courtesy of Rob Lavinsky, irocks.com; https://commons.wikimedia.org/wiki/File:Beryl-271349.jpg.)

Topaz is an aluminum fluoro-hydroxyl-silicate that can be colorless, blue, pale green, yellow, red, orange, and brown (Figure 15.26). It is usually found in granite pegmatites or gas cavities in rhyolite. Famous localities include Russia (Urals), Afghanistan, Sri Lanka, Argentina (Chivinar), the Czech Republic, Brazil, Italy, Germany, Norway, Sweden, Pakistan, Japan, Mexico, Australia (Flinders Island), Nigeria, and the United States (Utah).

Peridot is the mineral olivine, an iron or magnesium silicate. The gem variety of olivine is translucent green (Figure 15.27). It is found in mafic lavas and mantle-derived peridotite. Good quality gemstones are found in Australia, Brazil, China, Myanmar, Pakistan, Sri Lanka, Saudi Arabia, South Africa, Tanzania, Mexico, and the states of Arizona, New Mexico, Arkansas and Hawaii in the United States.

Opal is an **amorphous** (noncrystalline) **hydrated** (water-bearing) form of silica (Figure 15.28). It is deposited by groundwater in fissures and veins and can be found in any kind of rock. Opals that have weathered out of their host rock are considered **alluvial** (eroded, lose sediment) or placer deposits. The play of colors, called "fire," is attributed to light refracting through microscopic silica spheres within the opal. Finished stones are rounded and polished on one side and flat on the back. Historically, most opal has come from the state of South Australia, but recently good-quality gem opal has also come from Ethiopia and the state of Nevada.

Jade is a green gemstone valued for its warmth and beauty. Jade is actually two metamorphic rocks. The type known as **jadeite** is a sodium-rich aluminous pyroxene, whereas **nephrite** jade is a calcium-rich magnesium–iron–aluminous amphibole (Figure 15.29). Both types of jade range in color from white to light and dark green, yellow, brown, purple, and black. Nephrite is the more common form. It is found in low- to moderate-pressure and low- to moderate-temperature rocks such as dolomite and silicic rocks associated with serpentinite in subduction zones. Jadeite is restricted to low-temperature, high-pressure rocks and is exclusively found in serpentinite. Because it is tougher than the host rock, jade erodes to boulders and frequently has an iron oxidation rind. Important jade regions

FIGURE 15.26
Imperial Topaz from Ouro Preto, Minas Gerais, Brazil. (Courtesy of Madereugeneandrew; https://en.wikipedia.org/wiki/Topaz#/media/File:Imperial_Topaz.JPG.)

FIGURE 15.27
Peridot from Suppat, Kohistan District, Khyber Pakhtunkhwa, Pakistan. Size: 2.0 × 1.6 × 1.2 cm. (Courtesy of Rob Lavinsky/iRocks.com; https://commons.wikimedia.org/wiki/File:Forsterite-121354.jpg.)

FIGURE 15.28
Blue banded opal from Barco River, Queensland, Australia. (Courtesy of Aram Dulyan [User: Aramgutang] and the Natural History Museum, London; https://commons.wikimedia.org/wiki/File:Opal_banded.jpg.)

FIGURE 15.29
Two varieties of jade. (a) New Zealand nephrite, Museum of New Zealand Tepapa Tongarewa, Wellington. (Courtesy of Wmpearl; https://commons.wikimedia.org/wiki/File:New_Zealand_nephrite,_Museum_of_New _Zealand_Tepapa_Tongarewa,_Wellington.jpg.) (b) California jadeite. (Courtesy of Akos Kokai; https://commons .wikimedia.org/wiki/File:California_jadeite_(8044828151).jpg.)

include Myanmar, Turkestan, New Zealand, the Canadian province of British Columbia, and the US states of Wyoming and Alaska.

Tourmaline, a boron silicate, is a semiprecious gem. Gem varieties are either pink or green, and sometimes pink and green (see Tourmaline in section on Minerals). Occasionally, it is red, blue, or black. The colors are attributed to trace amounts of lithium, magnesium, aluminum, iron, sodium, and potassium. Tourmaline is found in granite pegmatites, granite, schist, and marble. The best gems come from Brazil, Afghanistan, Pakistan, Tanzania, Mozambique, Kenya, Nigeria, and Namibia.

Turquoise, another semiprecious stone, is a blue-green opaque mineral (Figure 15.30). It is a hydrated copper–aluminum phosphate that is deposited when slightly acidic water percolates through weathering copper sulfide deposits (chalcopyrite and bornite) or copper carbonates (**malachite** and **azurite**). Turquoise is usually found in arid climates in veins and cavities in granite porphyries and volcanic rocks. Important sources of turquoise include Iran, Egypt (Sinai), Chile, and the states of Arizona and New Mexico in the United States.

Lapis lazuli is a deep blue opaque stone that is a combination of three minerals: lazurite, calcite, and pyrite (Figure 15.31). It is a sodium- or calcium-bearing aluminum silicate usually found in metamorphic rocks or in **contact metamorphic** limestone (limestones that have been intruded by magma or hot fluids). The best deposits are in Afghanistan, but other sources are found in Chile, Russia (Siberia), Pakistan, Burma, India, Angola, Argentina, Canada, and the states of California and Colorado in the United States.

Garnets belong to a family of metamorphic silicate minerals that all grade into one another. All have the same physical properties and crystal form, but their compositions vary slightly. The color can be any shade of purple, red, green, and yellow, and is a result of varying amounts of calcium, magnesium, iron, manganese, aluminum, vanadium, or chromium (Figure 15.32). Some of the more common varieties are almandine (red), pyrope (red to purple), spessartine (orange to red), andradite (green and yellow), grossularite (green, yellow, red, and clear), and uvarovite (green). Garnets are usually found in gneiss and schist, but can also be found in mantle and lower crustal rocks such as eclogite and peridotite, and can result from contact metamorphism in limestone. Garnets are frequently found concentrated as placer deposits in stream channels and on beaches because of their high specific gravity and hardness.

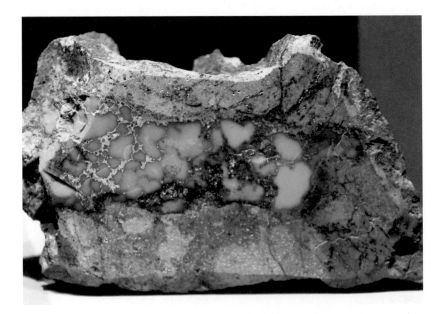

FIGURE 15.30
Turquoise from Cerillos, New Mexico, at the Smithsonian mineral collection, Washington D.C. (Courtesy of Tim Evanson; https://commons.wikimedia.org/wiki/File:Turquoise_Cerillos_Smithsonian.jpg.)

FIGURE 15.31
Lapis Lazuli with pyrite from Afghanistan. (Courtesy of Hannes Grobe; https://en.wikipedia.org/wiki/Lapis
_lazuli.)

FIGURE 15.32
Garnet crystals, Jeffery Mine, Quebec, Canada. (Courtesy of US Geological Survey; https://en.wikipedia.org/wiki
/Garnet.)

FIGURE 15.33
Citrine crystal from Ouro Preto, Minas Gerais, Brazil. Size: 3.5 × 3 × 2 cm. (Courtesy of Mike Keim. Copyright 2010–2017, Marin Mineral; http://www.marinmineral.com/quartz.html.)

Amethyst is purple quartz, or silicon dioxide. The color is the result of trace amounts of iron. Amethyst is found as crystals in open cavities and veins in igneous rocks. Good-quality stones come from Minas Gerais state in Brazil, Mexico, Uruguay, Russia, Austria, South Korea, India, Zambia, the provinces of Ontario and Nova Scotia in Canada, and the states of Arizona, Colorado, Texas, among others, in the United States.

Citrine is a transparent, pale yellow to reddish orange quartz (Figure 15.33). It is found in silica-rich volcanic and intrusive rocks in cavities and veins. The color is a result of trace amounts of iron. The best specimens come from Brazil, but it is also found in Russia, Madagascar, Congo, and Kazakhstan.

Rare Earth Minerals

Rare earth elements are essential components in high-tech electronics such as cell phones, silicon chips, light-emitting diodes (LEDs), compact fluorescent lamps, color monitors and television sets, batteries, superconductors, lasers, fiber optics, camera lenses, and x-ray machines as well as in special glass, drugs, and cleaning agents. Rare earth minerals contain one or more of 17 rare earth elements (Table 15.3).

Rare earth minerals are not particularly rare in the Earth's crust. They are, however, rarely concentrated in ore deposits, and are difficult to separate and process. They are considered "strategic" minerals because of their high-tech and military applications and because good deposits are few and far between.

TABLE 15.3

Rare Earth Elements and Some of Their Uses

Element	Use
Scandium	Strengthens aluminum for aerospace and sports (bikes, golf) applications, high-intensity street lamps, high-performance equipment
Yttrium	TV sets, cancer treatment drugs, strengthens aluminum and magnesium
Lanthanum	Camera lenses, battery electrodes, hydrogen storage
Cerium	Catalytic converters, colored glass, steel production
Praseodymium	Lasers, magnets, welding goggles
Neodymium	Extremely strong magnets, electric motors, lasers
Promethium	Nuclear batteries, luminous paint
Samarium	Nuclear reactor control rods, cancer treatment, x-ray lasers
Europium	Color TV screens, fluorescent glass, genetic screening
Gadolinium	Nuclear reactor shielding, nuclear marine propulsion, improves durability of metal alloys
Terbium	Sonar, TV sets, fuel cells
Dysprosium	Lighting, computer memory, transducers
Holmium	Lasers, colored glass, high-strength magnets
Erbium	Glass coloring, fiber-optics signal amplifiers, metallurgy
Thulium	Lasers, portable x-ray machines, high-temperature superconductors
Yttrium	Stainless steel, portable x-ray machines, lasers
Lutetium	Petroleum refining, LEDs, integrated circuits

Common rare earth minerals include apatite, fluorite, monazite, and zircon. Less common rare earths include aeschynite, allanite, bastnäsite, britholite, brockite cerite, fluocerite, gadolinite, parisite, stillwellite, synchysite, titanite, wakefieldite, and xenotime.

Rare earth minerals are usually found in **alkaline igneous rocks**, that is, rocks with more alkali metals (sodium and potassium) than silica. They are found in alkaline pegmatites and carbonatites as well as hydrothermal deposits associated with these rock types. Originally, most rare earths were mined as placer deposits in Brazil and India. Then, large vein deposits were found in South Africa and at the Mountain Pass mine in California. Now, most of the world's rare earth minerals come from Inner Mongolia in China.

Salt

Salt is essential for animal life. It is the mineral **halite**, and is abundant dissolved in the ocean. Salt is mined from evaporites, deposits formed by evaporation of seawater. Some of the major salt mines in the world are in Austria, Bosnia, Bulgaria, Canada, England, Germany, Italy, Morocco, Northern Ireland, Pakistan, Poland, Romania, Russia, and the United States (Figure 15.34).

Salt is used both as a food seasoning and as a food preservative. It is used in industry to make caustic soda, chlorine, plastics, and polyvinyl chloride; in road de-icing; in water conditioners; and in paper processing, among others.

Potash

Potash deposits contain water-soluble potassium, most commonly potassium chloride. Potassium is an essential element in fertilizers. Like salt and gypsum, potash is an

FIGURE 15.34
Salina Veche, Europe's largest salt mine, in Slănic, Prahova, Romania. The railing (lower middle) gives an idea of scale. (Courtesy of Andrei Stroe; https://commons.wikimedia.org/wiki/File:Slanic_Salt_Mine.jpg.)

evaporate deposit: it is found as layers rich in potassium salts that were deposited as a result of evaporating seawater.

Large deposits are found in Canada (Saskatchewan), Russia, Belarus, Brazil, Germany, and New Mexico and Michigan in the United States.

Phosphates

Phosphorus is required for photosynthesis and is a component of bones and teeth. **Phosphates** are rocks with high concentrations of phosphate minerals, mainly fluorapatite and hydroxyapatite. Fluorapatite is found in hydrothermal veins, pegmatites, and intrusive masses accompanying carbonatite and alkaline igneous rocks. Both fluorapatite and hydroxyapatite are found in phosphorites. **Phosphorites** are extensive dark brown to black marine mudstones and limestones enriched in phosphate minerals (Figure 15.35). For example, the Permian Phosphoria Formation in the states of Wyoming, Idaho, and Utah can be more than 400 m (1300 ft) thick and covers 350,000 km^2 (135,000 mi^2). Phosphate deposits also include organic **guano** (poop) deposited by bats and sea birds.

FIGURE 15.35
Phosphorite (4.6 cm across) from the Permian Phosphoria Formation, Simplot Mine, Bingham County, Idaho, USA. (Courtesy of James St. John; https://en.wikipedia.org/wiki/Phosphorite.)

The main use of phosphates is as an ingredient in fertilizer. It is also employed in water treatment, metallurgy, food preservatives, anticorrosion, cosmetics, ceramics, and fungicides. The ocean is the main reservoir for phosphates, and in places where cold, phosphate-rich bottom waters come to the surface, they support lavish marine ecosystems.

Significant phosphorite deposits occur in South Africa, China, Brazil, United States, Russia, France, Belgium, Spain, Morocco, Tunisia, Algeria, and Australia.

Sulfur

Sulfur is both mined and is produced as a byproduct of natural gas production. Mined sulfur comes mainly from volcanic fumarole and hot springs deposits. Anaerobic bacteria generate sulfur by breaking down calcium sulfate (gypsum) in salt domes. It is used as a fertilizer additive, in matches, insecticides, fungicides, and medicine.

Silica

Properly known as **silicon dioxide**, **silica** is the mineral quartz. It is mainly used in the making of glass and in the production of Portland cement. Silica is also used to make transistors, solar cells, optic fibers, and microchips (as in "Silicon Valley"); in ceramics; in sand casting of metals; to absorb moisture in packing materials and food processing; as a desiccant; in pest control powder; as a filtering agent in swimming pools and brewing; as a flow agent in powdered foods and medicine; and as an abrasive in toothpaste.

Silica is the main constituent in most sand. Sand is found on beaches, in rivers, in sand dunes, and as sandstone. Industrial grade deposits are essentially pure quartz. Silica is also mined from veins.

Calcium Carbonate

Calcium carbonate, in the form of limestone, is a common industrial mineral. Its main uses are as aggregate for structural foundations and roads and as the feedstock for cement

FIGURE 15.36
Rip rap is used to stabilize a river bank. (Courtesy of US Army Corps of Engineers; https://commons.wiki
media.org/wiki/File:Riprap.jpg.)

FIGURE 15.37
Dimension stone on Chicago's Monadnock building, Dearborne Street façade, showing granite entryways.
(Courtesy of Zol87; https://en.wikipedia.org/wiki/Monadnock_Building#/media/File:Monadnock_Building
_East_Facade.jpg.)

and concrete. A pure limestone is crushed and heated to temperatures above 825°C to create **lime** (calcium oxide). The lime can then be used to make concrete by adding back water and allowing it to dry in air.

Other uses for calcium carbonate include as a flux in smelting and metal refining, and as a soil neutralizer in agriculture. Limestone is mined just about anywhere it is found.

Sand, Gravel, and Rock

Sand and gravel, collectively known as **aggregate**, is the most valuable mined material, worth more than all other mineral products combined. Why should humble **granular materials** be worth so much? Simply because we use so much of it. Aggregate is used to make concrete, in asphalt pavement, as a road base and railroad track base, as a stable foundation for buildings to prevent uneven settling, and to improve drainage. Large stones are used as **rip rap**, or rock armor, to stabilize shorelines, riverbanks, road embankments, and other steep slopes (Figure 15.36).

Aggregate is found as beach and river channel deposits, as sandstone and conglomerate, or is made by crushing rock. Crushed sand, gravel, and stone can come from sandstone, limestone, igneous rocks (granite, volcanic cinders), and metamorphic rock. Most aggregate and stone is found and consumed locally.

Building Stone (Dimension Stone)

Building stone is any rock that is cut and finished to face buildings, to use as flooring, roofing material, columns, countertops, and so forth (Figure 15.37). The important qualities are color, patterns, and durability. The stone can be igneous (such as granite countertops), metamorphic (slate roofing, marble for building facing, walls, columns, and countertops), or sedimentary (sandstone or travertine building facing).

Most building stone comes from Brazil, Italy, and China.

References

Columbia Riverkeeper. 2009. Hanford and the River. columbiariverkeeper.org, http://columbiariver keeper.org/wp-content/uploads/2011/09/hanford_and_the_river_final2.pdf

Durden, T. 2017. "Serious Situation" after Tunnel Collapse at WA Nuclear Facility; Evacuation Ordered, No-Fly Zone in Place. May 9th. http://www.zerohedge.com/news/2017-05-09 /emergency-alert-declared-hanford-nuclear-facility-washington-evacuation-ordered

Wikipedia. Ore. Accessed online June 12, 2016. https://en.wikipedia.org/wiki/Ore

16

Black Gold (Texas Tea)

… And up through the ground come a bubblin crude.
Oil that is, black gold, Texas tea.

Paul Henning
From *The Ballad of Jed Clampett*

Whoever controls oil controls much more than oil.

John McCain
US senator

We're not running out of oil. There is plenty of oil left in the ground to last us many
decades, if not longer. We are, however, running short of cheap oil…

Peter Tertzakian
Economist, author, investment strategist, in *A Thousand Barrels a Second*

Petroleum is used for everything from transportation fuel to lubricants, for power generation to
making plastics, for paving and heating, for synthetic fibers in clothing and carpets, for paints
and pesticides, and for chemical feedstocks, pharmaceuticals, and fertilizers. We use hydrocarbons because on an energy-unit basis they are, along with coal, the least expensive source of
energy. People will continue to use hydrocarbons until the cost of **alternative energy** (nuclear,
solar, wind, and tidal) becomes competitive with oil- and gas-based energy (Figure 16.1) (see,
e.g., Open Energy Information 2016; US Energy Information Agency 2016). Since hydrocarbons
are a finite resource, they become more expensive as they get harder to find. That encourages
conservation and fosters use of alternative energy. Petroleum could also become competitive
with alternative fuels through policy choices like increased taxation. Some uses are harder to
convert to alternative energy than others: for example, it is difficult to fly a passenger aircraft
using solar power or to make plastics and lubricants without petroleum.

People have been using petroleum products since the dawn of time. Weathered oil leaves
a tarry residue known as **bitumen**. Bitumen has been found on Neanderthal stone tools
dating back 40,000 years. As long as 5000 years ago, natural oil and bitumen seeps in the
Middle East were used to calk boats, to waterproof baskets and baths, as mortar in stone
buildings, for paving roads, and for medicinal purposes. Egyptians used petroleum for
embalming mummies, and Romans used bitumen for mortar.

Early on, oil was used to make weapons. In 480 BCE, the Persians used oil-soaked flaming arrows in the siege of Athens. North American natives used oil to secure arrowheads
to shafts and weapons to handles, as well as to waterproof canoes and baskets and as a

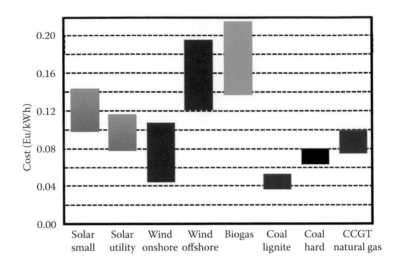

FIGURE 16.1
Comparison of the leveled cost of electricity (LCOE) for newly built renewable and fossil-fuel power stations in Euro/kWh in Germany, 2013. Calculation assumes a given annual global horizontal irradiation (GHI, in kilowatt-hours per square meter) for photovoltaics (solar) and a certain number of annual full load hours (FLH) for wind turbines, biogas, and conventional power plants. Investment-related factors are also taken into account. Small-scale solar, annual GHI of 1000 to 1200 kWh/m². Utility-scale solar, annual GHI of 1000 to 1200 kWh/m². Wind—onshore, annual FLH of 1300 to 2700 h. Wind—offshore, annual FLH of 2800 to 4000 h. Biogas, annual FLH of 6000 to 8000 h. Coal—lignite, annual FLH of 6600 to 7600 h. Coal—hard, annual FLH of 5500 to 6500 h. CCGT combined cycle—natural gas, annual FLH of 3000 to 4000 h (https://en.wikipedia.org/wiki/Cost_of_electricity_by_source).

medicine for colds, cuts, and burns. The Seneca tribe in upstate New York collected oil as body paint and for medicine: locally crude oil was referred to as "Seneca Oil" and the first settlers in the area bottled and sold Seneca Oil as a cure-all.

The Roman naturalist Pliny the Elder commented on the "**eternal fires**" in Azerbaijan and Persia. These were burning natural gas seeps. The "Pillars of Fire" near Baku in Azerbaijan are thought to have inspired the Zoroastrian religion.

The Chinese were drilling oil wells 240 m (800 ft) deep with bamboo poles in the third century CE. Using bamboo pipes to bring gas to their homes, they used natural gas for lighting. They burned oil for heat and to evaporate brine to make salt.

A milestone in the use of oil was the discovery of **distillation**. The Greeks of Alexandria had invented distillation in the first century CE. Naphtha, distilled from oil, may have been a component of "**Greek Fire**," an incendiary weapon developed by the Byzantine Empire around 672 CE. According to the Persian Muhammad ibn Zakarīya al Rāzi (854–925 CE), Arab and Persian chemists were distilling crude oil to create flaming weapons. Persian chemists also distilled kerosene to burn for light in the ninth or tenth century.

Marco Polo visited Baku in Azerbaijan during the 1200s and wrote that "Near the Georgian border, there is a spring from which gushes a stream of oil in such abundance that a hundred ships may load there at once." In the late 1300s, Yaqut al-Hamawi wrote that Baku had two oil fields that brought in 2000 silver dirhams a day from exports.

Russia drilled its first oil well and built a "rock oil" refinery in 1745 in Ukhta in the present-day Komi Republic. The kerosene produced was used in oil lamps in churches and monasteries. In the same year, the French were mining oil sands in Alsace. In 1846, the first well was drilled using a percussion tool drilling rig near Baku, reaching a depth of 21 m (69 ft). **Percussion tool rigs**, or cable tool drills, work by repeatedly raising and dropping a heavy chisel-like tool to create a hole. Modern **rotary drilling rigs** make holes by turning a drill bit at the end of a string of pipes. The bit grinds up rock in the well, and the **cuttings** are brought to the surface by a mixture of water and mud injected into the well.

In 1847, the Scottish chemist James Young began distilling oil that seeped from the roof of a coal mine near Alfreton, Derbyshire. He refined oil to use in lamps as well as a heavier type of oil to lubricate machinery. His oil was so popular that the supply was soon exhausted. Young began experimenting with creating oil from the coal. He found that, at low heat, the coal yielded a petroleum-like fluid that could be used to make paraffin wax, among other products. By 1851, he opened the first commercial refinery in the world producing lubricating oils, paraffin, and naphtha.

Abraham Gesner, a Canadian geologist, figured out how to refine kerosene from coal and oil shale. His Kerosene Gaslight Company, founded in 1850, provided lighting to Halifax, Nova Scotia, and by 1854, he had expanded to New York. The invention of the kerosene lamp in 1854 created even more demand for refined oil products.

Hand dug oil wells were excavated in Poland and Romania in the 1850s, and small refineries were built nearby. In 1858, James Williams dug a well at Oil Springs, Ontario (Canada) that touched off a local oil boom, as did the drilling of a well near Titusville, Pennsylvania, by Edwin Drake and his Seneca Oil Company in 1859. Drake's 21-m (69-ft) well is considered the beginning of the oil industry in the United States. The output of these early wells went to the production of kerosene for lamps, which was quickly replacing whale oil, and to natural gas street lighting. Even before the invention of the internal combustion engine for automobiles, the demand for kerosene outstripped supply and led to exploration and oil booms in Texas, Oklahoma, Louisiana, and California.

In the early years of the twentieth century, large oil discoveries were made in the Golden Lane near Tampico, Mexico, in Lake Maracaibo, Venezuela, on Sumatra in the Dutch East Indies, and in Persia (now Iran). The huge popularity of the automobile and the rapid changeover in trucking and railroads to oil spurred demand. The conversion of navies from coal to oil led to an international effort to find secure sources for national defense, especially after World War I. The US embargo of oil to Japan after they invaded Korea and China was a factor in their bombing Pearl Harbor. Japan wanted to keep the United States from contesting its invasion of Indonesia (to secure oil supplies). Germany's alliance with Romania and its North African and southern Russian campaigns were all based on securing sources of oil for their industry and the war effort. As Daniel Yergin states in *The Prize*, "Oil has meant mastery throughout the Twentieth Century."

By the 1940s, large discoveries had been made in Alberta, Canada, and the first offshore wells were drilled in the Caspian Sea off Baku.

We use a lot of oil and gas. In 2016, the world used about 96 million barrels per day (13.1 million metric tons/day) of oil, or 35 billion barrels (4.8 billion metric tons) per year. We used 9.6 billion m^3 (123.6 trillion ft^3) of natural gas during 2014. Much of the growth in demand comes from developing countries that are trying to replace wood and coal with cleaner sources of energy. There are 1.3 billion internal combustion engines in the world. Every year, another 90 million new engines are added while 40 million are junked, leaving a net increase of 50 million engines. About 1 million new electric engines are added each year. The electricity to run these engines comes primarily from coal-fired power plants. At this rate, unless something truly dramatic happens, it will be many, many years before alternative energy replaces carbon-based energy.

How Is Oil Formed?

Hydrocarbons are complex molecules of hydrogen and carbon. Contrary to popular belief, hydrocarbons are not formed from the decaying remains of dinosaurs. With the exception of methane, which can be an inorganic volcanic gas, hydrocarbons are the remains of mainly planktonic algae and woody material (Figure 16.2). These remains have been buried in the earth and cooked, like a dinner in a pressure cooker, by Earth's heat and pressure. Depending on the amount of heat and pressure, the hydrocarbons cook relatively quickly or slowly.

The **petroleum system** incorporates all of the essential components and processes that lead to oil and gas accumulations. These include the source rock, reservoir rock, seal rock, and overburden rock, and the processes include the generation, migration, and accumulation of hydrocarbons in a trap of some kind. All of the elements of this system must occur in a specific sequence in time and space in order for petroleum to accumulate. A sediment with large amounts of organic matter and the potential to generate hydrocarbons is a **source rock**. Regional depressions where large volumes of sediments accumulate are called **sedimentary basins** (Figure 16.3). Source rocks are deposited in these basins, usually under anoxic conditions, that is, there is not enough oxygen in the system to support the bacteria that would consume the organic matter. As the basin subsides, more sediments are deposited over the source rock and it gets warmer and is under more pressure. The sediments overlying the source rock are the **overburden**.

Source rock that has been buried in an environment too cool to generate hydrocarbons is said to be **immature**. Once the temperature is between 50°C and 130°C (120°F–266°F), the source rock will start to generate oil. When the source rock is generating oil, it is at **peak oil maturity**, or in the "oil window." If the heating continues, to between 130°C and 200°C (266°F–390°F), the organic matter will generate natural gases such as methane, ethane, propane, and butane, among others. At this point, the sediment is considered **past peak oil maturity** and in the **main gas maturity** phase, or in the "gas window" (Figure 16.4). If it gets hotter than this, the remaining organic matter is converted to graphite.

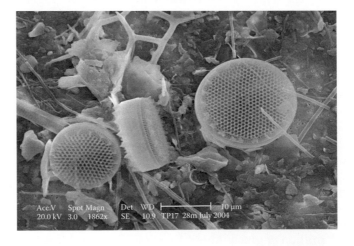

FIGURE 16.2
Marine diatoms, a type of algal plankton. Contrary to common belief, it is these critters, not dinosaurs, that create oil when they die. (Courtesy of Hellenic Centre for Marine Research; scanning electron microscope image by Kostas Tsobanoglou; https://commons.wikimedia.org/wiki/File:Diatoms-HCMR.jpg.)

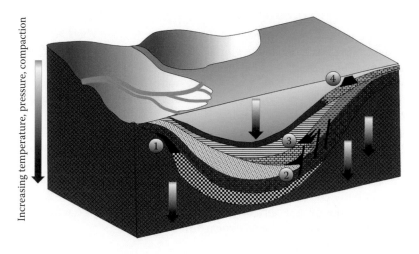

FIGURE 16.3
Schematic diagram of a sedimentary basin. Black indicates oil accumulations. (1) Updip sand pinchout trap. (2) Updip fault trap. (3) Updip facies change from porous limestone to shale. (4) Limestone reef trap. Brown, green, and purple units are organic-rich shale—potential source rock. Yellow units are sandstones—good reservoir rock. Blue units are limestones and reefs—also good reservoir rock. Rivers bring sand to the delta and beaches, both good environments to deposit reservoir rocks.

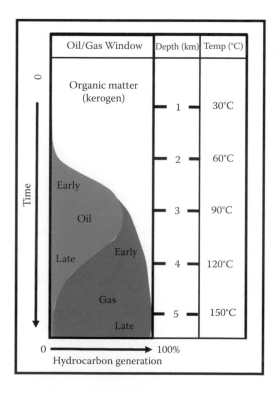

FIGURE 16.4
Hydrocarbon generation chart. Organic matter consists of insoluble **kerogen** and soluble **bitumen**. The deeper you go, the warmer it gets, and the more hydrocarbon (bitumen) is generated. Oil is generated first, then gas.

In the Gulf of California, diatoms, a major group of algal plankton, settle to the bottom when they die. The high heat flow over the East Pacific Rise in the Gulf cooks these diatoms almost instantly: crude oil has been discovered in recent sediments deposited over hydrothermal vents at temperatures between 116°C and 226°C (241°F–439°F). On the other hand, algal organic matter in the 742 to 800 Ma Chuar Group (black shales and dolomites in the Grand Canyon) has been subjected to low heat flow over all that time: the rocks are just now in the main oil generating window (Uphoff 1997).

Once oil is generated, it is literally under pressure to migrate out of the source rock. This process has been compared to a pot of soup on the oven: when the soup gets too hot, the pot boils over, lifting the lid momentarily to relieve the pressure inside the pot. So too newly generated hydrocarbons are expelled and migrate into adjacent **carrier beds**, usually sandstone, with the porosity and permeability that allows oil to migrate. The analogy to cooking continues: the area where oil is generated, usually in the warmer, deeper parts of a sedimentary basin, is called the **hydrocarbon kitchen**.

Source Rocks

Source rock typically has more than about 1% **total organic carbon**. This usually means organic-rich black shale, but it can also be organic-rich limestone or coal beds. The source rocks that generate oil usually have organic matter derived from algae or plankton. The source rocks that generate gas usually contain woody organic matter. Oil shale is a source rock that generated oil but never expelled it, so the oil is still trapped inside the rock.

Source rocks are able to preserve organic matter because they have little permeability: oxygenated groundwater is not able to penetrate the rock and degrade the organic matter. They are usually deposited as mud in quiet water, either in swamps on river deltas or in coastal lagoons in the case of coal, or deep anoxic lake or marine environments. They are preserved through burial.

Generation, Migration, and Entrapment

When the organic matter in a rock is mature and oil or gas is generated, it squeezes out of the pores in the source rock. This can be a slow movement from pore to pore, or the high pressure can create or open fractures that allow the oil to move through them to a carrier bed. Oil and gas are both lighter than water, so they **migrate**, or move upward along the bed by buoyancy, displacing the water already in the pores. They move up until the carrier bed changes laterally to a low-porosity or low-permeability sediment. This is a **stratigraphic trap**. Or the hydrocarbon migrates up until it encounters a fault or geologic structure such as a dome or anticline (Figure 16.5). This is a **structural trap**. An essential component of a trap is an impermeable **seal**. The seal is usually a shale or salt bed above the trap that prevents oil or gas from leaking upward out of the trap. Another requirement is the presence of good **reservoir rock**. This may be the carrier bed or another porous and permeable unit. It is usually a sandstone or limestone, and the hydrocarbons are found in the pores and voids (including open fractures) in the reservoir rock (Figure 16.6).

Methane hydrate is an unusual hydrocarbon occurrence in that the gas molecule is trapped in ice crystals. Hydrates occur under special temperature and pressure conditions found near the seafloor and under arctic permafrost. One cubic meter (35.3 ft^3) of hydrate contains 163 m^3 (5756 ft^3) of methane, so there are huge amounts of methane trapped in the near-freezing conditions of the hydrate stability zone. As yet, there is no economic way to produce hydrates.

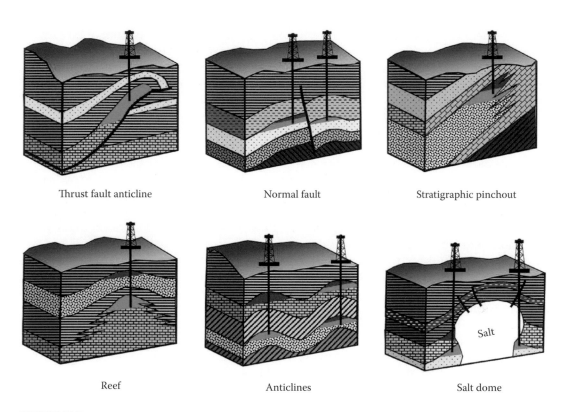

FIGURE 16.5
Some common hydrocarbon traps. Red indicates gas; green indicates oil accumulations.

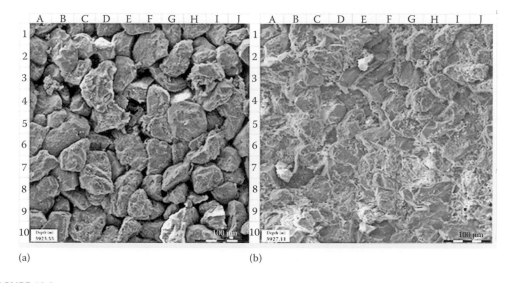

FIGURE 16.6
Oil does not occur in underground lakes. It occupies the pores between grains. (a) Scanning electron microscope image of pores in sandstone reservoir rock. (b) Nonreservoir sandstone has clay filling the pores. (Courtesy of Bien Co; https://www.linkedin.com/biz/6403984/feed?start=10.)

It may seem obvious, but not all traps have hydrocarbons. It may be that an area with good traps never generated any hydrocarbons. The oil may have been generated before the trap was formed and thus migrated right to the surface. Or the trap may have had hydrocarbons but then was faulted or tilted or eroded and the hydrocarbons escaped.

Preservation and Destruction of Hydrocarbons

Oil is generated when an oil source rock reaches the oil window. If a gas source rock reaches the gas window, only gas will be generated. If an oil source rock is heated through the oil window and into the gas window, then oil will be generated followed by gas. The hydrocarbons will accumulate if a trap exists along the migration path. A trap will then likely have both oil and gas. If enough gas is generated, it can displace all of the oil from the trap. The displaced oil continues to migrate up the bedding to the next trap. Migration under such conditions can lead to gas fields near the deep center of a sedimentary basin, oil and gas fields along the flanks of the basin, and oil fields near the shallow margins of a basin.

Traps are sometimes breached, allowing oil and gas to escape. Traps can be breached by later movement on the trapping fault, by tilting or folding, and by regional uplift and erosion. Faults with open fractures or faults that juxtapose permeable sandstones on both sides of the fault are the easiest way for traps to leak. Oil and gas then escape to the surface at **seeps** (Figure 16.7). Gas, and sometimes oil, can also leak directly upward if the low permeability of the seal is overcome by upward-directed buoyancy. This requires a thick enough accumulation of oil or gas, called the **hydrocarbon column**, to push upward until the seal is breached. When enough oil or gas has escaped, the column height and buoyant

FIGURE 16.7
A natural oil seep at La Brea tar pits, Los Angeles, California. The seep at La Brea was the first clue that a huge amount of oil exists in the Los Angeles basin. (Courtesy of bobjgalindo; https://en.wikipedia.org/wiki/La _Brea_Tar_Pits.)

pressure in the trap is reduced to the point where leakage stops. In such cases, only some of the oil or gas is preserved.

Sometimes, groundwater below an oil pool contains enough oxygen to maintain microbes that feast on hydrocarbons. When this happens, the oil is **biodegraded**. Since microbes generally consume the lighter parts of the oil, the result is a heavy oil or tar residue. The tar sands of Alberta are an example of biodegraded oil.

In other cases, where groundwater moves relatively quickly through a reservoir rock, the water can carry away the smaller oil molecules, the lighter hydrocarbons. This **water washing** also leaves behind tar or heavy oil.

Oil can also be destroyed through heat. Once it goes past the oil and gas windows, the oil can be cooked to a tar and even metamorphosed to graphite if the heat and pressure are intense enough.

Hydrocarbon accumulations are thus a result of just the right conditions occurring in time and space. An organic-rich source rock must exist and be heated enough to generate oil or gas. The petroleum must have a carrier bed to migrate along. A trap must exist along the migration path. The hydrocarbons must be preserved within the trap. If any of the components of the petroleum system are not in place or happen at the wrong time, the hydrocarbons will not generate or will migrate to the surface and be lost.

Exploring for Hydrocarbons

Originally, people happened upon natural oil seeps and then came up with uses for the tar-like hydrocarbons found there. All of the early oil wells were dug at or near seeps. Early in the history of oil exploration, drillers noticed that wells drilled on hills near seeps were more successful than those drilled elsewhere. When geologists mapped these hills, they learned that the layers were all inclined away from the high point. In other words, the oil had been trapped in anticlines or salt domes or reefs. So explorers began looking for rocky domes near seeps.

The drilling success rate in "**wildcat**" oil and gas exploration (exploring in frontier areas) is between 10% and 20%. Depending on where they are drilling, wells can cost a few hundred thousand dollars or several hundred million dollars to drill. Always on the lookout to improve their chances of success and keep costs down, the energy industry is always trying new techniques to locate hydrocarbons. These include seismic surveys, remote sensing surveys, geochemical surveys, and good old-fashioned field mapping.

Field mapping involves finding and defining geologic structure at the surface, primarily by measuring the strike and dip of layers. **Remote sensing** does the same thing by acquiring and examining airphotos, satellite images, and radar imagery, among others (Prost 2014; Figure 16.8). **Geochemical surveys** measure liquid or gaseous hydrocarbons in the soil or near-surface atmosphere and determine where they are concentrated at levels higher than normal. Geophysicists using **seismic surveys** investigate the subsurface by projecting sound waves into the ground and recording the reflections from the layers below (Figure 16.9). Seismic surveys can be processed to show either a two-dimensional cross section of a portion of the earth, or a 3D representation of the subsurface (Figures 16.10 and 16.11). It is then necessary to determine the best place to drill a well for the best chance to hit hydrocarbons.

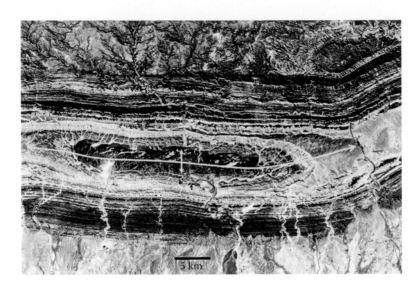

FIGURE 16.8
Landsat satellite false color image of Zinda Peer anticline (yellow symbol), Punjab, Pakistan. The oldest rock (dark) is exposed in the center of this fold. Remote sensing images such as this allow rapid mapping of surface structures. North is to the right. (Courtesy of NASA.)

FIGURE 16.9
Vibroseis seismic program layout and acquisition. Vibrating trucks send sound waves into the ground. These waves are reflected back from subsurface layers and are recorded by geophones planted in the surface. The geophones are connected by cables to a processing truck where the recorded signals are converted into seismic images that reveal the layers in the earth. (From Bott, R. D. 2004. *Our Petroleum Challenge: Sustainability into the 21st Century*, 7th edition. Canadian Centre for Energy Information. 131 pp.)

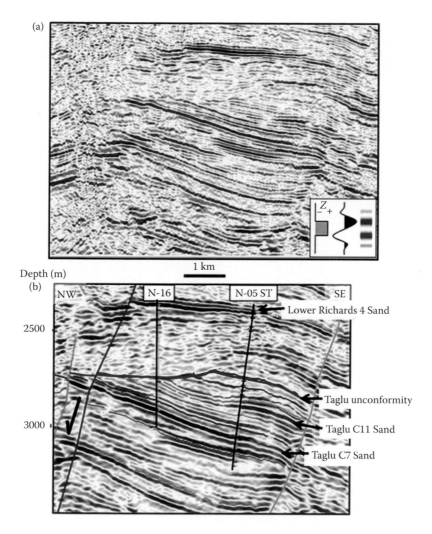

FIGURE 16.10
Seismic line through the Umiak field, Northwest Territories, Canada. (a) Uninterpreted line. (b) Interpreted line showing a tilted fault block below an angular unconformity (heavy red line) and a gentle fold in the overlying formations. The Umiak N-16 well and Umiak N-05 sidetrack (ST) well are shown. Depth is in meters. (From Prost, G. L., and G. M. Peasley. 2015. The Umiak field discovery, Northwest Territories, Canada. *American Association of Petroleum Geologists Bulletin* 99 (2), 273–292. Copyright American Association of Petroleum Geologists.)

What about the Oil Sands?

Oil sands are found around the world, most notably in Canada and Venezuela. Fort McMurray, Alberta, is in the heart of the Canadian oil sands. **Oil sands** (also called "tar sands") are sandstone with tar-like bitumen in the pores. The only difference between this oil and conventional oil is that bacteria in the groundwater have consumed the smaller, oilier hydrocarbon molecules and left behind the larger and heavier tar-like components. In this sense, the so-called **"dirty oil"** is no dirtier, or cleaner, than any other oil. Some of

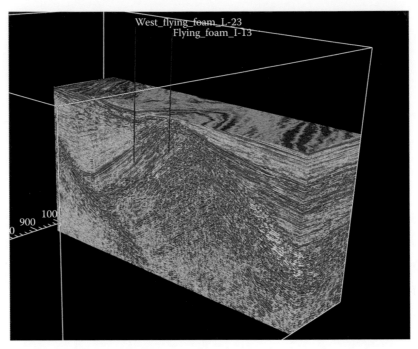

FIGURE 16.11
3D seismic cube showing two side views and a depth slice (top surface) of the Flying Foam structure, Jeanne d'Arc Basin, Canada. (Courtesy of Canadian Society of Exploration Geophysicists, Michael Enachescu, Memorial University, and WesternGeco. CSEG RECORDER, May 2007, Vol. 32, No. 05; http://csegrecorder.com/articles /view/digital-seismic-dilemma-ownership-and-copyright-of-offshore-data.)

this **heavy oil** is used to pave roads. Much of it is mixed with lighter hydrocarbons and shipped to refineries to be converted to gasoline and other products just like regular oil.

Oil sands are strip mined when they are close to the surface (Figure 16.12). When tar sand is too deep to mine, steam is injected through wells into the sand, the bitumen melts, and oil and water are pumped to a surface facility where the oil is separated and the water is cleaned and recycled. This process is called "Steam Assisted Gravity Drainage," or SAG-D.

Canadian oil sands development is subject to some of the most exacting environmental standards in the world. The Government of Alberta requires that companies reclaim 100% of the land after the oil has been extracted. Reclamation means that land is returned to a self-sustaining forest ecosystem with native vegetation and wildlife.

According to the government of Alberta (http://www.energy.alberta.ca/oilsands/791.asp),

- Alberta's oil sands are the third largest oil reserves in the world, after Venezuela and Saudi Arabia.
- Reserves shallow enough to mine (up to 75 m) are found only within the Athabasca oil sands area. Surface mineable area is equal to about 4800 km^2 and accounts for about 3.4% of total oil sands area.
- Disturbed oil sands surface minable area was about 895 km^2 in 2013, accounting for less than 1% of total oil sands area and about 0.2% of Alberta's boreal forest.

FIGURE 16.12
Suncor's Steepbank mine north of Fort McMurray, Alberta. (Courtesy of G. Prost.)

- The total area occupied by oil sands tailings ponds and associated structures (such as dikes) was 220 km² at the end of 2013.
- Alberta became the first jurisdiction in North America to legislate greenhouse gas (GHG) emissions reductions for large industrial facilities by passing the *Specified Gas Emitters Regulation*.
- Oil sands extraction currently accounts for approximately 8.5% of Canada's total GHG emissions and about 0.12% of global GHG emissions.
- Between 1990 and 2012, oil sands producers reduced per-barrel emissions by an average of 28%.
- Oil sands projects recycle 80%–95% of water used and use saline water whenever possible.
- *River Management Frameworks* from Alberta Environment and Parks impose strict limits on water usage.
- Mine operators are required to supply reclamation security bonds to ensure that reclamation requirements are met. Reclamation certificates are not issued until monitoring through time demonstrates that these particular lands meet the criteria for return to self-sustaining ecosystems. The first successful reclamation occurred in 2008 (Figure 16.13).

To put the surface disturbance into perspective, many cities have significantly larger footprints (Figure 16.14; Table 16.1).

FIGURE 16.13
Example of a terraced and partially reclaimed Suncor oil sands mine tailings near Tar Island as viewed from the Athabasca River. (Courtesy of G. Prost.)

FIGURE 16.14
Satellite image of Fort McMurray area and the Canadian oil sands mines (bright areas north of the city). Mines, tailings ponds, and structures occupy about 1115 km^2, or about the same area as Vancouver. Forest cover is green; lakes are dark blue to black. Yellow line is 100 km (60 mi). (Courtesy of Google Earth.)

TABLE 16.1

Sizes of Some of the Largest Cities in the World Relative to the 1115 km² Surface Disturbance of Canadian Oil Sands Operations

Rank	City	Metropolitan Land Area (km²)
1	London	9078
2	New York	8683
3	Los Angeles	8320
4	Tokyo	6993
5	Chicago	5498
6	Atlanta	5083
7	Philadelphia	4661
15	Paris	2723
19	Johannesburg	2396
22	Buenos Aires	2226

Source: Wikipedia, List of Largest Cities by Area. Accessed online July 10, 2016. https://en.wikipedia.org/wiki/List_of_largest_cities_by_area.

Production, Consumption, and the Price of Oil

Oil is a commodity traded around the world, and as such is no different than gold, wheat, and beef. World oil production and consumption are in a delicate balance (Figure 16.15). When production rises even one barrel above the level of consumption, the price falls because there is more oil than people need. When consumption exceeds production, the

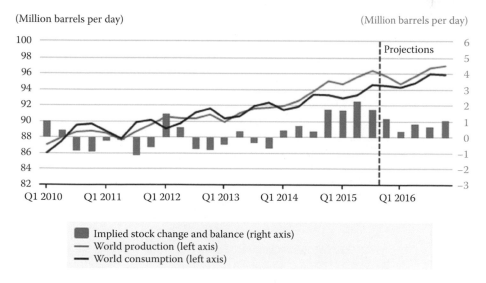

FIGURE 16.15

Oil production versus consumption over 6 years from 2010 to 2015. The price goes up when consumption exceeds production. (US Energy Information Agency, 2015. Short Term Energy Outlook. Accessed online July 15, 2016; http://www.eia.gov/outlooks/steo/.)

price goes up, because oil traders are willing to pay more to guarantee that they have enough oil for their customers. This is the dance of **supply and demand**.

An example of the delicate balance between supply and demand can be seen in the recent fall in oil prices. In an attempt to protect their market share, Saudi Arabia in 2014 increased production one half of one percent, to 9.6 million barrels/day. This small increase in production caused the benchmark crude known as West Texas Intermediate (WTI) to fall from an average of $107.45/barrel in June 2014 to $49.37 in January 2015. Other oil producers followed Saudi Arabia and raised production. This increased production, combined with a slowdown in demand from China, dropped the price of WTI to $29.01 in January 2016. The point is, price is *not* arbitrarily set by oil companies. While cartels like the Organization of Petroleum Exporting Countries try to control the price of oil, these efforts only work some of the time.

Consumption (and price) usually goes up when economies are doing well and industries are running at full tilt. Consumption (and price) usually goes down during an economic downturn because not as much fuel is needed for manufacturing and commerce. Production goes up and price comes down when new technologies like 3D seismic, horizontal drilling, hydraulic fracturing, or oil sands development make large new supplies of fuel available. Price goes up when war or political unrest makes deliveries uncertain and traders want to lock-in their sources of supply (Figure 16.16).

Local **price fluctuations** are the result of refinery shutdowns for maintenance or to change production runs for seasonal blends. In many areas, refiners are required to use a more expensive 10% **ethanol** (corn-derived alcohol) blend during the summer driving season. Seasonal changes are also the result of increases or decreases in driving: as gasoline demand goes up in the summer or over holidays, stations raise prices to ensure that

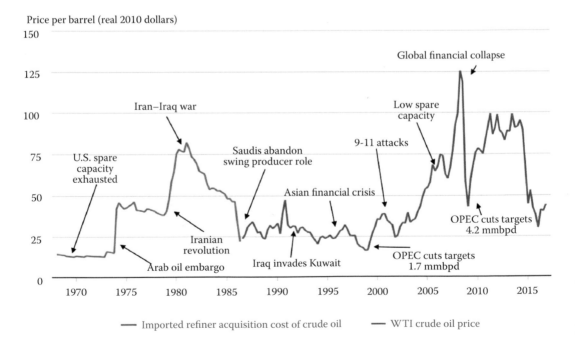

FIGURE 16.16
Oil price since 1970 in 2010 dollars. Key political and economic events are shown. (Courtesy of US Energy Information Agency, 2016. Cost and Performance Characteristics of New Generating Technologies, Annual Energy Outlook 2016. Accessed online December 8, 2016; http://www.eia.gov/finance/markets/crudeoil/reports _presentations/crude.pdf.)

there is enough fuel for those who really need it. If prices remained low, there would be the danger that stations run out of gasoline. Other price variations are attributed to different gasoline taxes from city to city or across state lines. Rarely, there is a refinery outage owing to a fire or storm that reduces supply locally and causes prices to rise temporarily.

The fact that low prices lead to less drilling, and eventually less oil supply, is the source of the oilfield quip that "the best cure for low prices is low prices."

Remaining Reserves and Peak Oil

Every now and then, we are told we are running out of oil. This is invariably followed in a few years by the statement that the world is awash in oil. What's going on?

In one sense, we *are* running out of oil, because there is only so much to be found and it takes millions of years to create more. However, the reality is that we are running out of cheap, easily found oil. When large fields are found near the surface and it only costs $500,000 to drill a well and produce the oil, that oil will be produced. When oil is found in 3 km of water and another 7 km below the seabed, wells can cost upward of $300 million each. Offshore platforms cost billions of dollars to build and install. Those wells won't be drilled unless the oil fetches $100/barrel or more. Likewise, when oil is $20/barrel, the oil sands are not economic to produce using steam injection. When oil is $50/barrel, many oil sands are barely breaking even. At $100/barrel, they are making good money and expanding output.

This does not just apply to the oil business: mining companies do the same thing. When gold was $32 an ounce, companies could only mine rich deposits. When the price of gold hit $2000 an ounce, they could mine much lower grades of ore, and consequently, there

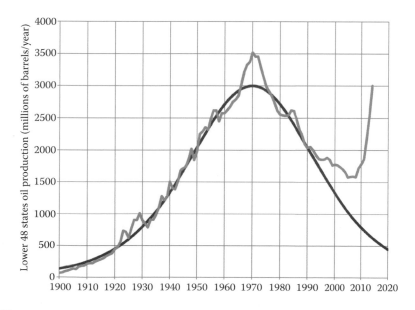

FIGURE 16.17
Hubbert's prediction for US crude oil production (red), and actual lower-48 states production through 2014 (green). The spike in actual production after 2010 is attributed to increased production of unconventional oil and gas. This is mainly attributed to frac'ing. (Courtesy of Plazak; https://en.wikipedia.org/wiki/Peak_oil.)

is much more ore (and gold) available. The point is, when prices are high, companies are willing to spend more to look for harder to produce or lower-quality resources.

Peak oil refers to the peak in worldwide production of conventional crude oil. As new oil is found, production rises to a peak. As oil becomes harder to find, production begins an irreversible decline. The theory of peak oil was proposed in 1956 by the geologist M. King Hubbert for production in the United States. He predicted that production would peak between 1965 and 1971. It peaked in 1970 (Figure 16.17). But there was a catch. Hubbert was referring to **conventional oil**, oil that is found in good sandstone or limestone reservoirs onshore or in shallow water and that is extracted using standard drilling practices. **Unconventional oil**, including deep water fields, oil sands, and shale oil, was either unknown at the time or the technology did not exist to extract it. The ability to produce unconventional oil is the reason for the recent uptick in the amount of oil produced in North America.

So, how much oil and gas is actually left? That's a moving target. We know how much oil and gas we have found (**proved reserves**), and we can guess how much might exist that has not been found (**undiscovered reserves**; Figures 16.18 through 16.20). But we won't really know how much is left until we look for it (that is, drill wells). And we won't look for the hard-to-find stuff till the price makes it worthwhile.

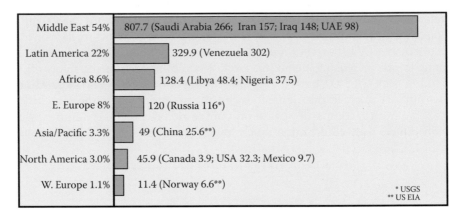

FIGURE 16.18
World proven conventional oil reserves (in billions of barrels) by region in 2016. (Modified from OPEC, 2017, Annual Statistical Bulletin tables 1.2 and 3.1, US Geological Survey, and US Energy Information Agency). Accessed online July 13, 2017; http://www.opec.org/opec_web/static_files_project/media/downloads/publications/ASB2017_13062017.pdf; https://en.wikipedia.org/wiki/List_of_countries_by_proven_oil_reserves; and https://en.wikipedia.org/wiki/Oil_reserves_in_Russia.)

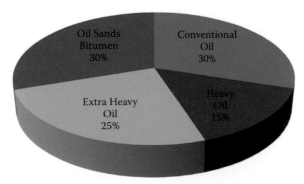

FIGURE 16.19
Total world oil reserves by type. (Courtesy of RockyMtnGuy; https://en.wikipedia.org/wiki/Peak_oil.)

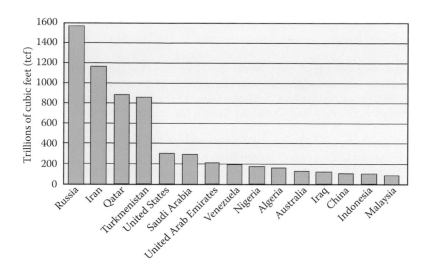

FIGURE 16.20
World natural gas reserves by major producing country in 2016. (Modified from Fossil Energy, World's Natural Gas Reserves. Accessed online July 10, 2016; http://www.trubacovs.eu/worlds-natural-gas-reserves/.)

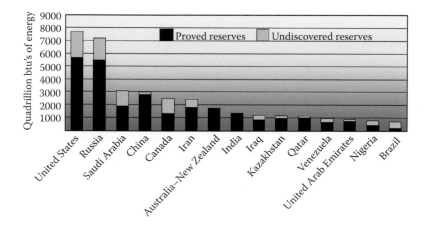

FIGURE 16.21
All fossil fuel energy reserves (oil, gas, coal), discovered and undiscovered. (Modified after O'Brien and Woolverton. 2009. World and U. S. Fossil Fuel Supplies. Copyright 2016, Agricultural Marketing Resource Center, all rights reserved. Accessed online December 8, 2016; http://www.agmrc.org/renewable-energy/energy/world-and-u-s-fossil-fuel-supplies/.)

How long will supplies last? We think we know, based on what we have found and what we guess might be found (Figure 16.21). But again, the answer is a moving target because we don't really know how much we haven't found, or what new technologies may come along. If we don't find any more resource, and continue producing and consuming current reserves at the 2010 rate, a prediction by British Petroleum in 2011 indicated that the world's proved oil reserves of 1382 billion barrels will last about 46 years. The same report suggests that world gas reserves of 187 trillion m³ will last another 59 years. But don't despair. The higher the price, the more oil that will be found. We're not running out of oil, we're just running out of cheap oil.

References

Agricultural Marketing Resource Center. World and U.S. Fossil Fuel Supplies. Accessed online July 10, 2016. http://www.agmrc.org/renewable-energy/energy/world-and-u-s-fossil-fuel-supplies/

Bott, R. D. 1999. *Our Petroleum Challenge: Exploring Canada's Oil and Gas Industry*, 6th edition. Petroleum Communication Foundation, Canadian Centre for Energy Information, College Park, MD, 101 pp.

Bott, R. D. 2004. *Our Petroleum Challenge: Sustainability into the 21st Century*, 7th edition. Canadian Centre for Energy Information, College Park, MD, 131 pp.

Fossil Energy, World's Natural Gas Reserves. Accessed online July 10, 2016. http://www.trubacovs.eu/worlds-natural-gas-reserves/

Ground Water Protection Council. 2011. State Oil and Gas Agency Groundwater Investigations and Their Role in Advancing Regulatory Reforms, A Two-State Review: Ohio and Texas.

Heisig, P. M., and T. Scott. 2012. Occurrence of Methane in Groundwater of South-Central New York State, 2012—Systematic Evaluation of a Glaciated Region by Hydrogeologic Setting. US Geological Survey Scientific Investigations Report 2013–5190.

Magoon, L. B., and W. G. Dow. 1994. The Petroleum System, *in* Magoon, L. B., and W. G. Dow, eds., *The Petroleum System—From Source to Trap*. American Association of Petroleum Geologists Memoir 60, pp. 3–24.

O'Brien, D., and M. Woolverton. 2009. World and U. S. Fossil Fuel Supplies. Agricultural Marketing Resource Center. Accessed online December 8, 2016. http://www.agmrc.org/renewable-energy/energy/world-and-u-s-fossil-fuel-supplies/

Open Energy Information. 2016. Transparent Cost Database. Accessed online December 8, 2016. http://en.openei.org/apps/TCDB/

Prost, G. L. 2014. *Remote Sensing for Geoscientists*. Taylor & Francis, 702 pp.

Prost, G. L., and G. M. Peasley. 2015. The Umiak field discovery, Northwest Territories, Canada. *American Association of Petroleum Geologists Bulletin* 99 (2), 273–292.

Sverrisson, H. January 9, 2015. Chevron's Value at Various Oil Prices. Accessed online July 15, 2016. http://seekingalpha.com/article/2810135-chevrons-value-at-various-oil-prices

Tertzakian, P. A. 2006. *A Thousand Barrels a Second—The Coming Oil Break Point and the Challenges Facing an Energy Dependent World*. McGraw-Hill, New York, 272 pp.

The Global Education Project. World Energy Supply. Accessed online July 10, 2016. http://www.theglobaleducationproject.org/earth/energy-supply.php

Uphoff, T. 1997. Precambrian Chuar Source Rock Play: An exploration case history in Southern Utah. *American Association of Petroleum Geologists Bulletin* 81 (1), 1–15.

US Energy Information Agency. October 2015. Short Term Energy Outlook. Accessed online July 15, 2016. http://seekingalpha.com/article/3561846-longer-term-world-oil-supply-demand-provide-thesis-oil-exploration-production-investment?page=2

US Energy Information Agency. 2016. Cost and Performance Characteristics of New Generating Technologies, Annual Energy Outlook 2016. Accessed online December 8, 2016. http://www.eia.gov/forecasts/aeo/assumptions/pdf/table_8.2.pdf

US Geological Survey. 2005. Natural Gases in Ground Water near Tioga Junction, Tioga County, North-Central Pennsylvania—Occurrence and Use of Isotopes to Determine Origins. US Department of the Interior, US Geological Survey Scientific Investigations Report 2007-5085.

Wikipedia. List of Largest Cities by Area. Accessed online July 10, 2016. https://en.wikipedia.org/wiki/List_of_largest_cities_by_area

Wikipedia. Peak Oil. Accessed online July 10, 2016. https://en.wikipedia.org/wiki/Peak_oil

Woodward, C. M. 2002. *Britain's Offshore Oil and Gas*. UK Offshore Operators Association and The Natural History Museum, 56 pp.

Yergin, D. 1991. *The Prize—The Epic Quest for Oil, Money, & Power*. Free Press, New York, 885 pp.

17

Climate Change

The debate is over! There's no longer any debate in the scientific community about this. But the political systems around the world have held this at arm's length because it's an inconvenient truth, because they don't want to accept that it's a moral imperative.

Al Gore
US Vice President

Essentially, all models are wrong, but some are useful.

George E. P. Box
Statistician

We are headed to a radically new Earth, at least from our perspective. But from the planet's perspective, this is nothing new. As the geologist Peter Ward is fond of pointing out, we are actually heading back to a time kind of like the Miocene. The Miocene ended 5.5 million years ago, and it was the last time that the planet had no icecaps.

Annalee Newitz
Journalist

Let's be clear: climate changes. And the earth has been warming.

What follows is an attempt to provide an impartial view of global warming from a geologic perspective. First, a disclaimer. The senior author (GP) has worked for the US Geological Survey mapping coal, for the mining industry looking for mineral deposits, and for the energy industry exploring for and developing oil and gas fields. You are welcome to consider him a "denier" or "shill for the fossil fuel industry," but keep in mind that labels do not refute any of the statements that follow. Evidence should always be evaluated on its own merit.

Debate is never over among scientists. To claim that it is, as Al Gore did, contradicts the scientific method. The scientific method requires that all propositions be testable and falsifiable. If a theory passes the test of prediction hundreds of times but fails just once, the theory must be revised.

Conventional wisdom is frequently wrong. There once was a consensus that Earth was the center of the universe. Deniers were imprisoned (Galileo) or burned (Bruno). Conventional wisdom once held that people were created 6000 years ago in their present form. Evolutionists who proposed that humans evolved from "lower" forms were mercilessly ridiculed. Conventional wisdom once held that continents cannot move about the face of the earth. Those who proposed otherwise were relegated to the fringes of science.

Geologists have a unique perspective on climate change, by which most people mean global warming. Rather than dealing in years or decades, the geologist's time frame encompasses millennia to millions of years. There is ample evidence that Earth has several times been much warmer, as during Late Cretaceous (65–100 Ma), and during the Paleocene–Eocene Thermal Maximum (56 Ma). There is also indisputable evidence that

Earth has been much cooler than now, as during the Precambrian Snowball Earth episode (650 million years ago and earlier), during Permian glaciation (250–300 Ma), and during the recent ice ages. Climate is constantly changing.

What We Know

- Earth started cooling around 15 Ma, during the middle Miocene (Figure 17.1).
- There have been several ice ages in the past 120,000 years (Figure 17.2).
- The last ice age, the Wisconsinan, lasted from about 80,000 to 12,500 years ago. Ice sheets between 1 and 4 km thick (between 0.5 and 2.5 mi) covered Toronto, Chicago, New York, Edinburgh, Berlin, and Moscow. In North America, the ice sheets extended south as far as Iowa and southern Illinois (Figures 17.3 and 17.4).
- At the peak of the last ice age, sea level was as much as 110 m (350 ft) lower than it is today. Large parts of the continental shelves were grasslands grazed by herds of giant mammals. Fishermen off the east coast of the United States have dredged up the remains of mastodons from the ocean floor—300 km (180 miles) from the present coastline.
- In the past 12,000 years, average global temperature has risen between 5°C and 8°C (between 9°F and 15°F).

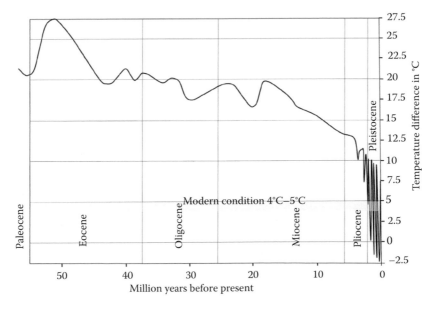

FIGURE 17.1
Temperatures in central Europe during the Cenozoic, from the Paleocene–Eocene Thermal Maximum 50 million years ago to the Pleistocene ice ages and interglacial periods. Earth began cooling significantly around 15 Ma. (From Gerhard et al. 2001. *Geological Perspectives of Global Climate Change*. American Association of Petroleum Geologists, Studies in Geology #47, 372 pp. Modified after Anderson and Borns. 1997. *The Ice Age World*. Oslo, Scandinavian University Press.)

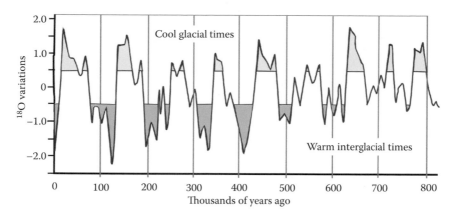

FIGURE 17.2
Cool glacial epochs (ice ages) and interglacial periods over the past 800,000 years based on oxygen isotopes. (From Lang, 2002. *Ice Age Mammals of North America*. Mountain Press Publishing, Missoula, 226 pp. Modified after Morrison, 1991. Introduction, Quaternary Nonglacial Geology: Conterminous U.S. In *The Geology of North America*, K-2, Geological Society of America, pp. 1–12. Imbrie et al., 1984. The orbital theory of Pleistocene climate: Support from a revised chronology of the marine $\delta^{18}O$ record. In *Milankovitch and Climate*, Part I, A. L. Berger and others (eds.), Reidel Publishing Company. pp. 269–305.)

- Cities built on subsiding deltas or coastal plains, like New Orleans, Amsterdam, and Venice, are doomed. The waterlogged sediments they are built on are compacting. They will eventually drown whether sea levels rise or not.
- Carbon dioxide (CO_2) is a greenhouse gas. The amount of carbon dioxide in the atmosphere is rising. In the past 1000 years, it has increased about 125 parts per

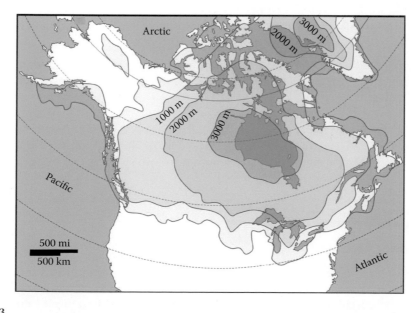

FIGURE 17.3
Extent and thickness of glacial ice in North America about 18,000 years ago. (Based on work by McIntyre 1981. CLIMAP Project, Lamont-Doherty Earth Observatory; http://written-in-stone-seen-through-my-lens.blogspot.com/2014/03/climbing-geology-and-tectonics-of.html.)

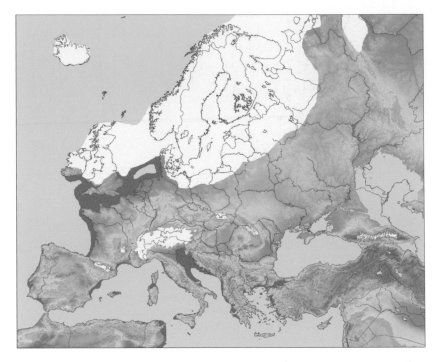

FIGURE 17.4
Extent of ice sheets over Europe about 20,000 years ago. (Courtesy of Ulamm; https://commons.wikimedia.org/wiki/File:Weichsel-W%C3%BCrm-Glaciation.png?uselang=en-gb.)

million (ppm), to about 400 ppm. About 64% of that increase occurred in the past 50 years, and is largely attributed to the burning of fossil fuels (coal, oil, and gas).

- Water vapor and methane are also greenhouse gases.
- The sun is the source of most of the heating of Earth's surface. The amount of sunlight arriving at Earth varies over time as a result of solar processes and wobbles in Earth's orbit.

Few would argue with the observation that Earth has been warming since the end of the last ice age some 12,000 years ago. What is being debated is whether the human contribution to global warming is significant and whether we can stop the warming.

A Word about Models

Models allow us to make predictions. A good example of a model is a retirement calculator. You enter things you know, or have control over, like your expenses, your income, and how long until you retire. You also enter things you don't know and have little control over, like the rate of inflation, how the stock market will perform, and when you might die. You then run the program many times, changing the inputs slightly, and the program gives you a distribution of likely outcomes. Every outcome has a specific probability of happening: you would like at least a 90% chance of having money until the end of your life. The model is useful, but no single run of the program is likely to be correct.

Another example of modeling that's familiar to most people is the prediction of hurricane storm tracks. Emergency planners need to know where and when a hurricane will make landfall. The tracks are generated by computer models where the inputs are fairly well known. Inputs include wind speed and direction, atmospheric pressure, air temperature at various elevations, ocean temperature, and terrain elevation. This information is constantly being updated with data from satellites, hurricane hunter planes, and buoys. Multiple models are used to come up with a range of possible outcomes extending a week or more into the future (Figure 17.5).

The farther the prediction extends into the future, the farther apart possible storm tracks diverge. One model predicted the correct landfall of Hurricane Sandy 5 days in advance, but there was an 800-km-wide (500-mi-wide) range of potential outcomes. This large range of outcomes tells us that the storm track models have a hard time predicting accurately more than a few days in advance.

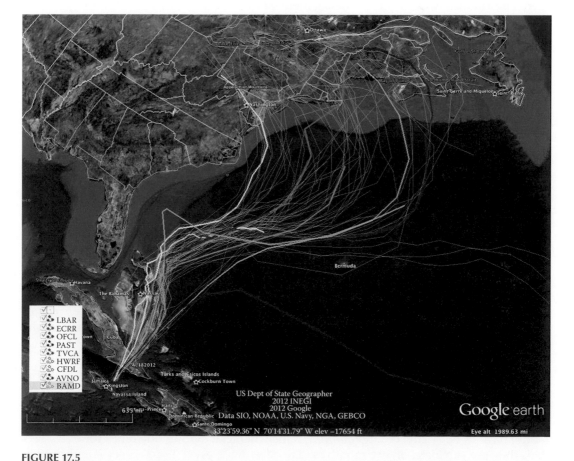

FIGURE 17.5
Range of possible storm tracks predicted by weather models for Hurricane Sandy, October 25th, 2012. The storm made landfall on October 29th. (Courtesy of US Department of State Geographer. Copyright Google, INEGI.)

Unlike retirement calculators, hurricane path prediction models are highly complex, have multiple known inputs, and predict up to 8 days into the future. "Searching for the... single best model is misguided. Complex phenomena require multi-pronged research attacks. A track model that does well for part of the lifetime of a storm could really tank for another part of it.... We cannot tell in advance which track models are going to work best...." (Johnson 2012).

Now, consider climate models. They attempt to show what the climate will be like 100 years or more from now. These models have hundreds of variables. We kind of know how these variables behave, but we don't always understand how they interact and change over time.

To give you some idea of the complexity that goes into climate models, here are some variables that climate scientists have to account for:

- Magnitude of solar radiation over time
- Earth's orbital variations
- The amount and timing of degassing of volcanoes
- Global and local atmospheric composition
- The amount of CO_2 and methane generated by humans
- The amount of CO_2 and methane generated naturally
- Absorption of atmospheric CO_2 by plants
- The amount of carbon taken out of the atmosphere by burial of plants
- The amount of forest burned each year
- The amount of cooling owing to reflection of incoming radiation by clouds, aerosols, and icecaps
- The melting rates of glaciers and ice caps
- Shifting of ocean currents
- The amount of CO_2 used by sea creatures to build their shells
- The amount of CO_2 absorbed or released by oceans
- The numbers of tropical storms in a given year
- The amount of positive and negative feedback from water vapor
- The amount and rate of melting seasonal sea ice

Take just one input, the amount of CO_2 generated by humans in the next 100 years. This requires projecting the world population, standards of living, new sources of energy, the amount of nuclear and renewable energy being used, and the cost of the energy. And the model may even miss important inputs because we may not *know* all the important inputs.

There is also the issue of **resolution**. The smallest "unit" of resolution in a model is called a cell. The size of the cell has to be large, or the computer running the program would grind to a halt. Most current models use cells 100 to 250 km on a side with 10 to 30 atmospheric layers and another 10 to 30 ocean layers. The time intervals used in the model have to be relatively large as well: rather than calculating outcomes for every second, they are calculated every month or day or perhaps every hour. Every simplification adds uncertainty.

The mathematician David Orrell describes the problems inherent in modeling complex systems (Orrell 2007). Complex systems are the result of a large number of choices, and there may exist no set of equations that tell how the system is going to behave. Even if the system always develops or plays out the same way, you cannot tell the result at step 1000 without actually running the model. Because unpredictability is the result of the complexity of the system itself rather than initial conditions, the model cannot be improved by inputting better data. He states "the climate system therefore consists of a nested series of non-linear feedback loops that are in a kind of dynamic balance.... Weather and climate predictions are directly linked, and to believe that models can fail at the former but succeed at the latter is nothing but wishful thinking...."

In other words, models simplify reality. They contain incomplete data and data that are not well understood. They miss important inputs or contain unconscious biases. Uncertainty increases the farther you look into the future.

This brings us to the United Nations **Intergovernmental Panel on Climate Change** (IPCC), the internationally accepted authority on global warming. The IPCC feels that models are useful to estimate greenhouse gas (GHG) emissions and their impact on Earth's climate. About every 5 years, the panel produces a summary of current work for policymakers. IPCC predicts global temperature rise over the next 100 years. Having reviewed the uncertainties in retirement calculators and storm track models (which cover days to decades), consider now the 100-year prediction from the 2014 IPCC Summary for Policymakers:

> Without additional efforts to reduce GHG emissions beyond those in place today, emissions growth is expected to persist driven by growth in global population and economic activities. Baseline scenarios... result in global mean surface temperature increases in 2100 from 3.7°C to 4.8°C compared to pre-industrial levels (range based on median climate response; the range is 2.5°C to 7.8°C when including climate uncertainty...) (high confidence).

Lots of Moving Parts

In the 1970s, scientists warned us about global cooling and the beginning of a new ice age (Time 1974; Newsweek 1975). Today, we hear news stories about the warmest day, month, or year ever. Then, you may hear a story about how the climate has *not* warmed significantly over the past 10 years. You don't know who or what to believe.

Climate is a result of complex interactions between the sun, atmosphere, oceans, land masses, ice sheets, and biosphere. To get some background, let's look at the systems involved.

Role of the Sun

Ultimately, the sun drives Earth's climate (Scafetta and West 2008). Changes in the intensity of solar radiation cause changes in global climate. These changes occur in cycles that span tens or hundreds of thousands of years. The cycles depend on what's going on inside the sun, as well as wobbles in Earth's orbit. It's those changes in Earth's orbit that cause ice ages and interglacial periods, or **interglacials**.

Climate change is amplified by **feedback** mechanisms—for example, solar warming causes the release of methane and carbon dioxide from the oceans, and these gases in turn cause more warming.

The sun's radiation is either reflected or absorbed. Absorbed radiation is emitted as heat. Ultraviolet radiation is mostly absorbed by the upper atmosphere and warms it. Visible light is absorbed by the lower atmosphere, oceans, and land. Surface reflectance, or **albedo**, is important: dark surfaces, such as plowed fields, absorb light and re-emit heat, while bright surfaces like snow-covered ground and ice sheets reflect incoming radiation and contribute to cooling.

How do we know if the climate change we see today is in line with these cycles, or if it's something abnormal? To know that, you have to look at data from the past. However, accurate records only go back a few hundred years. The longest continuous temperature records are from central England and go back to 1659. To discern temperatures from before then, scientists use evidence from tree rings, ice cores, ocean sediments, and oxygen isotopes in sea shells to measure past changes in solar radiation. These **temperature proxies** tell us that radiation from the sun has varied a lot over time. An increase in **sunspots** may be responsible for as much as half the global warming of the last 110 years (NASA Earth Observatory; IPCC 2001). Eleven-year sunspot cycles appear to have a role in global warming and cooling, depending on whether they are increasing or decreasing (Figure 17.6).

The total amount of sunlight hitting Earth, called **total solar irradiance**, or TSI, also changes week to week and even day to day. These rapid fluctuations are the result of season, time of day, and latitude.

Studies indicate that the earth has warmed between 0.3°C and 0.6°C (between 0.5°F and 1.0°F) over the past 100 years (Figure 17.7). Why? There had been a slight increase in solar radiation since 1750, but it was offset by natural cooling—the result of the volcanic eruptions

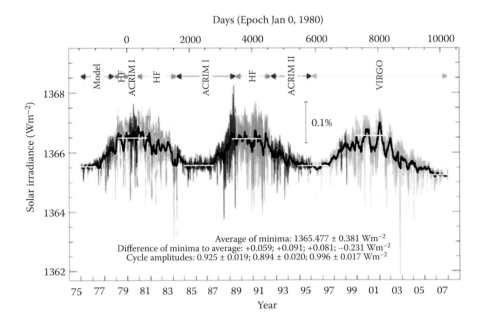

FIGURE 17.6

Two-and-a-half sunspot cycles of TSI. Compiled by the Physikalisch-Meteorologisches Observatorium/World Radiation Center, Davos, Switzerland. TSI given as daily values; different colors indicate different measuring projects. (Courtesy of the SOHO consortium [European Space Agency and NASA]; https://sohowww.nascom.nasa.gov/publications/ESA_Bull126.pdf.)

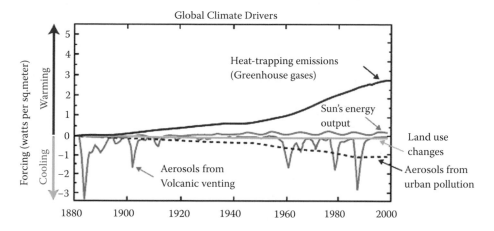

FIGURE 17.7
Climate drivers. (From Hansen et al. 2005. Earth's energy imbalance: Confirmation and implications. *Science* 308, 1431–1435. Figure adapted by Union of Concerned Scientists, The Sun-Climate Connection. Copyright Union of Concerned Scientists; http://www.ucsusa.org/global_warming/science_and_impacts/science/global -warming-faq.html#.WEnkhLIrKUk.)

that dumped reflective particles into the atmosphere. Heating as a result of solar radiation has been decreasing during the late twentieth century, mainly around urban areas, owing to increasing amounts of pollution reflecting light back into space. This has been accompanied by an increase in atmospheric heat-trapping gases since the beginning of the industrial revolution in the early 1800s. Many scientists believe that this increase is mostly attributed to the burning of fossil fuels (coal, oil, and gas), farming, and deforestation.

There are indications that the sun today may be moving to a minimum of radiation similar to that of the **Maunder Minimum**, a time when there were few sunspots (Abdussamatov 2009). The Maunder Minimum lasted from 1645 to 1715 and roughly coincides with the middle of the **Little Ice Age**. During the Little Ice Age, between 1300 and 1850, the Northern Hemisphere was colder than average (Figure 17.8). Is there a link between sunspots and Earth's temperature? Whether a link exists is controversial, but the possibility of one does raise questions. For example, if a period of fewer sunspots results in global cooling, could it cancel out warming caused by greenhouse gases?

The other factor that affects global warming is changes in Earth's orbit. There are regular cycles in the orbit that repeat every 20,000, 40,000, and 100,000 years. These are known as **Milankovitch cycles**, after the Serbian geophysicist who postulated that orbital variations have an effect on climate. There are three overlapping cycles—one caused by changes in the shape of Earth's orbit, one caused by the tilt of Earth's axis, and one caused by the wobble of this axis. (Think of a spinning top that wobbles as it spins.)

These cycles affect the total amount of sunlight hitting Earth. When all of these factors combine to bring Earth closer to the sun, climate gets warmer. When these factors cause Earth to be farther from the sun, it gets cooler.

About 2 million years ago, these orbital variations initiated the current cycle of ice ages, which last 80,000 to 100,000 years and are separated by warm periods that last 20,000 to 40,000 years.

Needless to say, the relationship between Earth's climate and changes in solar radiation is complex. Whereas we can predict orbital cycles, we have no way of predicting whether TSI will increase or decrease.

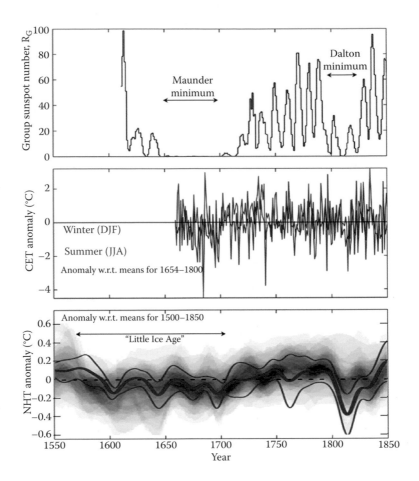

FIGURE 17.8
Comparison of sunspots (top), Central England Temperature (CET) observations (middle), and reconstructions and modeling of Northern Hemisphere Temperatures (NHT). NHT in gray are from paleoclimate reconstructions (darker gray showing higher probabilities) and those in red are from model simulations. (CET data are from the UK Met Office. NHT data are from Figure 1 of the Intergovernmental Panel on Climate Change 2013. Climate Change 2013, The Physical Science Basis, WG1, 5th Assessment Report, IPCC; https://en.wikipedia.org /wiki/Maunder_Minimum.)

The Atmosphere, Aerosols, and Greenhouse Gases

The atmosphere contains both greenhouse and non-greenhouse gases (Table 17.1; Figure 17.9). **Greenhouse gases** absorb light and emit heat. This is a good thing for Earth: without the greenhouse effect, Earth would be about 33°C (59°F) cooler and life as we know it would not exist. The problem is that too much greenhouse gas might raise the temperature to a point where it becomes harmful to life. An example of a runaway greenhouse effect is Venus, where surface temperatures are 462°C (864°F). That's hot enough to melt lead.

The major natural greenhouse gas is water vapor, which accounts for about 50% of Earth's greenhouse effect (Table 17.2). Clouds contribute 25%, carbon dioxide contributes 20%, and the minor greenhouse gases (ozone and nitrous oxide) and aerosols account for the remaining 5% (Schmidt et al. 2010). The role of clouds is more complex: they reflect incoming radiation, but also trap heat radiated from the ground.

TABLE 17.1

Composition of Dry Air by Volume

Gas	Volume
Nitrogen (N_2)	780,840 ppmv (78.084%)
Oxygen (O_2)	209,460 ppmv (20.946%)
Argon (Ar)	9,340 ppmv (0.9340%)
Carbon dioxide (CO_2)	383 ppmv (0.0383%)
Neon (Ne)	18.18 ppmv
Helium (He)	5.24 ppmv
Methane (CH_4)	1.745 ppmv
Krypton (Kr)	1.14 ppmv
Hydrogen (H_2)	0.55 ppmv
Not included in above dry atmosphere:	
Water vapor (H_2O)	typically 1% to 4% (highly variable)

Source: https://en.wikipedia.org/wiki/Atmosphere_of_Earth.
Abbreviation: ppmv: parts per million by volume.

Greenhouse gases are said to "**force**" heating. **Positive feedback mechanisms** are effects that cause more of the same (Figure 17.10), as when a warming Earth causes more water to evaporate; more water vapor in the atmosphere causes more heating.

Processes that remove gases like CO_2 from the atmosphere are **sinks**; those that release gases are **sources**. Plants, including grasslands, forests, and marine plankton, absorb CO_2 and generate oxygen (O_2) as a result of their metabolism. It is important to know whether global biomass is increasing or decreasing in order to gauge the effect on CO_2. Another major CO_2 sink is the ocean: seawater contains large amounts of dissolved gases, and all animals that form calcium carbonate shells derive the carbon from CO_2 dissolved in seawater (Figure 17.11).

Aerosols are minute particles suspended in the atmosphere. They include dust, sulfates, smoke, and other particulates that contribute to global cooling by reflecting incoming radiation back into space. A major contributor of aerosols is explosive volcanism such as the eruption of Tambora in 1815 and Pinatubo in 1991. Several years of global cooling followed each eruption.

Carbon Dioxide

Most attention is focused on the role of carbon dioxide in raising global temperatures. Major sources of CO_2, in addition to burning of fossil fuels, include natural and human-caused forest fires (as in slash-and-burn agriculture), animal respiration, and volcanic gas emissions.

Carbon dioxide concentrations are discussed in parts per million. The current concentration, around 400 ppm, sounds like a lot. In reality, it is a small part of the atmosphere, less than half of 1%. Having said that, even this amount adds to global warming. CO_2 in the atmosphere is estimated at 730 to 770 billion tons (gigatons). The human contribution of CO_2 as a result of burning fossil fuels is between 6 and 30 gigatons annually. This is 1% to 4% of the total, a small amount overall. If we include water vapor, the most abundant greenhouse gas, the annual human contribution is about 1% of total greenhouse gases. And some of that is absorbed naturally by plants and the oceans. But the human contribution adds to the total nonetheless.

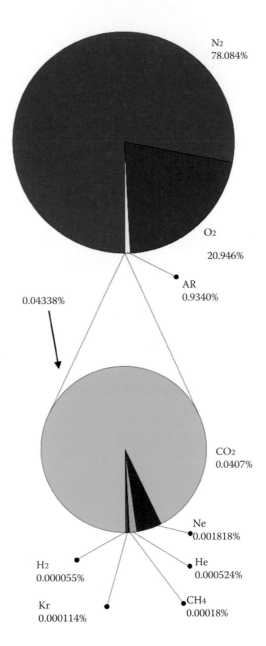

FIGURE 17.9
Composition of the atmosphere by volume. The lower pie chart represents trace gases that together make up about 0.043% of the atmosphere. (Courtesy of Mysid; https://en.wikipedia.org/wiki/Atmosphere_of_Earth.)

With human greenhouse gas emissions effectively overwhelmed by natural sources, it is worth debating whether limiting the burning of hydrocarbons will have any effect on global climate. Even if we all agree that human-derived CO_2 will increase global temperature, it is fair to ask *"how much?"* Is it significant, or negligible?

Carbon dioxide concentrations in the atmosphere have been much higher as well as lower in the past (Figure 17.12). The correlation to temperature appears tenuous.

TABLE 17.2

Major Atmospheric Greenhouse Gases and Their Contribution to Warming

Compound	Formula	Concentration in Atmosphere (ppm)	Contribution to Warming (%)
Water vapor and clouds	H_2O	10–50,000[a]	36–72
Carbon dioxide	CO_2	~400	9–26
Methane	CH_4	~1.8	4–9
Ozone	O_3	2–8[b]	3–7

Source: https://en.wikipedia.org/wiki/Greenhouse_gas.

[a] Water vapor strongly varies locally.

[b] The concentration in stratosphere. About 90% of the ozone in Earth's atmosphere is contained in the stratosphere.

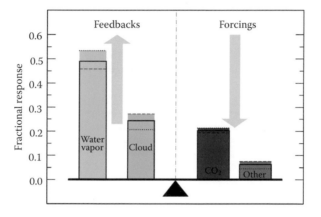

FIGURE 17.10
Schmidt et al. (2010) analyzed relative contributions of various individual components of the atmosphere to the total greenhouse effect. (Courtesy of NASA [Lacis 2010]; https://en.wikipedia.org/wiki/Greenhouse_gas.)

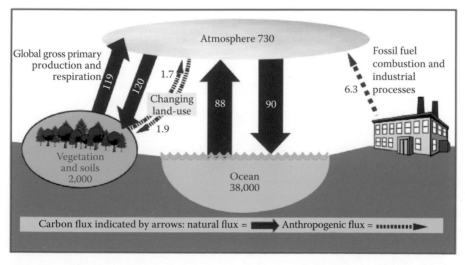

FIGURE 17.11
Movement of carbon to and from the atmosphere in gigatons of CO_2/year. **Anthropogenic** refers to human-generated carbon dioxide. **Flux** refers to the flow from one source to another. (Data from Intergovernmental Panel on Climate Change, *Climate Change 2001: The Scientific Basis [U.K. 2001].* Figure from National Oceanic and Atmospheric Administration [NOAA]; http://www.esrl.noaa.gov/research/themes/carbon/.)

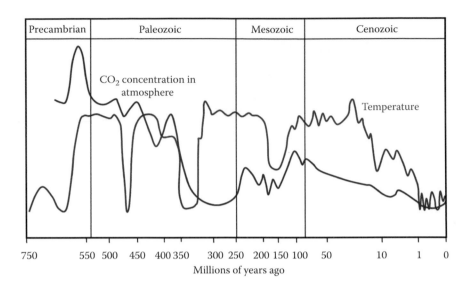

FIGURE 17.12

CO_2 concentration and temperature over geologic time. Vertical scale is relative. Time scale is not linear. (From Scotese, 2002. Analysis of temperature oscillations in geological eras; Ruddiman, 2001. *Earth's Climate: Past and Future*. W. H. Freeman and Sons, New York; Pagani et al. 2005. Marked decline in atmospheric carbon dioxide concentrations during the Paleocene. *Science* 309 (5734), 600–603. Based on compilation by Nahle, 2007. Cycles of global climate change. *Biology Cabinet Journal Online*. Article no. 295.)

In 2009, the US Environmental Protection Agency (EPA) stated that CO_2 should be considered a pollutant (Johnson 2009). For plants, CO_2 is not a pollutant. It is food, an essential source of carbon needed to grow and prosper. Plants need CO_2 the way animals need oxygen: without a minimum amount of CO_2, plants suffocate and die. An increase in atmospheric CO_2 invariably causes an increase in plant biomass. Up to half of Earth's vegetation-covered land has shown a significant increase in the amount of leaf cover, or **greening**, over the past 35 years as a result of increased levels of CO_2 (Zhu et al. 2016). The increase, measured by satellites, is equivalent to an area twice the size of the continental United States. Roughly half of the 10 billion tons of carbon emitted by humans each year gets stored in plants and the oceans. "Studies have reported an increasing carbon sink on land since the 1980s, which is entirely consistent with the idea of a greening Earth," according to Shilong Piao of the College of Urban and Environmental Sciences, Beijing University (NASA Goddard Space Flight Center 2016).

There is some discussion as to whether heating of the atmosphere follows a rise in CO_2, or whether a rise in CO_2 follows heating of the atmosphere. Evidence from Antarctic ice cores suggests that a rise in CO_2 follows a rise in temperature. This makes sense because warmer temperatures cause seawater to release dissolved CO_2 into the atmosphere. Recent climate models indicate that initial warming was a result of Milankovitch cycles. But the CO_2 released by warming oceans resulted in further warming (Shakun et al. 2012). Thus, although the CO_2 increase lagged behind the initial temperature increase, the initial warming accounted for only about 7% of the overall global temperature rise. Increased CO_2 accounted for the remainder.

Methane

Methane (CH_4) is another important greenhouse gas. With atmospheric concentration currently around 1.75 ppm and rising, methane traps 25 times more heat than the same amount of CO_2. Methane comes from natural sources such as wetlands (swamp gas), volcanos, coal beds, natural gas seeps, termites, and wildfires. Wetlands are the largest natural source, emitting methane from bacteria that decompose organic material in the absence of oxygen. According to the EPA (2010), more than 60% of global methane emissions come from human activities. At between 50 and 100 million metric tons of methane a year, rice paddies are perhaps the largest source of man-made methane emissions. The warm, waterlogged soil of the paddies provides ideal conditions for methane generation. The belches, farts, and manure of billions of livestock around the world contribute methane to the air. Garbage dumps and landfills contain rotting organic matter that generates methane. Oil industry activity—drilling, flaring, and leaky pipelines—contribute small amounts of methane, too.

Water Vapor

Water vapor is the most abundant greenhouse gas, making up between 1% and 4% of the atmosphere. It accounts for between 36% and 66% of greenhouse warming under clear skies, and for 66% to 85% of warming when clouds are included.

NASA satellites have been measuring humidity levels in the lower atmosphere since 2002. We know that water vapor increases warming, and warmer air holds more water vapor. Scientists use the humidity measurements to estimate the amount of positive feedback attributed to water vapor. It is not clear how strong this feedback is, but some think it may be enough to almost double the warming due to CO_2 alone (Dessler et al. 2008).

No one, however, is talking about reducing the amount of water vapor in the atmosphere.

Complex Interactions

The slight increase in TSI since 1750 has been offset by natural cooling—the result of the volcanic eruptions ejecting reflective particles into the atmosphere. Solar heating actually decreased slightly in the late twentieth century, mainly around urban areas, because increased air pollution reflects light back into space.

On the other hand, there has been an increase in heat-trapping gases since the beginning of the industrial revolution in the early 1800s. Because of, or perhaps despite these multiple, complex feedbacks, Earth has warmed 0.3°C to 0.6°C (0.5°F and 1.0°F) over the past 100 years.

Role of the Oceans

Oceans influence global warming. The top few meters contain as much heat energy as the entire atmosphere, and ocean water contains several atmospheres worth of dissolved gas. Much of this gas would be released to the atmosphere if the ocean were to warm appreciably. This would have a positive feedback effect on Earth's climate and increase warming.

Currents

Temperature and salinity drive global currents. This **thermohaline circulation** is also called the **Meridional Overturning Circulation** or the "**global conveyor belt**." Freezing of seawater in the North Atlantic causes the remaining cold water to become saltier and denser. This cold, dense water sinks to the ocean floor and flows south, eventually warming and rising to the surface in the northern Indian Ocean and in the North Pacific (Figure 17.13). These currents transfer heat and cause mixing of deep and surface waters.

Air near the equator receives more solar radiation than air near the poles. As the air warms and rises, it moves toward the poles and draws in cooler air below. This convective air circulation, combined with Earth's rotation, causes wind. Wind patterns drive **surface currents**. Surface currents, too, are a means of distributing water across the ocean and heat around the planet.

In the 1950s, oceanographers began collecting **deep sea cores** and bringing back samples of mid-ocean sediments. Analysis of deep marine shells suggested that climate could change abruptly. As we have seen, Milankovitch proposed that Earth's orbital variations trigger ice ages. Although incident sunlight varied only slightly, something was amplifying the effect. Speculation was that it had to do with ocean circulation and salinity. If melting ice sheets caused seawater in the North Atlantic to be less salty and dense, the water would not sink and currents might stop. Without movement of warm equatorial waters northward, the pole would cool and a new ice age would begin (American Institute of Physics 2015).

Study of Greenland and Antarctic **ice cores** began in the 1970s. Ice core work supported studies of deep sea sediments that indicated climate could change abruptly. Bubbles trapped in ice provided atmospheric composition going back 800,000 years or more. There appeared to be an abrupt increase in atmospheric CO_2 at the end of the last ice age. Could rapid changes in atmospheric CO_2 be related to ocean circulation changes?

Ice cores indicated that the Southern Hemisphere was often warm at times when the Northern Hemisphere was cool, and vice versa. During a time known as the **Younger**

FIGURE 17.13
Thermohaline circulation. Blue arrows show the path of deep, cold, dense, saltier water currents. Red arrows indicate the path of warmer, less dense surface water. It takes about 1000 years for a "parcel" of water to complete the global journey. (Courtesy of NOAA; http://oceanservice.noaa.gov/education/tutorial_currents/05conveyor2.html.)

Dryas, around 11,000 years ago, a shift in circulation had cooled the Northern Hemisphere abruptly while the Southern Hemisphere warmed.

Oceanographers suggested that there might be two stable ocean conditions: one with currents, and one without. They feared that Atlantic circulation is now slowing and about to shut down. The 8°C average global warming over the past 12,000 years, however, had not shut down ocean currents. In 2007, the IPCC concluded that a dramatic shift in ocean circulation was "very unlikely" in the twenty-first century (IPCC 2007).

The Ocean–Atmosphere Linked System

Currents and the atmosphere form a linked system for distributing heat around the globe. Climate scientists now recognize connected ocean–atmosphere climate cycles. The **El Niño–Southern Oscillation** describes related ocean surface temperatures and atmospheric pressure changes that occur every 2 to 7 years. "El Niño" causes weather effects from torrential rains in California to droughts in Kansas. The **Atlantic Multidecadal Oscillation** (AMO), seen mainly in the North Atlantic, affects Atlantic hurricanes and weather in North America, Europe, northeast Brazil, and western North Africa.

The Atlantic carries a lot of heat from the equator to the Arctic. Measuring the temperature of seawater indicates that atmospheric heat is absorbed fairly quickly into the top few meters, and much more slowly in deeper water. Near the ocean bottom, seawater is close to freezing. By 2007, a program called Argo had established a global network of 3000 sensors that measure temperature and salinity down to depths of 2000 m (6600 ft) (University of California San Diego, Argo Program, online). Argo revealed that an apparent pause in global warming seen since the late 1990s did not apply to the deep oceans. While atmospheric warming may have paused, deep seawater appears to be getting warmer (American Institute of Physics 2015).

Ocean Models

Marine scientists build ocean models to understand the relationship between temperature, salinity, currents, and climate. As soon as computers became available, oceanographers began to build simple models. The first 3D ocean models that coupled ocean circulation to the atmosphere were created in the 1970s. "Modelers had not yet fully grasped ... global ocean circulation. Their grid boxes were still too large to realistically represent giant eddies or narrow currents like the Gulf Stream.... It turned out that the ocean system, like most features of climate, was more complex than the first models had supposed" (American Institute of Physics 2015, online). By the start of the new millennium, modelers were fairly confident in results showing that Earth is warming. They just weren't sure how much.

We still don't understand all of the components and interactions that go into ocean and current models. For example, how does the uptake of CO_2 by plankton affect climate? Plankton are everywhere near the ocean surface, but thrive in warm surface water and where bottom water upwelling brings nutrients to the surface. Both surface water temperatures and upwelling depend on currents. Hence, the amount of CO_2 taken out of the atmosphere by plankton ultimately depends on currents. Ocean modelers don't know why atmospheric warming has paused since the late 1990s, or the reason for the 60- to 80-year oscillations in North Atlantic currents. They don't understand the effects of a rough sea bottom on currents and vertical mixing of water, or how changes in trade winds affect currents. Until modelers identify all the drivers for ocean circulation and understand their impacts and interactions, these unknowns will lead to uncertainties in global warming models.

The Land Surface

The surface of the land mostly absorbs sunlight and reradiates heat. The land contributes aerosols in the form of dust and smoke. It is where most human activities generate greenhouse gases. Some urban areas absorb and reradiate more heat than rural areas owing to the large amount of concrete and asphalt as opposed to vegetation. This is known as the **urban heat island** effect.

Ice Sheets

Ice sheets, including continental glaciers and sea ice, contribute to cooling by reflecting sunlight back into space. Ice sheets are primarily in Greenland and Antarctica, but smaller areas exist in Canada, Russia, Scandinavia, and the United States (Alaska). The Antarctic ice shelves are extensions of Antarctic continental glaciers. Sea ice is mainly seasonal; it is permanent (for now) only around the North Pole.

Melting ice sheets decrease solar reflection and contribute to warming. Melting ice adds freshwater to the oceans, thereby changing salinity and density and influencing global currents. Melting of ice sheets also raises sea levels somewhat. Melting of the Greenland ice sheets would contribute 2 cm (0.79 in) rise in the next 100 years, and melting Antarctic ice would cause an 8-cm (3-in) rise over the same time (Lomborg 2001).

The Biosphere

Biomass, primarily plankton but also forests, swamps, marshes and grasslands, remove large amounts of CO_2 from the atmosphere. Forests, swamps, and marshlands are carbon sinks that lock up carbon when organic matter is buried and becomes peat (the first stage in forming coal). There are indications that global vegetation cover has been increasing over the past few decades (global greening). Carbon is also stored in oil and gas deposits.

Plants exhale both water vapor and aerosols. As we have seen, water vapor is a greenhouse gas except when it is in clouds: the role of clouds as a net absorber or radiator of heat is still unclear and may change depending on circumstances. Aerosols generally, but not always, reflect solar radiation. **Forest fires**, whether natural or man-made, add carbon particulates to the atmosphere. This carbon is characterized as **black carbon** when it absorbs radiation and adds to warming and as **organic carbon** when it reflects radiation and contributes to cooling. The amount of each type of carbon depends on what is being burned and how complete the combustion is.

To Summarize

Earth has warmed 5°C to 8°C since the end of the last ice age around 12,000 years ago, and 0.3°C to 0.6°C over the past 100 years. Carbon dioxide is a greenhouse gas, and humans have been adding CO_2 to the atmosphere since the beginning of the Industrial Revolution in the early 1800s. There are many climate drivers, from changing solar radiation to orbital variations and changing atmospheric composition. Model-based predictions are only as good as the data that go into the model. The farther into the future one predicts, the greater the uncertainty and range of possible outcomes.

The debate continues as to whether current warming is primarily a result of human burning of fossil fuels. The human contribution may be significant—or not. While the weight of public opinion has ruled in favor of human-caused warming, opinion does not establish facts. Climate changes over centuries and millennia: we may not know for sure whether there is a human contribution to warming for another century or more. Despite that, many think there is no time to waste responding to this crisis. What should be done?

The Worst-Case Scenario?

Rather than apportion blame, which does nothing to solve the problem, we should have a backup plan for the worst-case scenario. If warming is man-made, we might be able to do something about it. The worst case would be that we do everything to stop warming, but the warming is completely natural and there is nothing humans can do about it. What would this outcome look like?

It turns out Earth has seen this before. Sea levels have come up over 100 m (330 ft) since the last ice age and many early human archaeological sites are now under water (Fleming 2004). Sea level has risen between 10 and 25 cm (4 to 10 in) since 1900, and may rise another 30 to 50 cm by the year 2100. Three quarters of this rise is attributed to warming of the oceans and expansion of seawater. The cheapest and easiest response would be building dikes around low-lying cities and farms, as has been done in the Netherlands and New Orleans. But this merely delays the inevitable (McPhee 1989). Other proposed responses include geoengineering solutions such as fertilizing the oceans so that more planktonic algae absorb CO_2 and take it out of circulation when they die and fall to the bottom. Adding sulfur aerosols to the stratosphere would cool the earth. Carbon capture and storage takes smokestack emissions and pumps them into deep underground formations. Yet each of these proposed solutions has potentially damaging or unknown side effects.

Ultimately, rising sea level will require moving to higher ground. Low-lying islands and coastal plains will be surrendered to the sea. Tropical rainforests will expand into areas that are now desert, and mid-latitude deserts will migrate away from the equator. Temperate grasslands will move toward the poles. Boreal forests will move into areas now covered by tundra. But it's not all bad news. As temperate climates migrate toward the poles, good agricultural land will move as well. An estimated 4°C to 5°C temperature increase (the extreme high end of predictions) would increase worldwide rainfall 10% to 15% (Crosson 1997). Most plants do better with more CO_2 in the atmosphere, and warmer temperatures enhance the fertilization effect of CO_2. There would also be a longer growing season. Agricultural production might actually improve. During the **medieval warm period** (roughly the years 950 to 1250), vineyards were common in England and the Norse were able to successfully colonize Greenland (English-Wine.com, online). Over the coming century, global biomass may increase over 40% owing to a combination of warmer temperatures, increasing CO_2 fertilization, a longer growing season, and more precipitation (Cramer et al. 2001).

There is much talk about global warming causing extreme weather: more storms, more severe storms, more flooding, more destruction, and greater loss of life. As we have seen, climate is always changing. This includes the El Niño–La Niña events in the Pacific, and the weaker AMO. While damage and loss of life have increased, this is a result of more people and more development in risk-prone areas, not because of more severe storms.

Observational evidence of the overall trend in the North Atlantic indicates that there are fewer storms (Karl et al. 1997). While there has been an increase in the number of cyclones over the past 50 years in the North Pacific, the number of tropical storms in the North Indian Ocean has decreased, there has been virtually no change in the southwest Indian Ocean and South Pacific, and there has been a decrease in storms around Australia (Landsea 2000). In other words, there is no good evidence that global warming is causing more storms.

While Earth is getting warmer, the warming has not been uniform. Night temperatures have increased more than daytime temperatures, and there has been more warming in winter than in summer. Winter temperatures are warming most in the Arctic. This means more frost-free days for agriculture. Only Australia and New Zealand have seen maximum (daytime) temperature increases: there has been virtually no change in the United States, and maximum temperatures reportedly have declined in China. The Central England temperature data indicate fewer cold days, but not more hot days (Jones et al. 1999). Thus, average global warming is the result of warmer minimum (nighttime and winter) temperatures and only slight increases in maximum (daytime, summer) temperatures.

Twice as many people die from exposure to cold in winter as die from heat in the summer (IPCC 2001). A warming Earth would decrease the number of deaths overall. There is concern that a warming Earth means more areas will be subject to tropical diseases. In fact, diseases like malaria were endemic to northern Europe and northern North America until just after World War II. These areas brought mosquito-borne diseases under control by eradicating mosquitos and their breeding grounds, not by controlling climate.

Economic analyses indicate that it is far more expensive to cut CO_2 emissions than to adapt to warmer temperatures (Lomborg 2001, p. 318). Scientists admit that cutting emissions may not significantly decrease warming, at least in the short term. It is worth discussing whether the money we are putting into controlling climate would be better spent building dikes or moving people to higher ground. Humans are successful at least partly because they adapt to changing circumstances. As it has been, so shall it be.

References

Abdussamatov, H. 2009. The Sun defines the climate. *Nauka I Zhizn* (Science and Life), 1, 34–42. Translated by Lucy Hancock.

American Institute of Physics. 2015. The Discovery of Global Warming—Ocean Currents and Climate. Accessed online July 14, 2016. https://www.aip.org/history/climate/oceans.htm

Anderson, B. G., and H. W. Borns. 1997. *The Ice Age World*. Scandinavian University Press, Oslo.

Cramer, W., A. Bondeau, F. I. Woodward, I. C. Prentice, R. A. Betts, V. Brovkin, P. M. Cox, V. Fisher, J. A. Foley, A. D. Friend, C. Kucharik, M. R. Lomas, N. Ramankutty, S. Sitch, B. Smith, A. White, and C. Young-Molling. 2001. Global response of terrestrial ecosystem structure and function to CO_2 and climate change: Results from six dynamic global vegetation models. *Global Change Biology* 7 (4), 357–373.

Crosson, P. 1997. Impacts of Climate Change on Agriculture. Climate Issues Brief No. 4, Resources for the Future. Accessed online July 19, 2016. http://www.climatecaucus.net/images/Climate_Agr.pdf

Dessler, A., Z. Zhang, and P. Yang. 2008. Water-vapor climate feedback inferred from climate fluctuations, 2003–2008. *Geophysical Research Letters* 35 (20). Bibcode: 2008GeoRL..3520704D. doi:10.1029/2008GL035333.

English-Wine.com. The History of English Wine. Accessed online July 14, 2016. http://www.english-wine.com/history.html

EPA. 2010. Methane and Nitrous Oxide Emissions from Natural Sources. US Environmental Protection Agency, Washington, D.C. Accessed online July 14, 2016. http://nepis.epa.gov/Exe/ZyNET.exe/P100717T.TXT?ZyActionD=ZyDocument&Client=EPA&Index=2006+Thru+2010&Docs=&Query=&Time=&EndTime=&SearchMethod=1&TocRestrict=n&Toc=&TocEntry=&QField=&QFieldYear=&QFieldMonth=&QFieldDay=&IntQFieldOp=0&ExtQFieldOp=0&XmlQuery=&File=D%3A%5Czyfiles%5CIndex%20Data%5C06thru10%5CTxt%5C00000017%5CP100717T.txt&User=ANONYMOUS&Password=anonymous&SortMethod=h%7C-&MaximumDocuments=1&FuzzyDegree=0&ImageQuality=r75g8/r75g8/x150y150g16/i425&Display=p%7Cf&DefSeekPage=x&SearchBack=ZyActionL&Back=ZyActionS&BackDesc=Results%20page&MaximumPages=1&ZyEntry=1&SeekPage=x&ZyPURL

Fleming, N. 2004. *Submarine Prehistoric Archaeology of the North Sea*. Council for British Archaeology Book 141, 220 pp.

Gerhard, L. C., W. E. Harrison, and B. M. Hanson. 2001. *Geological Perspectives of Global Climate Change*. American Association of Petroleum Geologists, Studies in Geology #47, 372 pp.

Gray, L. J. et al. 2010. Solar influences on climate. *Reviews of Geophysics*, 48, RG4001, doi:10.1029/2009RG000282. http://solar-center.stanford.edu/sun-on-earth/2009RG000282.pdf

Hansen, J., L. Nazarenko, R. Ruedy, M. Sato, J. Willis, A. Del Genio, D. Koch, A. Lacis, K. Lo, S. Menon, T. Novakov, J. Perlwitz, G. Russell, G.A. Schmidt, and N. Tausnev. 2005. Earth's energy imbalance: Confirmation and implications. *Science* 308, 1431–1435.

Imbrie, J. et al. 1984. The orbital theory of Pleistocene climate: Support from a revised chronology of the marine $\delta^{18}O$ record. In *Milankovitch and Climate*, Part I, A. L. Berger and others (eds.), Reidel Publishing Company, Dordrecht, Netherlands, pp. 269–305.

Intergovernmental Panel on Climate Change. 2001. Climate Change 2001. Accessed online July 10, 2016. https://www.ipcc.ch/ipccreports/tar/

Intergovernmental Panel on Climate Change. 2007. Climate Change 2007: Climate Change Impacts, Adaptation and Vulnerability. Contribution of Working Group II to the Fourth Assessment Report of the IPCC. Accessed online July 10, 2016. http://www.ipcc.ch/

Intergovernmental Panel on Climate Change. 2013. Climate Change 2013, The Physical Science Basis, WG1, 5th Assessment Report, IPCC.

Intergovernmental Panel on Climate Change. 2014. Summary for policymakers. In *Climate Change 2014: Impacts, Adaptation, and Vulnerability. Part A: Global and Sectoral Aspects*. Contribution of Working Group II to the Fifth Assessment Report of the Intergovernmental Panel on Climate Change. Field, C. B., V. R. Barros, D. J. Dokken, K. J. Mach, M. D. Mastrandrea, T. E. Bilir, M. Chatterjee, K. L. Ebi, Y. O. Estrada, R. C. Genova, B. Girma, E. S. Kissel, A. N. Levy, S. MacCracken, P. R. Mastrandrea, and L. L. White (eds.). Cambridge University Press, Cambridge, United Kingdom and New York, pp. 1–32.

Johnson, K. 2009. How carbon dioxide became a 'pollutant.' *The Wall Street Journal*, April 18. https://www.wsj.com/articles/SB124001537515830975

Johnson, M. E. 2012. Hurricane Sandy: Big Data Predicted Big Power Outages. Information Week. Accessed online July 26, 2016. http://www.informationweek.com/big-data/big-data-analytics/hurricane-sandy-big-data-predicted-big-power-outages/d/d-id/1107329?

Jones, P. D., E. B. Horton, C. K. Folland, M. Hulme, D. E. Parker, and T. A. Basnett. 1999. The use of indices to identify changes in climatic extremes. *Climatic Change* 42 (1), 131–149.

Karl, T. R., N. Nicholls, and J. Gregory. 1997. The coming climate. *Scientific American* 276 (5), 78–83.

Kiehl, J. T., and K. E. Trenberth. 1997. Earth's annual global mean energy budget. *Bulletin of the American Meteorological Society* 78 (2), 197–208. Bibcode:1997BAMS...78..197K. doi:10.1175/1520-0477(1997) 078 <0197:EAGMEB> 2.0.CO;2.

Lacis, A. 2010. NASA *GISS: CO₂: The Thermostat that Controls Earth's Temperature,* New York: NASA GISS.

Landsea, C. W. 2000. Climate variability of tropical cyclones: Past, present, and future, In *Storms,* Pielke and Pielke (eds.), Routledge, New York, pp. 220–241. http://www.aoml.noaa.gov/hrd /Landsea/climvari/

Lange, I. M. 2002. *Ice Age Mammals of North America.* Mountain Press Publishing, Missoula, 226 pp. (This book and a number of other fine books on geology and natural history may be found at www.mountain-press.com or a print catalog may be requested by email at info@mtnpress.com or by calling 1-800-234-5308.)

Liu, J., Mao, Y. Pan, S. Peng, J. Peñuelas, B. Poulter, T. A. M. Pugh, B. D. Stocker et al. 2016. Greening of the Earth and its drivers. *Nature Climate Change.* doi:10.1038/nclimate3004. Published online April 25, 2016. http://www.nature.com/nclimate/journal/vaop/ncurrent/full/nclimate3004 .html

Lomborg, B. 2001. *The Skeptical Environmentalist.* Cambridge, The Cambridge University Press, pp. 258–324.

McIntyre, A. 1981. CLIMAP Project, Lamont-Doherty Earth Observatory. http://written-in-stone -seen-through-my-lens.blogspot.com/2014/03/climbing-geology-and-tectonics-of.html

McPhee, J. 1989. *The Control of Nature.* "Atchafalaya." Noonday Press, Farrar, Straus and Giroux, New York, pp. 3–94.

Met Office. 2016. Hadley Centre Central England Temperature dataset. Accessed online July 16, 2016. http://www.metoffice.gov.uk/hadobs/hadcet/

Morrison, R. B. 1991. Introduction, Quaternary Nonglacial Geology: Conterminous U.S. In *The Geology of North America,* K-2, Geological Society of America, Boulder, CO, pp. 1–12.

Nahle, N. 2007. Cycles of global climate change. *Biology Cabinet Journal Online.* Article no. 295. Accessed online August 9, 2016. http://www.biocab.org/Climate_Geologic_Timescale.html and http://www.biocab.org/Carbon_Dioxide_Geological_Timescale.html.

National Oceanic and Atmospheric Administration. Carbon Cycle Science. Accessed online July 12, 2016. http://www.esrl.noaa.gov/research/themes/carbon/

NASA Earth Observatory. The Sun and Global Warming. Accessed online July 10, 2016. http:// earthobservatory.nasa.gov/Features/SORCE/sorce_04.php

NASA Goddard Space Flight Center, 2016. Carbon dioxide fertilization greening Earth, study finds. PUBLIC RELEASE: 26-APR-2016. Accessed online July 10, 2016. http://www.eurekalert.org /pub_releases/2016-04/nsfc-cdf042616.php

Newitz, A. AZ Quotes. Accessed online August 9, 2016. http://www.azquotes.com/quote/1473981

Newsweek. 1975. The Cooling World. Peter Gwynne. April 28, 1975.

Orrell, D. 2007. *Apollo's Arrow.* Harper Collins Publishers Ltd., Toronto. 449 pp.

Pagani, M. et al. 2005. Marked decline in atmospheric carbon dioxide concentrations during the Paleocene. *Science* 309 (5734), 600–603.

Relfe, S. 2007. Global Warming Will Cause—The Coming Ice Age. Accessed online July 10, 2016. http://www.metatech.org/07/ice_age_global_warming.html

Ruddiman, W. F. 2001. *Earth's Climate: Past and Future.* W. H. Freeman and Sons, New York.

Scafetta, N., and B. J. West. 2008. Is climate sensitive to solar variability? *Physics Today,* 2 p. accessed online September 11, 2016. www.physicstoday.org

Schmidt, G. A., R. Ruedy, R. L. Miller, and A. A. Lacis. 2010. The attribution of the present-day total greenhouse effect. *Journal of Geophysical Research.* 115Bibcode:2010JGRD..11520106S, doi:10.1029/2010JD014287, D20106.

Scotese, C. R. 2002. Analysis of temperature oscillations in geological eras. http://www.scotese .com/climate.htm

Shakun, J. D., P. U. Clark, F. He, S. A. Marcott, A. C. Mix, Z. Liu, B. Otto-Bliesner, A. Schmittner, and E. Bard. 2012. Global warming preceded by increasing carbon dioxide concentrations during the last deglaciation. *Nature* 484, 49–54 (April 5, 2012). doi:10.1038/nature10915.

Share, J. 2014. Written in Stone … seen through my lens. Accessed online July 10, 2016. http://writ ten-in-stone-seen-through-my-lens.blogspot.com/2014/03/climbing-geology-and-tectonics-of .html

skepticalscience.com. Shakun et al. Clarify the CO2-Temperature Lag. Accessed online July 10, 2016. http://www.skepticalscience.com/print.php?n=1391

Stanford Solar Center. 2008. Global Warming. http://solar-center.stanford.edu/sun-on-earth/glob -warm.html

Time. 1974. Science: Another Ice Age? June 24. Accessed online July 10, 2016. http://content.time .com/time/magazine/article/0,9171,944914,00.html

Union of Concerned Scientists. The Sun–Climate Connection. Accessed online July 10, 2016. http:// www.ucsusa.org/global_warming/science_and_impacts/science/effect-of-sun-on-climate -faq.html#bf-toc-0

University of California San Diego, Argo Program. Accessed online July 10, 2016. http://www.argo .ucsd.edu/About_Argo.html

Wikipedia. Atmosphere of Earth. Accessed online July 10, 2016. https://en.wikipedia.org/wiki /Atmosphere_of_Earth

Wikipedia. Greenhouse Gas. Accessed online July 12, 2016. https://en.wikipedia.org/wiki/Greenhouse _gas#cite_note-kiehl197-17

Wikipedia. Maunder Minimum. Accessed online July 12, 2016. https://en.wikipedia.org/wiki/Maunder _Minimum

Wild, M., H. Gilgen, A. Roesch, A. Ohmura, C. N. Long, E. G. Dutton, B. Forgan, A. Kallis, V. Russak, and A. Tsvetkov. 2005. From dimming to brightening: Decadal changes in solar radiation at Earth's surface. *Science* May 6; 308 (5723), 847–850.

Zhu, Z., S. Piao, R. B. Myneni, M. Huang, Z. Zeng, J. G. Canadell, P. Ciais, S. Sitch, P. Friedlingstein, A. Arneth, C. Cao, L. Cheng, E. Kato, C. Koven, Y. Li, X. Lian, Y. Liu, R. Liu, J. Mao, Y. Pan, S. Peng, J. Penuelas, B. Poulter, T. A. M. Pugh, B. D. Stocker, N. Viovy, X. Wang, Y. Wang, Z. Xiao, H. Yang, S. Zaehle, and N. Zeng. 2016. Greening of the Earth and its drivers. *Nature Climate Change* DOI: 10.1038/NCLIMATE3004 http://www.nature.com/nclimate/journal/v6/n8/full /nclimate3004.html and http://luc4c.eu/public_files/findings_and_downloads/publications /MS_with_figs_Pugh%202016.pdf

18

So You Want to Be a Geologist

No Geologist worth anything is permanently bound to a desk or laboratory, but the charming notion that true science can only be based on unbiased observation of nature in the raw is mythology. Creative work, in geology and anywhere else, is interaction and synthesis: half-baked ideas from a bar room, rocks in the field, chains of thought from lonely walks, numbers squeezed from rocks in a laboratory, numbers from a calculator riveted to a desk, fancy equipment usually malfunctioning on expensive ships, cheap equipment in the human cranium, arguments before a road cut.

Stephen Jay Gould
Author, paleontologist, in *An Urchin in the Storm*

Geologists are never at a loss for paperweights.

Bill Bryson
Author, humorist, in *A Short History of Nearly Everything*

Geologists study the earth and its history. This benefits society by providing natural resources such as metals, building stone, energy, and water. They work with engineers to build dams that don't collapse and roads that won't buckle. They work to stabilize slopes and help developers avoid faults, flooding, and subsidence. They investigate the processes of erosion, flooding, landslides, earthquakes, and volcanic eruptions to minimize catastrophic damage from these events. Geologists change the way we look at the world by observing and crafting a history of the earth. They propose theories of evolution, mountain building, ice ages, moving continents, and changing climate. Then, they test those theories.

Geology is not a philosophy, but it does emphasize a certain worldview. This geologic perspective stresses the awesome power of nature acting over unimaginable spans of time and, as a result, tends to minimize the influence of humans on the world.

What Does a Geologist Do?

Many people become geologists because they love the natural beauty of the landscape and want to work outdoors. While some geologist careers involve working outdoors in scenic and remote areas, many more involve office work on computers or in laboratories. Some geologists work on their own, but most jobs require teamwork in a setting that integrates multiple specialties. The senior author's last work team consisted of geologists, geophysicists, reservoir engineers, and an intern. The team was almost equally divided between male and female and was truly international: we had members originally from the United States, Canada, China, New Zealand, Russia, and Nigeria.

The **average starting salary** for a geologist in the United States in 2016 is around $60,000/year. Salaries range from around $40,000 to more than $200,000 depending on location, industry, education, and experience. Profit sharing and performance bonuses can increase this amount.

Geologists' **work location**, in the sense of where they live, may be limited. For example, the mining industry in the United States is concentrated in Spokane, Reno, and Tucson. In Canada, it is primarily in Toronto and Vancouver. The oil industry in the United States is located in Houston, with outliers in Denver, Dallas, Midland, Oklahoma City, Tulsa, and Anchorage. In Canada, the oil industry is concentrated in Calgary with outliers in Fort McMurray and St. John's. This doesn't mean geologists don't get to see other areas. Whereas much work can be done in an office, collecting samples and making maps take geologists around the world, frequently to areas far from the comforts of home (Figures 18.1 through 18.9).

There are many subcategories of geology. They include mineral exploration and mining geology, petroleum exploration and development, environmental geology, groundwater hydrology, engineering geology, geomorphology (the study of landscapes), and mapping for federal or state geological surveys. You can also be a paleontologist (study fossils) or a seismologist (study earthquakes) or teach geology. Most of these have been addressed in various sections of this book.

Each of these fields is further subdivided into specializations. Within petroleum geology, for example, you can be a petrophysicist, rock mechanics expert, micropaleontologist (or "biostratigrapher"), geochemist, photogeologist/remote sensing specialist, petroleum systems modeler, sequence stratigrapher, or structural geologist. A structural geologist might focus their career on fault seal analysis, interpreting dipmeter data, or balancing cross sections.

There are several closely related fields like geophysics, petroleum engineering, civil engineering, oceanography, evolutionary biology, climatology, agriculture/soil science, ecology, and geography.

FIGURE 18.1
The senior author (GP) "on the road" mapping in Abu Dhabi. (Courtesy of G. Prost.)

FIGURE 18.2
Encounter with a Bedouin shepherd while exploring for oil in Jordan. (Courtesy of G. Prost.)

FIGURE 18.3
Stuck on a tidal flat in Abu Dhabi with the tide coming in. And no shovel. (Courtesy of G. Prost.)

So what, exactly, can you do with a background in geology? The primary **employers** of geologists are environmental firms, followed by government agencies, the oil and gas industry, universities and high schools, and mining. Some geologists work for nonprofit agencies like the Peace Corps to help underdeveloped countries understand and catalog their mineral and water resources. Others work as freelance consultants, preparing maps or monitoring wells or prospecting for minerals. With this in mind, some of the more common geologic careers are discussed in the following sections.

FIGURE 18.4
Making a friend while mapping in the Gobi basin, Mongolia. (Courtesy of G. Prost.)

FIGURE 18.5
Examining the Paclele mud volcanos, Berca, Romania. (Courtesy of G. Prost.)

FIGURE 18.6
Posing by war wreckage while mapping in the Sinai, Egypt. (Courtesy of G. Prost.)

FIGURE 18.7
Sometimes, the job involves negotiating leases or data acquisition agreements. Negotiating a work agreement with Russia required a trip to Moscow. (Courtesy of G. Prost.)

FIGURE 18.8
Mapping structure along the Mekong River near Vientiane, Laos. (Courtesy of G. Prost.)

FIGURE 18.9
The senior author (crouching) while mapping in the Sierra Madre Oriental, Mexico, with "los leones de Pemex." (Courtesy of G. Prost.)

Resource Geology

Those who explore for and develop metals and nonmetallic mineral resources including coal, granite, aggregate, and dimension stone are known as **economic geologists** (Figure 18.10). **Petroleum geologists** explore for and produce oil and natural gas (Figure 18.11). **Hydrologists** and **hydrogeologists** monitor groundwater and surface water and are concerned with finding supplies of freshwater, monitoring water pollution, understanding groundwater flow and replenishment, studying streamflow and flooding, locating water wells and springs, and monitoring stream erosion.

Environmental Geology

Environmental geologists are concerned with soil and water pollution such as that caused by mine tailings, natural acid drainage, or leaking pipelines. They monitor and try to predict natural disasters such as earthquakes, landslides, volcanic eruptions, floods, and tsunamis, particularly as they affect humans and wildlife. **Seismologists** studying earthquakes and **volcanologists** studying eruptions may be included in this category.

Geological Engineering

Mining engineers design mines, calculate reserves, and plan the extraction of mineral deposits. **Petroleum engineers** design drilling programs and plan oil and gas field development. They are usually responsible for calculating reserves. **Geotechnical engineers** and **civil engineers** determine the mechanical properties of rock and soil and apply them to building of roads, dams, bridges, power plants, and other types of infrastructure development.

FIGURE 18.10
Exploring for copper-moly porphyries in the Belmont Mountains, Arizona. (Courtesy of G. Prost.)

FIGURE 18.11
Mapping petroleum reservoir rocks in the Richardson Mountains, Northwest Territories, Canada. (Courtesy of G. Prost.)

Marine Geology

Marine geologists study the geology of the ocean floor and its sediments, including submarine erosion and sediment transport, sedimentation, and mineral deposits. They study reefs and currents, submarine volcanos and spreading centers, deep sea trenches and faulting and tsunamis.

Geochemistry

Geochemists determine the composition of rocks and soils, the constituents of Earth's interior, and determine the age, provenance, and burial history of sediments. They help find metal deposits and oil fields by measuring and interpreting trace amounts of metals, minerals, and organic matter. Geochemists monitor soil and water pollution, and reconstruct ancient atmospheres, oceans, and environments from trace chemical indicators.

Geophysics

Geophysicists monitor earthquakes and calculate their magnitude and epicenters. They help find petroleum deposits and monitor magma chambers by providing acoustic imagery of the interior of the earth. They generate gravity and magnetic maps to help locate metallic mineral deposits. They monitor groundwater pollution. Geophysicists reveal the structure of Earth's interior.

Paleontology

The study of fossil animals and plants and their behavior and evolution falls to **paleontologists**. Combined with the study of sediments (stratigraphy), they are able to reconstruct ancient environments and determine the age of sediments using fossils. Paleontologists survey construction sites to locate and preserve important fossils.

Geological Surveys and Government Agencies

Much of the fundamental mapping and research in geology is done by federal and state geological or mineral surveys. The basic functions of these groups are to identify and quantify the resources that belong to the country or state and to help regulate exploration and extraction so that it is safe and environmentally responsible. An objective of geological surveys is seeing that the public benefits from resource extraction by means of royalties and user fees.

In the United States, for example, federal agencies that hire geologists include the following:

US Geological Survey (USGS) (https://www2.usgs.gov/ohr/, http://geology.com /groups.htm)

Bureau of Land Management (BLM) (http://www.blm.gov/wo/st/en.html)

Bureau of Ocean Energy Management, Regulation and Enforcement (BOEMRE) (http://www.boem.gov/employment/)

US Army Corps of Engineers (http://www.usace.army.mil/Careers/How-to-Apply/)

US Environmental Protection Agency (EPA) (https://www.epa.gov/careers/science -careers-epa)

There are also state geological surveys. See http://geology.com/groups.htm.

Academia

Geologists teach Earth Science at the high school level. They also teach all aspects of geology at the university level.

Geology professors teach geology courses and do geologic research. Universities frequently require them to bring in federal or corporate funding to help support graduate students. They may be required to publish the results of their research. Professors are usually hired as "tenure track," meaning they are on probation for a period of years until they complete their PhD and are determined to be a good fit. **Tenure** is a form of job security that is meant to allow expression of novel or controversial ideas free of intimidation and retaliation. Tenured professors can only be terminated for serious misconduct or severe funding problems.

Once they have been given tenure, they become full professors. They can also be hired as "non-tenure track" professors, meaning they are lecturers, visiting professors, or adjunct professors.

Other Geologic Specialties

Mineralogists study the chemical and crystalline structure of minerals and their resulting properties.

Petrologists investigate igneous, metamorphic, and sedimentary rocks. They determine the origin, history, composition, and structure of these rocks.

Sedimentologists research sedimentary rocks. They describe the characteristics of sediments and determine their origins and history.

Stratigraphers examine the age relationships among sedimentary layers using fossils, radiometric dating, and composition (Figure 18.12). They determine which layers are continuous across regions and how a formation's composition and depositional environment changes in space and time. They determine "which way is up?" in a section of rock.

Structural geologists probe the structure of the earth, including folds and faults (Figure 18.13). They determine the deformation history of regions, such as how and when mountains and basins formed and the movement of continents over time.

Geomorphologists establish the origin and evolution of landscapes by studying landforms and the erosional and depositional processes that shape them. They determine what areas are prone to landslides, floods, and subsidence.

Glacial geologists study present-day and ancient glaciers and the sediments, soils, and landscapes associated with ice sheets. They help us understand the ice age history of the earth and the ever-changing climate.

Regardless of which branch of geology you decide on, the rewards of the work and the people you work with make this an enjoyable and fulfilling career (Figure 18.14).

FIGURE 18.12
Ancient stratigraphy. Red Rock Canyon exposes Precambrian sediments of the Uinta Mountain Group, Uinta Mountains, Utah. (Courtesy of G. Prost.)

FIGURE 18.13
A company class examines rock deformation in the Canadian fold-thrust belt, Banff National Park, Alberta. (Courtesy of G. Prost.)

FIGURE 18.14
Parched geologists socialize over a pint after a hard day's work examining turbidite deposits. This watering hole is at Punta San Carlos, Baja California.

Education Requirements

All it takes to be a rock hound is an interest in minerals, fossils, and the earth in general. You can read popular books about geology (Busbey et al. 1997; Dixon and Bernor 1992; Eyles and Miall 2007; Lambert 2007), or you can learn by observing on your own. There are many clubs, societies, and guides to mineral collecting as well: see *Guidebooks* in the section on Minerals and Mineral Identification.

If you are looking for a career in geology or a related field, you will spend several years in a university and get a science degree. It is never too early to get started.

High School

Geoscience education has traditionally been given short shrift in the kindergarten through 12th grade (K-12) curriculum, at least in the United States. Few secondary schools offer courses in geology. That may be changing. In 2011 the National Research Council developed a set of standards for K-12 science education (http://www.nextgenscience.org /framework-k%E2%80%9312-science-education). The Next Generation Science Standards (NGSS) call for specific learnings in Earth and space science because of the role these play in the economy and daily life. Topics such as earthquakes, volcanic activity, plate tectonics, energy and natural resources, the role of water on Earth's surface, and global warming are addressed. These standards are not federally mandated: it is up to the states to adopt the NGSS.

High school-level resources on Earth science are highlighted during Earth Science Week, usually in October each year. Check out http://ww3.earthscienceweek.org/ for classroom activities, Earth science videos, news, resources, Earth science organizations, and photo and essay contests.

If your high school offers a course in geoscience, take it. Otherwise, the best background to prepare for university is taking as much math and science as possible. Geology is an integrative discipline: it uses bits and pieces from all other branches of science. Geometry and algebra provide a background for understanding structural geology. Statistics and computer science help analyze data. Physics prepares you for petrology, structural geology, and optical mineralogy. Chemistry is needed to understand mineralogy and geochemistry. Physical geography, the study of natural processes that shape the earth, provides an excellent background for physical geology. The parts of biology that pertain to evolution, skeletal structure, and classification of life are helpful in historical geology and paleontology. Last, but not least, any class that emphasizes critical thinking and that teaches the ability to write and communicate clearly and logically is essential.

A list of schools that offer geology degrees can be found at http://geology.com/colleges .htm. Early in your high school senior year, you should arrange to visit a university and talk to the professors if at all possible.

University

Any career in geology requires, at a minimum, a Bachelor of Science (BSc) degree from a recognized and accredited university. The BSc provides an introduction to most aspects of geology. In many countries, the BSc is all that is needed for a career in geology. In the United States, a Master of Science (MSc) is needed for many professional careers. In addition to classwork, the MSc requires writing a **thesis**, personal research into some aspect of geology. A few careers, notably teaching and research at the university level, require a

Doctor of Philosophy (PhD) degree. In addition to classwork, the PhD requires a written **dissertation** that comprises extensive research and original work in geology.

Undergraduate School

Each school has its own requirements for graduating with a four-year geology Bachelor of Science degree. After the introductory classes, most schools offer similar core programs that include, at a minimum, courses in stratigraphy, structural geology, mineralogy, petrology, and paleontology. An essential part of geological training is learning to prepare and deliver clear presentations. This should be part of your classwork.

In addition to classes in geology, the student is usually required to take supporting classes in physical chemistry, physics, calculus, basic computer programming, and technical report writing.

Field Camp

Most degree programs require a **field class** where students learn how to make geologic maps and how to apply what was learned in class to solve geologic problems. Not all schools offer these classes, but those that do often accept qualified students from other schools. Learning how to observe, collect samples, take measurements with a Brunton compass, take notes, measure stratigraphic sections, correlate stratigraphy, compile maps, and write summary reports is where everything you learned in class comes together. **Field school** is the highlight of most undergraduate programs (Figures 18.15 and 18.16).

FIGURE 18.15
Mapping in the Grand Canyon, Northern Arizona University geology field school, 1972. (Courtesy of G. Prost.)

FIGURE 18.16
The senior author (far left), at underground mapping field camp, Colorado School of Mines, 1974. (Courtesy of G. Prost.)

You can find a list of field programs at Geology.com (Table 18.1). Check with the school to see that their program offers the experience you want and to ensure the credits will transfer.

Graduate School

Before going to graduate school, know what your goals are. Do you want to teach, do research, or work in industry? Do you want to specialize in some aspect of geology? The answer will determine which degree you go for and which school you apply to, since each school has strengths in different areas.

Graduate school in Geology can be a 2-year Master of Science (sometimes stretches to 3 years), a 4-year PhD program (sometimes stretches to 7 years or more), or a program where you go straight from your BSc to the PhD, typically in 5 to 6 years (Table 18.2). Although each school is different, the first year of a graduate program is usually spent taking classwork and writing a thesis/dissertation proposal. The proposal requires you to set out a problem that you want to solve and lay out a program of research and classwork to resolve the problem.

Graduate programs are monitored by **degree committees** consisting of professors from your department as well as other related departments. The Masters degree is fairly well supervised by your degree committee. The work does not have to be original research, though it may be. If you are in a 2-year program, you will do field work during your first summer. You may be required to go through a thesis defense where you explain what you did and justify the results. You will come out of this program with the ability to work on your own with minimal supervision, minimal guidance, and minimal training.

A PhD program must contain original work and is essentially unsupervised. Your faculty advisors are available, but you are essentially on your own to figure out what you need to do and how to do it. You have several summers to do any necessary field work. PhD

TABLE 18.1

US and Canadian Geology Field Camp Directory

School	Field Location
Canadian Schools	
McGill	Nevada–Arizona–California; Massif Central, France
Memorial	Newfoundland, International
Queen's University	Quebec, United States
University of Alberta	Northwest Territories
University of British Columbia	Oliver, British Columbia
University of Calgary	Alberta, USA
University of Toronto	Various locations worldwide
US Schools	
Adams State University	Colorado/New Mexico
Albion College	Wyoming/Montana
Ball State University	South Dakota/Wyoming/Montana
Boise State University	Italy
Bowling Green State University	New Mexico/Colorado
California State University, Sacramento	California
Central Washington University	Oregon/Idaho
Clemson University	Kentucky/Tennessee
Colorado School of Mines	Colorado
Concord University	Colorado/Utah
Cornell University	Argentina
East Carolina University	New Mexico/Colorado
Eastern Washington University	Montana
Fort Hays State University	Colorado/Utah
Idaho State University	Idaho
Illinois State University	South Dakota/Wyoming
Indiana University	Montana
Iowa State University/University of Nebraska Lincoln	Wyoming
James Madison University	Ireland
Kent State University	Idaho/South Dakota/Wyoming
Lehigh University	Ohio/Wisconsin/Minnesota/South Dakota/Wyoming/Idaho
Louisiana State University	Two different courses (freshmen and senior): Colorado
Massey University	New Zealand—Late December to Early February
Miami University-Ohio	Montana/Wyoming/Idaho/Alberta
Michigan Technological University	South Africa
Montana State University	Montana
New Mexico Tech	New Mexico/Colorado
North Carolina State University	New Mexico
Northern Arizona University	Arizona
Northern Illinois University	Wyoming/South Dakota
Oklahoma State University	Colorado
Oregon State University	Two courses: California/Oregon
Pennsylvania State University	Utah/Wyoming/Idaho/Montana

(*Continued*)

TABLE 18.1 (CONTINUED)

US and Canadian Geology Field Camp Directory

School	Field Location
Salem State College	Montana
South Dakota School of Mines and Technology	California—December/January
South Dakota School of Mines and Technology	Multiple courses: Arizona/California/Hawaii/Iceland/India/ Montana/South America/Spain/Turkey/Wyoming
Southern Illinois University	Montana/Wyoming
Southern Utah University	Utah
Stephen F. Austin State University	Texas/New Mexico/Arizona/Utah
Sul Ross State University	Texas
SUNY Buffalo	Colorado/Utah/Wyoming
SUNY Cortland	Three courses: New York (offered in even-numbered years)
Texas A&M	Texas
University of Akron	South Dakota/Wyoming
University of Alaska-Fairbanks	Alaska
University of Arizona	Wyoming/Utah/Nevada
University of Arkansas	Montana
University of Arkansas Little Rock	Colorado
University of Florida	New Mexico
University of Houston	Montana
University of Kentucky	Colorado
University of Louisiana-Lafayette	Texas/Oklahoma/South Dakota/Wyoming
University of Memphis	South Dakota/Montana/Wyoming
University of Michigan	Wyoming
University of Minnesota	Three different courses: Montana/Colorado/Minnesota
University of Minnesota-Duluth	Minnesota
University of Missouri-Columbia	Wyoming
University of Montana	Montana
University of Nevada-Las Vegas	Utah/Nevada/Idaho/Texas/California
University of Nevada-Reno	Nevada
University of North Dakota	Multiple courses: Arizona/California/Hawaii/Iceland/India/ Montana/South America/Spain/Turkey/Wyoming
University of Oklahoma	Colorado
University of Oregon	Montana/Oregon
University of South Florida	Multiple sessions: Florida/Georgia/South Carolina/Idaho
University of St Andrews	Scotland
University of Texas-Arlington	New Mexico/Texas
University of Texas-Austin	Multiple courses: Texas/New Mexico/Utah/Wyoming
University of Texas-Dallas	Two courses: New Mexico/Colorado
University of Washington	Montana
University of Wisconsin-Eau Claire	Multiple courses: South Dakota/Arkansas/Wisconsin/ Michigan/Wisconsin
University of Wisconsin-Milwaukee	Multiple courses: Arizona/California/Hawaii/Iceland/India/ Montana/South America/Spain/Turkey/Wyoming

(Continued)

TABLE 18.1 (CONTINUED)

US and Canadian Geology Field Camp Directory

School	Field Location
University of Wisconsin-Oshkosh	Utah
University of Wyoming—Laramie	Wyoming/Utah
West Virginia University	South Dakota/Wyoming/Montana
Western Illinois University	Montana/South Dakota/Wyoming
Western Michigan University	Michigan
Wichita State University	Wyoming and Montana
Wright State University	Multiple field courses

Source: From Geology.com. Copyright 2005–2016 Geology.com. All Rights Reserved (http://geology.com /field-camp.shtml). See also http://education.usgs.gov/nagt/geofieldcamps.html.

candidates have to jump several hurdles before they get their degree. They are expected to take and pass **comprehensive exams** ("comps") where their faculty committee asks oral or written questions regarding their research and classwork. In addition, any geologic or scientific question can be asked. Some of the questions asked of the senior author include "Why is the sky blue?", "How far from a convergent plate margin can you detect the tectonic effects of convergence?", and "Explain the geologic history of the rock I just handed you." You may have to pass a foreign language exam. You will have to defend your dissertation. You will come out of this program needing little guidance and supervision when starting your career.

Internships

As an undergraduate or graduate student, it is highly desirable to get work experience in geology. Many companies offer internships, usually paid, where you can work in a professional setting and start making contacts and building a network (Figure 18.17). Check with your school, take advantage of interview opportunities, and contact local companies to see if you can get this experience.

Professional Geologist Certification

Many states and provinces require you to get professional certification in order to call yourself a geologist or to work on projects that involve public safety. Certification means the state has certified that you know your profession and can work without harming the public or environment and that you are aware of the professional code of ethics. It is the modern equivalent of belonging to a medieval guild. Instead of apprentice–journeyman– master, there is "geologist-in-training" and "professional geologist" (P. Geol.). The difference is the amount of work experience you have.

In addition to submitting your university transcripts, there will be a background check to ensure you are not a felon, and you are usually required to take an exam or exams, register with the state, and pay annual license fees. Most jurisdictions require you to take continuing education classes to keep current in the profession. Learn more about it at the National Association of State Boards of Geology (http://www.asbog.org/) or from your local state or provincial board.

Professional groups work to maintain standards and to certify competence and ethical conduct. The American Institute of Professional Geologists is one such group:

TABLE 18.2

Some US and Canadian Geology Schools

Canadian Universities	California State University—Monterey Bay
Alberta	California State University—Northridge
University of Alberta	California State University—Sacramento
University of Calgary	California State University—San Bernardino
British Columbia	California State University—Stanislaus
University of British Columbia	Cerritos College
Newfoundland	College of Marin
Memorial University	College of the Redwoods
Toronto	College of the Siskiyous
Queens University	Cypress College
University of Toronto	Humboldt State University
Quebec	Loma Linda University
McGill University	Modesto Junior College
US Universities	Moorpark College
Alabama	Occidental College
Auburn University	Pomona College
Jacksonville State University	Saddleback College
University of Alabama	San Diego State University
University of Alabama—Huntsville	San Francisco State University
University of South Alabama	San Jose State University
Alaska	Santa Barbara City College
Alaska Pacific University	Scripps Institution of Oceanography
University of Alaska, Anchorage	Stanford University
University of Alaska, Fairbanks	University of California—Berkeley
Arizona	University of California—Davis
Arizona State University	University of California—Irvine
Northern Arizona University	University of California—Los Angeles
University of Arizona	University of California—Riverside
Arkansas	University of California—Santa Barbara
Arkansas Tech University	University of California—Santa Cruz
University of Arkansas—Fayetteville	University of the Pacific
University of Arkansas—Little Rock	University of Southern California
California	Whittier College
American River College	Colorado
California Institute of Technology	Adams State College
California Lutheran University	Colorado College
California State Polytechnic University, Pomona	Colorado Mesa University
California Polytechnic, San Luis Obispo	Colorado School of Mines
California State University—Bakersfield	Colorado State University—Fort Collins
California State University—Chico	Fort Lewis College
California State University—Dominguez Hills	University of Colorado at Colorado Springs
California State University—Fresno	University of Colorado at Boulder
California State University—Fullerton	University of Denver
California State University—East Bay	University of Northern Colorado
California State University—Long Beach	Western State College of Colorado
California State University—Los Angeles	

(Continued)

TABLE 18.2 (CONTINUED)

Some US and Canadian Geology Schools

Connecticut
    Central Connecticut State University
    Eastern Connecticut State University
    University of Connecticut
    Wesleyan University
    Yale University
Delaware
    University of Delaware
District of Columbia
    George Washington University
Florida
    Florida Atlantic University
    Florida Institute of Technology
    Florida International University
    Florida State University
    University of Florida
    University of Miami
    University of South Florida—St. Petersburg
    University of South Florida—Tampa
Georgia
    Columbus State University
    Emory University
    Georgia Institute of Technology
    Georgia Southern University
    Georgia Southwestern State University
    Georgia State University
    Skidaway Institute of Oceanography
    University of Georgia
    University of West Georgia
    Valdosta State University
Hawaii
    University of Hawaii at Hilo
    University of Hawaii at Manoa
Idaho
    Boise State University
    BYU-Idaho
    Idaho State University
    Lewis-Clark State College
    University of Idaho
Illinois
    Augustana College
    Eastern Illinois University
    Illinois State University
    Northeastern Illinois University
    Northern Illinois University

    Northwestern University
    Olivet Nazarene University
    Southern Illinois University, Carbondale
    University of Chicago
    University of Illinois—Chicago
    University of Illinois at Urbana-Champaign
    Western Illinois University
    Wheaton College
Indiana
    Ball State University
    DePauw University
    Earlham College
    Hanover College
    Indiana State University
    Indiana University
    Indiana University Northwest
    Indiana University—Purdue at Fort Wayne
    Indiana University—Purdue at Indianapolis
    Purdue University
    Saint Joseph's College
    University of Indianapolis
    University of Notre Dame
    University of Southern Indiana
Iowa
    Cornell College
    Drake University
    Iowa State University
    Simpson College
    University of Iowa
    University of Northern Iowa
Kansas
    Emporia State University
    Fort Hays State University
    Kansas State University
    University of Kansas
    Wichita State University
Kentucky
    Eastern Kentucky University
    Morehead State University
    Murray State University
    Northern Kentucky University
    University of Kentucky
    University of Louisville
    Western Kentucky University

(Continued)

TABLE 18.2 (CONTINUED)

Some US and Canadian Geology Schools

Louisiana	University of Michigan
Centenary College of Louisiana	Wayne State University
Louisiana State University	Western Michigan University
Tulane University	Minnesota
University of Louisiana at Lafayette	Bemidji State University
University of Louisiana at Monroe	Carleton College
University of New Orleans	Concordia College
Maine	Macalester College
Bates College	Minnesota State University Mankato
Bowdoin College	Minnesota State University Moorhead
Colby College	Southwest Minnesota State University
University of Maine at Orono	St. Cloud State University
University of Southern Maine	University of Minnesota—Minneapolis
Maryland	University of Minnesota—Duluth
Frostburg State University	University of Minnesota—Morris
Johns Hopkins University	University of St. Thomas
Towson University	Winona State University
University of Maryland—College Park	Mississippi
Massachusetts	Mississippi State University
Amherst College	University of Mississippi
Boston College	University of Southern Mississippi
Boston University	Missouri
Bridgewater State College	Missouri State University
Fitchburg State College	Missouri University of Science & Technology
Harvard University	Northwest Missouri State University
Massachusetts Institute of Technology	Saint Louis University
Mount Holyoke College	Southeast Missouri State University
Northeastern University	University of Central Missouri
Salem State College	University of Missouri—Columbia
Smith College	University of Missouri—Kansas City
Tufts University	Washington University
University of Massachusetts—Amherst	Montana
University of Massachusetts—Lowell	Montana Tech of the University of Montana
Wellesley College	Montana State University—Bozeman
Williams College	University of Montana
Michigan	University of Montana—Western
Albion College	Nebraska
Calvin College	Chadron State College
Central Michigan University	University of Nebraska at Kearney
Eastern Michigan University	University of Nebraska at Lincoln
Grand Valley State University	University of Nebraska at Omaha
Hope College	Nevada
Lake Superior State University	University of Nevada—Las Vegas
Michigan State University	University of Nevada—Reno
Michigan Technological University	

(Continued)

TABLE 18.2 (CONTINUED)

Some US and Canadian Geology Schools

New Hampshire	State University of New York—Plattsburgh
Dartmouth College	State University of New York—Stony Brook
Keene State College	Syracuse University
University of New Hampshire	The College of Saint Rose
New Jersey	Union College
Montclair State University	University of Rochester
New Jersey City University	Utica College
Princeton University	Vassar College
Richard Stockton College of New Jersey	York College CUNY
Rider University	North Carolina
Rutgers, The State University of New Jersey	Appalachian State University
William Paterson University	Duke University
New Mexico	East Carolina University
Eastern New Mexico University	Guilford College
New Mexico Institute of Mining & Technology	North Carolina Central University
New Mexico State University	North Carolina State University- Raleigh
University of New Mexico	University of North Carolina at Chapel Hill
Western New Mexico University	University of North Carolina at Charlotte
New York	University of North Carolina at Wilmington
Alfred University	Western Carolina University
Brooklyn College CUNY	North Dakota
City College CUNY	Minot State University
Colgate University	North Dakota State University—Fargo
Columbia University—Lamont-Doherty	University of North Dakota
Cornell University	Ohio
Hamilton College	Antioch College
Hartwick College	Ashland University
Hobart and William Smith Colleges	Bowling Green State University
Hofstra University	Case Western Reserve University
Hunter College CUNY	Cedarville University
Queens College CUNY	Cleveland State University
Rensselaer Polytechnic Institute	College of Wooster
Skidmore College	Denison University
St. Lawrence University	Kent State University
SUNY, College at Brockport	Marietta College
SUNY, College at Buffalo	Miami University
SUNY, College at Cortland	Mount Union College
SUNY, College at Fredonia	Muskingum College
SUNY, College at New Paltz	Oberlin College
SUNY, College at Oneonta	Ohio State University
SUNY, College at Oswego	Ohio University
SUNY, College at Potsdam	Ohio Wesleyan University
State University of New York—Albany	Otterbein College
State University of New York—Binghamton	University of Akron
State University of New York—Buffalo	University of Cincinnati
State University of New York—Geneseo	

(Continued)

TABLE 18.2 (CONTINUED)

Some US and Canadian Geology Schools

University of Dayton	Rhode Island
University of Toledo	Brown University
Wittenburg University	University of Rhode Island—Kingston
Wright State University	University of Rhode Island—Naragansett Bay
Youngstown State University	South Carolina
Oklahoma	Clemson University
Oklahoma State University	College of Charleston
University of Oklahoma	Furman University
University of Tulsa	University of South Carolina
Oregon	University of South Carolina—Aiken
Oregon State University	South Dakota
Portland State University	South Dakota School of Mines & Technology
Southern Oregon University	University of South Dakota
University of Oregon	Tennessee
Western Oregon University	Austin Peay State University
Pennsylvania	Middle Tennessee State University
Allegheny College	Tennessee Technological University
Bloomsburg University of Pennsylvania	University of Memphis
Bryn Mawr College	University of Tennessee—Chattanooga
Bucknell University	University of Tennessee—Knoxville
California University of Pennsylvania	University of Tennessee—Martin
Clarion University of Pennsylvania	University of the South—Sewanee
Dickinson College	Vanderbilt University
Drexel University	Texas
Edinboro University of Pennsylvania	Baylor University
Franklin & Marshall College	Hardin-Simmons University
Gannon University	Lamar University
Indiana University of Pennsylvania	McMurry University
Juniata College	Midwestern State University
Kutztown University of Pennsylvania	Rice University
La Salle University	Sam Houston State University
Lafayette College	Southern Methodist University
Lehigh University	Stephen F. Austin State University
Lock Haven University of Pennsylvania	Sul Ross State University
Mansfield University of Pennsylvania	Tarleton State University
Millersville University of Pennsylvania	Texas A&M University
Pennsylvania State University	Texas A&M University—Corpus Christi
Shippensburg University of Pennsylvania	Texas A&M University—Kingsville
Slippery Rock University of Pennsylvania	Texas Christian University
Temple University	Texas State University
University of Pennsylvania	Texas Tech University
University of Pittsburgh	Trinity University
University of Pittsburgh at Johnstown	University of Houston
West Chester University of Pennsylvania	University of Texas at Austin
Wilkes University	University of Texas at Arlington

(Continued)

TABLE 18.2 (CONTINUED)

Some US and Canadian Geology Schools

University of Texas at Dallas	Pacific Lutheran University
University of Texas at El Paso	University of Puget Sound
University of Texas at San Antonio	University of Washington
University of Texas of the Permian Basin	Washington State University
Utah	Western Washington University
Brigham Young University	Whitman College
Southern Utah University	West Virginia
University of Utah	Concord College
Utah State University	Marshall University
Utah Valley University	West Virginia University
Weber State University	Wisconsin
Westminster College	Beloit College
Vermont	Lawrence University
Lyndon State College	Northland College
Middlebury College	St. Norbert College
University of Vermont	University of Wisconsin—Eau Claire
Virginia	University of Wisconsin—Green Bay
College of William and Mary	University of Wisconsin—Madison
George Mason University	University of Wisconsin—Milwaukee
James Madison University	University of Wisconsin—Oshkosh
Old Dominion University	University of Wisconsin—Parkside
Radford University	University of Wisconsin—Platteville
University of Mary Washington	University of Wisconsin—River Falls
University of Virginia	University of Wisconsin—Stevens Point
University of Virginia's College at Wise	Wyoming
Virginia Polytechnic Institute & State University	Casper College
Washington and Lee University	University of Wyoming
Washington	
Central Washington University	
Eastern Washington University	

Source: From Geology.com. Copyright 2005–2016 Geology.com. All Rights Reserved (http//geology.com/colleges .htm).

http://www.aipg.org/. Specialized groups also have certification processes. For example, if you meet certain requirements, the American Association of Petroleum Geologists will endorse you as a Certified Petroleum Geologist, Petroleum Geophysicist, or Coal Geologist. Advantages include improved skills, higher credibility, and a stronger voice in influencing industry and government decisions and regulations.

Starting Your Career

Check with your school placement office and go to as many interviews as you can. Additionally, there are websites that provide career information.

FIGURE 18.17
The senior author's undergraduate internship included two summers on a team collecting geochemical stream-bed samples as part of a mineral exploration program in central Idaho. One of the areas worked was the Sawtooth Range, shown here viewed from the Salmon River Valley.

Sources of Information

https://www.sokanu.com/careers/geologist/	What geologists do.
http://www.earthworks-jobs.com/	Jobs in Earth science.
http://geology.com/	Earth science news and information.

Even with a good education and some experience under your belt, the best help in getting a job, any job, is a good attitude, curiosity, and love of what you do.

References

Bryson, B. 2003. *A Short History of Nearly Everything.* Broadway Books, New York. 560 pp.

Busbey III, A. B., R. R. Coenraads, P. Willis, D. Roots. 1997. *Rocks & Fossils. The Nature Company Guides,* Time Life Books, Sydney. 288 pp.

Dixon, D., and R. L. Bernor. 1992. *The Practical Geologist.* Fireside Book, Simon & Schuster, Inc. New York. 160 pp.

Eyles, N., and A. Miall. 2007. *Canada Rocks: The Geologic Journey.* Fitzhenry & Whiteside, Markham. 512 pp.

Gould, S. J. 1987. *An Urchin in the Storm: Essays about Books and Ideas.* W.W. Norton and Company, New York. 260 pp.

Lambert, D. 2007. *The Field Guide to Geology.* Checkmark Books, Infobase Publishing, New York. 304 pp.

Index

Page numbers followed by f and t indicate figures and tables, respectively.